The discovery of coherent structures in turbulence has fostered the hope that the study of vortices will lead to models and an understanding of turbulent flow, thereby solving or at least making less mysterious one of the great unsolved problems of classical physics. Vortex dynamics is a natural paradigm for the field of chaotic motion and modern dynamical system theory.

The emphasis in this monograph is on the classical theory of inviscid incompressible fluids containing finite regions of vorticity. The effects of viscosity, compressibility, inhomogeneity, and stratification are enormously important in many fields of application, from hypersonic flight to global environmental fluid mechanics. However, this volume focuses on those aspects of fluid motion which are primarily controlled by the vorticity and are such that the effects of the other fluid properties are secondary.

Cambridge Monographs on Mechanics and Applied Mathematics

General Editors
G. K. BATCHELOR, *Professor Emeritus of Applied Mathematics, Department of Applied Mathematics, University of Cambridge*
L. B. FREUND, *Professor of Engineering, Division of Engineering, Brown University*

VORTEX DYNAMICS

Vortex Dynamics

P. G. SAFFMAN

California Institute of Technology

CAMBRIDGE UNIVERSITY PRESS

PUBLISHED BY THE PRESS SYNDICATE OF THE UNIVERSITY OF CAMBRIDGE
The Pitt Building, Trumpington Street, Cambridge CB2 1RP, United Kingdom

CAMBRIDGE UNIVERSITY PRESS
The Edinburgh Building, Cambridge CB2 2RU, United Kingdom
40 West 20th Street, New York, NY 10011-4211, USA
10 Stamford Road, Oakleigh, Melbourne 3166, Australia

First published 1992
Reprinted 1993
First paperback edition with corrections 1995
Reprinted 1995, 1997

Printed in the United States of America

Typeset in Times

A catalogue record for this book is available from the British Library

Library of Congess Cataloguing-in-Publication Data is available

ISBN 0-521-42058-X hardback
ISBN 0-521-47739-5 paperback

CONTENTS

PREFACE

In the past three decades, the study of vortices and vortex motions – which originated in Helmholtz's great paper of 1858, 'Uber Integrale der hydrodynamischen Gleichungen welche den Wirbelbewegungen entsprechen' (translated by Tait [1867]), and continued in the brilliant work of Lord Kelvin and others in the nineteenth century, and Prandtl and his Göttingen school in the first half of this century – has received continuing impetus from problems arising in physics, engineering and mathematics. As aptly remarked by Küchemann [1965], vortices are the 'sinews and muscles of fluid motions'. Aerodynamic problems of stability, control, delta wing aerodynamics, high lift devices, the jumbo jet wake hazard phenomenon, among other concerns, have led to a myriad of studies. Smith [1986] reviews some of this work. The realisation that many problems involving interfacial motion can be cast in the form of vortex sheet dynamics has stimulated much interest. The discovery (rediscovery?) of coherent structures in turbulence has fostered the hope that the study of vortices will lead to models and an understanding of turbulent flow, thereby solving or at least making less mysterious one of the great unsolved problems of classical physics. Vortex dynamics is a natural paradigm for the field of chaotic motion and modern dynamical system theory. It is perhaps not well known that the father of modern dynamics and chaos wrote a monograph on vorticity (Poincaré [1893]). The theory of line vortices and vortex rings is a part of modern macroscopic treatments of liquid helium II, as is described by Donnelly and Roberts [1974]. Even Kelvin's [1867a] vortex theory of matter may one day have some topical interest.

Much of the modern work has been made possible by the remarkable advances in experimental techniques and the development of powerful computers, coupled with progress in numerical methods. Experiments and computations beyond the dreams of previous generations of students of vorticity are now routine, as we have tools with a power far greater than those available to the great scientists who founded the subject. Of course, this is not an unmixed blessing, as data and information are produced at a rate greater than the capability of the average scientist to absorb it. Indeed, the need for deep insight and analysis is more crucial than ever, in order to channel the energy available into productive research. As Kelvin [1880] remarked over 100 years ago, 'Crowds of exceedingly interesting cases present themselves'. We do not have the time, energy or resources for all members of this crowd, and the choice is overwhelming. Fortunately, analysis is still alive and helps to keep order, besides producing some of the important recent discoveries, such as the finite time vortex sheet singularity formation demonstrated by Moore [1979]. A powerful symbiosis between numerics and analysis has resulted in synergistic effects. Lamb's [1932 §159a] pessimistic remark, 'The motion of a solid in a liquid endowed with vorticity is a problem of considerable interest, but is unfortunately not very tractable', has lost much of its force. Vortex dynamics is a field of active research.

During the past twenty years, I have been teaching a course to graduate students in applied mathematics, aeronautics and engineering at Caltech on the fundamentals of vortex dynamics. An overview was published by Saffman and Baker [1979]. The present monograph is based on the notes of this course, and I am grateful to Prof Ari Glezer for providing me with a set of class notes more complete and in better order than those from which I lectured, and to Prof Hans Hornung who emphasised the pedagogical need for a text. In a field as well studied as vorticity, with a wealth of applications and a plethora of topics to investigate, it is difficult if not impossible to be both a scholar and an active researcher. If research is given priority over scholarship, then regretfully but inevitably the references to the corpus of work will be incomplete. This applies even more to papers and books not in the English language. In particular, there is outstanding research in the Soviet literature which is not properly acknowledged. A recent bibliography by Soviet workers (Akhmetov et al. [1988]) for the period 1975–87 contains 852 papers and is still not complete.

The emphasis in this monograph is on the classical theory of inviscid incompressible fluids containing finite regions of vorticity. Effects of viscosity, compressibility, inhomogeneity, stratification, and the like are enormously important in many fields of application, from hypersonic flight to global environmental fluid mechanics. But the discussion here focuses on those aspects of fluid motion which are primarily controlled by the vorticity and are such that the effects of the other fluid properties are secondary. It was my hope to present an up-to-date version of Chapter VII of Lamb's *Hydrodynamics* [1932], which is in my opinion (but not, I admit, in everybody's) unequalled in its combination of scholarship and research (although even Lamb omits some significant topics), but Lamb is not easily imitated or followed. There is no discussion here of the numerical methods for the computation of unsteady vortex flows such as the so-called vortex methods; this is more appropriately left to works on numerical analysis. Also omitted is any consideration of boundary layer theory, which can be and is treated as a strong interaction between the convection by external flow fields and its own self-induced velocity of the vorticity produced by viscous diffusion from a wall; boundary layer separation is then the highly non-trivial matter of determining where and under what conditions vorticity is convected away from the boundaries.

I should like to take this opportunity to thank the individuals and agencies who have supported my research on vorticity during the past twenty years and are therefore to some extent responsible for this monograph. The concern of Milt Rogers of the Air Force Office of Scientific Research over the jumbo jet, wake-hazard problem was instrumental in awakening much of the interest in vorticity in the 1960s and in starting a research programme at Caltech. Subsequently, support was provided by Ralph Cooper and Bob Whitehead of the Office of Naval Research and Milt Rose and Don Austin of what is now the Department of Energy. I owe a debt to all the scientific colleagues who participated in arguments and discussions and suggested many ideas; I have tried to acknowledge specific contributions. In particular, I wish to acknowledge an extensive, fruitful collaboration during the past twenty years with Prof Derek Moore.

Equations are numbered consecutively starting with 1 in each section. If the equation being referred to is in a different section of the same chapter, the section number is given first in arabic numerals separated by a dot from the equation number. If it is in a different chapter, the chapter number is given first in arabic numerals, followed, after a dot, by the section number. Sections are referred to by number if in the same chapter, and prefixed by the chapter number in arabic numerals if in a different chapter.

1

FUNDAMENTAL PROPERTIES OF VORTICITY

1.1 Relation between velocity and vorticity

The motion of a fluid is described by a vector field $\mathbf{u}(\mathbf{x}, t)$.[1] The curl of the velocity is called the vorticity $\boldsymbol{\omega}(\mathbf{x}, t)$. In the various notations

$$\boldsymbol{\omega}(\mathbf{x}, t) = \omega_i \equiv \operatorname{curl} \mathbf{u} = \epsilon_{ijk} \frac{\partial u_k}{\partial x_j} = \left(\frac{\partial w}{\partial y} - \frac{\partial v}{\partial z}, \frac{\partial u}{\partial z} - \frac{\partial w}{\partial x}, \frac{\partial v}{\partial x} - \frac{\partial u}{\partial y} \right)$$

$$= (\xi, \eta, \zeta). \tag{1}$$

It follows from the definition that the vorticity is solenoidal, that is,

$$\operatorname{div} \boldsymbol{\omega} = \frac{\partial \omega_i}{\partial x_i} = \frac{\partial \xi}{\partial x} + \frac{\partial \eta}{\partial y} + \frac{\partial \zeta}{\partial z} = 0. \tag{2}$$

The significance and importance of vorticity for the description and understanding of fluid flow stems from the facts, first, that (1) may be inverted to give the velocity field as an integral over the vorticity field, and, second, that when the viscous diffusion of vorticity is negligible, the fluid is barotropic (i.e., the density ρ is a single-valued function of the pressure p) and the external forces are conservative, then the vorticity satisfies conservation principles known as the Helmholtz laws (see §5), which allow the vorticity to be 'followed'. In particular, vorticity is not created and a compact distribution of vorticity remains compact,[2] so that the structure

---

[1] It is convenient to use various notations. The velocity will be denoted by a vector $\mathbf{u}$, or by Cartesian components (u, v, w), or by tensor components (u_1, u_2, u_3), or by unit vector decomposition $u\mathbf{i} + v\mathbf{j} + w\mathbf{k}$, according to circumstances. The position vector $\mathbf{x}$ will be denoted by (x, y, z), or $\mathbf{x} = (x_1, x_2, x_3)$ or $\mathbf{r} = (r_1, r_2, r_3)$, and similarly for other vectors. The summation convention will be implied when tensor notation is used.

[2] We do not use 'compact' in its pure mathematical sense, but in the ordinary sense of being localised to a finite region, outside of which the vorticity is zero or at most exponentially small.

and evolution of the fluid flow is more economically described in terms of the vorticity field than in terms of the velocity field.

We need to know conditions for which the inversion is possible and determines **u** uniquely. Sufficient but not necessary are the following six statements:

(i) The velocity field is solenoidal, that is,

$$\text{div } \mathbf{u} = 0. \tag{3}$$

(This condition is satisfied if the fluid is incompressible, that is, the material derivative $D\rho/Dt \equiv \partial\rho/\partial t + \mathbf{u} \cdot \nabla\rho = 0$; it is not necessary that the fluid be homogeneous, that is, $\rho =$ constant.)

(ii) The region occupied by the fluid is singly connected.

(iii) The normal component of fluid velocity, U_n, is given on all bounding surfaces S.

(iv) The velocity vanishes at infinity when the fluid is unbounded.

(v) The normal component of vorticity vanishes on S.

(vi) The vorticity field is compact.

The velocity is then given uniquely by the sum of a solenoidal vector potential component and an irrotational scalar component

$$\mathbf{u}(\mathbf{x}, t) = \mathbf{u}_v(\mathbf{x}, t) + \nabla\Phi \tag{4}$$

where

$$\mathbf{u}_v = \frac{1}{4\pi} \int \frac{\omega(\mathbf{x}', t) \times (\mathbf{x} - \mathbf{x}')}{|\mathbf{x} - \mathbf{x}'|^3} \, d\mathbf{x}' = -\frac{1}{4\pi} \int \omega(\mathbf{x}', t) \times \nabla \frac{1}{|\mathbf{x}' - \mathbf{x}|} d\mathbf{x}', \tag{5}$$

and $\nabla\Phi$ is determined uniquely (e.g., see Lamb [1932 §35]) by the classical potential problem

$$\nabla^2\Phi = 0, \quad \frac{\partial\Phi}{\partial n} = U_n - \mathbf{u}_v \cdot \mathbf{n} \quad \text{on } S, \quad \Phi \to 0 \text{ as } \mathbf{x} \to \infty. \tag{6}$$

In (4), $\mathbf{u}_v$ is a solenoidal field satisfying

$$\text{curl } \mathbf{u}_v = \omega, \tag{7}$$

and $\nabla\Phi$ is the irrotational, solenoidal field which can be added to satisfy a single boundary condition on the velocity field on S. If there are no bounding surfaces, $\Phi = 0$.

The proof of (4) is a standard exercise in vector analysis. A brief heuristic derivation is given here in order to demonstrate how the six conditions enter. For brevity, we write $\omega' = \omega(\mathbf{x}', t)$, $r = |\mathbf{x} - \mathbf{x}'|$, $\partial/\partial\mathbf{x} = \nabla$,

$\partial/\partial x' = \nabla'$, and so on. The following two vector identities are used repeatedly;

$$\text{div}(\mathbf{a} \times \mathbf{b}) = \mathbf{b} \cdot \text{curl}\,\mathbf{a} - \mathbf{a} \cdot \text{curl}\,\mathbf{b}, \tag{8}$$

$$\text{curl}(\mathbf{a} \times \mathbf{b}) = \mathbf{a}\,\text{div}\,\mathbf{b} - \mathbf{b}\,\text{div}\,\mathbf{a} - (\mathbf{a} \cdot \nabla)\mathbf{b} + (\mathbf{b}.\,\nabla)\mathbf{a}. \tag{9}$$

The following results of generalized function theory are also applied:

$$\text{curl grad}(1/r) = 0, \tag{10}$$

$$\nabla^2(1/r) = -4\pi\,\delta(\mathbf{x} - \mathbf{x}'). \tag{11}$$

First we consider $\mathbf{u}_v$. Taking the divergence of (5) and using (8) and (10), we have

$$\text{div}\,\mathbf{u}_v = -\frac{1}{4\pi}\int \text{div}(\boldsymbol{\omega}' \times \nabla(1/r))\,d\mathbf{x}' = \frac{1}{4\pi}\int \boldsymbol{\omega}' \cdot \text{curl grad}(1/r)\,d\mathbf{x}' = 0. \tag{12}$$

Taking the curl of (5), and applying (9), we have

$$\text{curl}\,\mathbf{u}_v = -1/4\pi \int \boldsymbol{\omega}' \nabla^2(1/r)\,d\mathbf{x}' + 1/4\pi \int (\boldsymbol{\omega}' \cdot \nabla)\nabla(1/r)\,d\mathbf{x}'$$
$$= \boldsymbol{\omega} - 1/4\pi \int (\boldsymbol{\omega}' \cdot \nabla')\nabla(1/r)\,d\mathbf{x}', \tag{13}$$

on using (11), the property of the delta function ($\int f(x)\delta(\mathbf{x})d\mathbf{x} = f(0)$), and $\nabla(1/r) = -\nabla'(1/r)$. Since $\text{div}'\,\boldsymbol{\omega}' = 0$, we can rewrite the last term in (13) as

$$(\boldsymbol{\omega} \cdot \nabla')\nabla(1/r) = \nabla' \cdot (\boldsymbol{\omega}'\nabla(1/r)),$$

and then from the divergence theorem,[3]

$$\int (\boldsymbol{\omega}' \cdot \nabla')\nabla(\frac{1}{r})\,d\mathbf{x}' = -\int (\boldsymbol{\omega}' \cdot \mathbf{n})\nabla(\frac{1}{r})\,dS' = 0 \tag{14}$$

by virtue of conditions (v) and (vi). Hence (7) follows. Conditions (i), (ii), (iii) and (iv) then specify a unique solenoidal irrotational velocity field $\nabla\Phi$ given by (6), which ensures that $\mathbf{u}$ is uniquely determined by $\boldsymbol{\omega}$. Note that only one boundary condition on $\mathbf{u}$ may be applied on S. Thus, in general, a no-slip boundary condition cannot be applied.

An alternative procedure employs the vector potential $\mathbf{A}(\mathbf{x}, t)$, which exists when (3) is satisfied, defined by

$$\mathbf{u}_v = \text{curl}\,\mathbf{A}. \tag{15}$$

---

[3] We shall apply the convention that $\mathbf{n}$ is the unit normal vector on a bounding surface S directed out of the fluid.

To determine $\mathbf{A}$, take the curl of (15), giving

$$\omega = \operatorname{curl} \operatorname{curl} \mathbf{A} = \operatorname{grad} \operatorname{div} \mathbf{A} - \nabla^2 \mathbf{A}. \tag{16}$$

The vector potential is arbitrary to the addition of the gradient of a scalar. Choose the scalar (i.e., fix the gauge of $\mathbf{A}$) by requiring that $\operatorname{div} \mathbf{A} = 0$. Then (16) is the Poisson equation

$$\nabla^2 \mathbf{A} = -\omega, \tag{17}$$

with solution

$$\mathbf{A} = \frac{1}{4\pi} \int \frac{\omega'}{r} \, d\mathbf{x}'. \tag{18}$$

It is now necessary to verify that this $\mathbf{A}$ is solenoidal. By arguments similar to those employed above (see also Batchelor[1967 §2.4.]), we find that

$$\operatorname{div} \mathbf{A} = \frac{1}{4\pi} \int \omega'.\nabla(\frac{1}{r}) \, d\mathbf{x}' = \frac{1}{4\pi} \int \frac{\omega' \cdot \mathbf{n}'}{r} \, dS' = 0,$$

when conditions (v) and (vi) are satisfied. Taking the curl of (18), we obtain (5). An alternative form is

$$\mathbf{u}_v = \frac{1}{4\pi} \int \frac{1}{r} \operatorname{curl} \omega' \, dV' - \frac{1}{4\pi} \int \frac{\omega' \times \mathbf{n}'}{r} \, dS'. \tag{19}$$

Instead of using the Green's function (18), the inversion can be carried out using only solutions of the scalar Laplace and Poisson equations with boundary conditions on S. There is no unique recipe, and the best procedure depends upon the application. For example, take $\mathbf{B}$ as any solution of the three scalar equations for its Cartesian co-ordinates,

$$\nabla^2 \mathbf{B} = -\omega. \tag{20}$$

Note that $\operatorname{div} \mathbf{B}$ is harmonic because of (2), but not necessarily zero. Let g denote the value of $\operatorname{div} \mathbf{B}$ on S. Now solve the equation for the Cartesian components of $\mathbf{C}$, defined by

$$\nabla^2 \mathbf{C} = 0, \quad \operatorname{div} \mathbf{C} = -g \quad \text{on } S. \tag{21}$$

This equation is also underdetermined and can be solved in an infinity of ways. For example, one can take $C_2 = C_3 = 0$, and $\partial C_1/\partial x = -g$ on S. Let f denote the solution of

$$\nabla^2 f = 0, \quad f = -g \quad \text{on } S. \tag{22}$$

Then

$$C_1 = \int_0^x f(\xi, y, z)d\xi + h(y, z), \quad \nabla^2 h = -\frac{\partial f}{\partial x}(0, y, z) \tag{23}$$

is a solution of (21), since $\partial C_1/\partial x = f$ and $\nabla^2 C_1 = \nabla^2 h + \partial f(0, y, z)/\partial x$.

Regardless of how **B** and **C** are determined in detail, $\mathbf{A} = \mathbf{B} + \mathbf{C}$ satisfies (17). Further, div **A** is harmonic and vanishes on S by construction. Hence, div $\mathbf{A} = 0$ inside S, and **A** is a vector potential.

As a simple example, consider the case when $\omega = (1, 0, 0)$ inside a sphere of unit radius. Then

$$\mathbf{B} = -\frac{x^2 + y^2 + z^2}{6}\mathbf{i}$$

is a solution of (20), where **i** is the unit vector in the x-direction. Then $g = -x/3$, and $f = x/3$ is the solution of (22). We find from (23) that $h = -\frac{1}{12}(y^2 + z^2)$, and

$$\mathbf{C} = \left(\frac{x^2}{6} - \frac{1}{12}(y^2 + z^2)\right)\mathbf{i}.$$

Thus

$$\mathbf{A} = -\frac{1}{4}(y^2 + z^2)\mathbf{i}$$

is solenoidal and satisfies (17). The corresponding velocity field is

$$\mathbf{u}_v = (0, -\tfrac{1}{2}z, \tfrac{1}{2}y).$$

Hirasaki and Hellums [1970] consider a three-dimensional rectangular geometry and use as boundary conditions on the vector potential **A** that satisfies (17)

$$\frac{\partial}{\partial n}(\mathbf{n} \cdot \mathbf{A}) = 0, \quad \mathbf{n} \times \mathbf{A} = 0, \tag{24}$$

which leads to three scalar Poisson equations with part Dirichlet and part Neumann conditions.

An arbitrary value U_n for the normal velocity is satisfied as before by adding the gradient of a harmonic function ϕ to curl **A** such that $\partial\phi/\partial n = U_n - \mathbf{n} \cdot \text{curl} A$.

Note that with direct inversion, which does not use the free-space Green's function (5) or (18), it is not necessary to require that conditions (v) and/or (vi) are satisfied. We describe later (§2.4) how to extend the vorticity field so that the Green's function can be used when (v) is not satisfied.

1.2 Vorticity and rotation

A kinematical interpretation of vorticity is provided by analysing the relative motion near a point. The relative velocity δu_i of two fluid particles separated by δx_i can be written

$$\delta u_i = \partial u_i/\partial x_j \, \delta x_j = e_{ij}\delta x_j + \Omega_{ij}\delta x_j, \tag{1}$$

where

$$e_{ij} = \tfrac{1}{2}(\partial u_i/\partial x_j + \partial u_j/\partial x_i) \tag{2}$$

and

$$\Omega_{ij} = \tfrac{1}{2}(\partial u_i/\partial x_j - \partial u_j/\partial x_i) = -\tfrac{1}{2}\epsilon_{ijk}\omega_k \tag{3}$$

are the rate of strain tensor and 'vorticity tensor', respectively. Note the converse relation

$$\omega_i = -\epsilon_{ijk}\Omega_{jk}. \tag{4}$$

The two contributions to the right-hand side of (1) constitute a pure straining motion and rigid body rotation. In the pure straining motion, line elements are extended or contracted, and spheres are deformed into quadrics with principal axes along those of the rate of strain tensor. In the rigid body rotation, line elements stay of constant length and spheres remain spheres while rotating with angular velocity

$$\Omega = \tfrac{1}{2}\omega. \tag{5}$$

Regions of fluid in which the vorticity is identically zero are said to be in irrotational motion. Solid particles do in general rotate when suspended in fluid in irrotational motion, even if the fluid is assumed to be ideal (i.e., zero viscosity). This is because a body of general shape is subject to a torque when suspended in irrotational flow of an ideal fluid (Landau and Lifshitz [1959 §11]). The particular feature of irrotational flow is that the torque vanishes for non-spherical bodies which have rotational symmetry about three perpendicular axes[4] (e.g., cubes or regular polyhedra), but does

---

[4] The same is true for microscopic particles in a real fluid. An irrotational solution of the Euler equation for a uniform fluid is also a solution of the Navier–Stokes equations, and bodies of the appropriate symmetry will experience no torque when sufficiently small for the motion around them to be described by the creeping flow equations. Consider, for example, a dumb-bell with one end at the origin and the other at the point (x, y) in the two-dimensional plane strain $u = \alpha x$, $v = -\alpha y$. Assuming that the force on the ends is proportional to the relative velocity of ball and fluid, there will be a torque $\propto 2\alpha xy$ tending to rotate the body. If the body consists of two equal

not vanish in general when such bodies are suspended in rotational flow, that is, $\omega \neq 0$. In a two-dimensional irrotational flow, a square or cross will not rotate when carried along with the flow, but a rod or diagonal cross or ellipse will.

The vorticity is twice the average angular velocity around an infinitesimal circle since

$$\frac{1}{2\pi l} \oint \frac{\mathbf{u}}{l} \cdot d\mathbf{s} = \frac{1}{2\pi l^2} \int \omega \cdot d\mathbf{A} \quad \text{by Stokes's theorem} \tag{6}$$

$$\to \tfrac{1}{2} \omega \quad \text{as} \quad l \to 0.$$

A dynamical interpretation in terms of angular momentum density follows from the calculation of the angular momentum about its centroid of an infinitesimal fluid particle. Suppose the fluid particle has volume δV and centroid at $\mathbf{x}$. Its angular momentum $\delta \mathbf{A}$ about its centroid is

$$\delta A_i = \rho \int_{\partial V} \epsilon_{ijk} \, \delta x_j \, \delta u_k \, dV. \tag{7}$$

Substituting (1) for δu_k, we obtain

$$\delta A_i = \epsilon_{ijk} \, e_{kl} \, I_{jl} + \tfrac{1}{2} (\delta_{ij} I_{kh} - I_{ij}) \omega_j, \tag{8}$$

where

$$I_{ij} = \rho \int_{\partial V} \delta x_i \, \delta x_j \, dV \tag{9}$$

is the inertia tensor of the fluid particle. The first term on the right-hand side of (8) vanishes if the body has spherical symmetry, as is the case for cubes, in which case the inertia tensor is diagonal. The second term is the angular momentum of the fluid particle rotating as a solid body with angular velocity $\tfrac{1}{2} \omega$. Vorticity can therefore be identified as proportional to the angular momentum of fluid particles whose inertia tensors have spherical symmetry.[5]

---

dumbbells joined at right angles with ends at (x, y) and (x', y'), the sum of the torques is proportional to $2\alpha(xy + x'y') = 0$.

[5] It can be speculated that a gyroscopic interpretation can be provided for vorticity and that the motion of fluids endowed with vorticity can be better understood by examining gyroscopic motion. For example, Coles [1967] pointed out that the Rayleigh criterion for the stability of flow between rotating cylinders is equivalent to the criterion that vorticity and angular rotation (precession) are in the same sense, which is the condition for stability of a spinning top.

1.3 Circulation

A scalar functional of considerable importance in the description of vortex flows is the circulation Γ around a simple closed curve C, defined as the line integral of the velocity.

$$\Gamma = \oint_C \mathbf{u} \cdot d\mathbf{s}. \tag{1}$$

We shall say that a curve is reducible if it can be shrunk continuously to a point without going outside the fluid. It follows from Stokes's theorem that the circulation around a reducible curve is equal to the flux of vorticity through an open surface A bounded by the curve, that is,

$$\int_A \omega \cdot \mathbf{n} \, dS = \oint_C \mathbf{u} \cdot d\mathbf{s} = \Gamma. \tag{2}$$

Some convention is required for the sense in which the circulation is taken and the direction of the normal to the surface. We shall suppose that the relation is that described by a right-hand-threaded screw.

The vanishing of the circulation for all closed curves implies that the vorticity is zero and the flow is irrotational. The converse is true if the fluid is contained in a simply connected region, but may be false if the region is multiconnected. Irrotational flow about a torus may have non-zero circulation around curves which thread the hole.

The circulation is important for its conservation principles (Kelvin's circulation theorem §6), its relation to the forces on bodies (the Kutta lift §3.1) and its use in vortex sheet dynamics as a variable for the parametrization of the sheet shape (the Birkhoff–Rott equation §8.1).

1.4 Vortex lines and tubes

In a region of fluid where the vorticity does not vanish identically, curves drawn parallel to the vorticity vector at each point of the curve are known as vortex lines. They are the solution families of the differential equations

$$\frac{dx}{\xi} = \frac{dy}{\eta} = \frac{dz}{\zeta}. \tag{1}$$

The vortex lines passing through the points of a reducible curve define a cylindrical-like volume called a vortex tube. This term commonly implies

that the cross-sections are small and oval, or even infinitesimal. The tube
has the property that the vorticity is everywhere parallel to its surface, that
is, $\omega \cdot \mathbf{n} = 0$ on the surface of a vortex tube. It follows that the flux of
vorticity through any cross-section of the tube is constant, since if A_1 and
A_2 are two cross-sections, the divergence theorem gives

$$\int_{A_1} \omega \cdot \mathbf{n} \, dS - \int_{A_2} \omega \cdot \mathbf{n} \, dS = \int_{A_1 - A_2} \text{div} \, \omega \, dV = 0. \tag{2}$$

From the relation (3.2) between flux and circulation, the flux of vorticity
along the tube is equal to the circulation round any closed curve on the
tube wall which encloses the tube once. This quantity is called the strength
of the tube and is the most natural measure of its intensity.

Because the strength of the tube does not vary with position along the
tube, it follows that vortex tubes are either closed, go to infinity or end
on solid boundaries.[6] If the vorticity field is compact, the tubes must be
closed or begin and end on boundaries. It is sometimes stated erroneously
that vortex lines cannot begin or end in the fluid and therefore either form
closed curves or begin and end on boundaries (e.g., Lamb [1932 §145]). In
general, a vortex line is infinitely long and passes infinitely often infinitely
close to itself even if the field is compact. However, symmetry of the flow
field, which is often the case for flows studied analytically, may produce
closed vortex lines.[7]

The vorticity field can vanish at an isolated point (a vortex null) without contradicting the statement that vortex lines do not end in the fluid.
Examination of the line structure for the case $\omega_i = \alpha_{ij} x_j$, $\alpha_{ii} = 0$, shows
that lines either cross or are closed ovals about the isolated point. In the
former case, we define the direction at the isolated point as the limit as
the point is approached.

In a real fluid satisfying the no-slip boundary condition, vortex lines
must be tangential to the surface of a body at rest, except at isolated points
of attachment and separation, because the normal component of vorticity
vanishes (since the circulation is zero for any circuit on the body). In this
case, vortex tubes cannot end on the body and must either be closed or go
to infinity at both ends.[8] Vortex tubes can end on the surface of a rotating
body. In this case, constraint (v) of §1 is violated; see §2.4 for a discussion

[6] The solenoidal structure of the vorticity field also precludes the tube ending on a
singularity in the fluid.

[7] See Truesdell [1954] for a thorough discussion of the geometry of vortex lines.

[8] See Lighthill [1963 Chap. II] for a discussion of the shape of vortex lines in a
three-dimensional boundary layer.

of how the vorticity field can be extended to handle this difficulty (see also Batchelor [1967 §2.4]).

When the vortex tube is immediately surrounded by irrotational fluid, it will be referred to as a vortex filament.[9] Since there is no vorticity outside the filament, it is not necessary to restrict the calculation of the filament strength to curves lying upon its walls; any curve threaded by the filament can be used.

A vortex filament is often just called a vortex, but we shall use this term to denote any finite volume of vorticity immersed in irrotational fluid (with obvious modifications when the flow is two-dimensional). Of course, the vortex filament and vortex require that the fluid is ideal to make strict sense, because viscosity diffuses vorticity, but they are useful approximations for real fluids of small viscosity.

1.5 The laws of vortex motion

Three laws of vortex motion were given by Helmholtz [1858]. For the motion of an ideal barotropic fluid under the action of conservative external body forces,[10] they can be expressed as follows:

  I. Fluid particles originally free of vorticity remain free of vorticity.
 II. Fluid particles on a vortex line at any instant will be on a vortex line at all subsequent times. Alternatively, it can be said that vortex lines and tubes move with the fluid.
III. The strength of a vortex tube does not vary with time during the motion of the fluid.

The first law implies Lagrange's theorem on the persistence of irrotational motion. Alternative formulations can be given.[11] There are several ways of proving these laws. We shall start with the Euler equations for the conservation of mass and momentum in an ideal fluid moving in the presence of an external force $\mathbf{F}$ per unit mass.

$$\rho_t + \mathbf{u} \cdot \nabla \rho + \rho \, \mathrm{div} \, \mathbf{u} = 0, \tag{1}$$

$$\mathbf{u}_t + \mathbf{u} \cdot \nabla \mathbf{u} = -(1/\rho) \, \nabla p + \mathbf{F}. \tag{2}$$

---

[9] The usage of this expression is neither uniform nor consistent. It is often used to denote an infinitesimal vortex tube. Helmholtz [1858] used it in both senses.

[10] External body forces $\mathbf{F}$ per unit mass are conservative when $\mathbf{F}$ is the gradient of a single-valued scalar, that is, $\mathrm{curl}\,\mathbf{F} = 0$ and $\oint \mathbf{F} \cdot d\mathbf{r} = 0$ for all closed, not necessarily reducible, curves in the fluid.

[11] For a historical discussion, see Lamb [1932 §146].

We take the curl of (2), employing (1.9) and the identity

$$\mathbf{u} \cdot \nabla \mathbf{u} = \operatorname{grad} \tfrac{1}{2} \mathbf{u}^2 - \mathbf{u} \times \omega, \tag{3}$$

obtaining a general evolution equation for the vorticity

$$\omega_t + \mathbf{u} \cdot \nabla \omega = \omega \cdot \nabla \mathbf{u} - \omega \operatorname{div} \mathbf{u} + (1/\rho^2)\nabla \rho \times \nabla p + \operatorname{curl} \mathbf{F}. \tag{4}$$

(This is referred to as Fridman's equation by Kochin, Kibel and Roze [1964].) In a barotropic fluid, the surfaces of constant density (isopycnics) are parallel to the surfaces of constant pressure (isobars) and $\nabla \rho \times \nabla p = 0$. The last term vanishes when the external forces are conservative, giving in this case Helmholtz's equation

$$\omega_t + \mathbf{u} \cdot \nabla \omega = \omega \cdot \nabla \mathbf{u} - \omega \operatorname{div} \mathbf{u}. \tag{5}$$

The noteworthy feature of this equation, responsible in part for the deep importance of vorticity, is the absence of the pressure and the dependence of the rate of change of the vorticity on the instantaneous local values of the velocity and vorticity fields alone. (In contrast, the pressure field for a uniform fluid of constant density is given by the Poisson equation $\nabla^2 p = -\rho \partial^2 (u_i u_j)/\partial x_i \partial x_j$ with not completely certain but probably Neumann boundary conditions.) Substituting from (1) for $\operatorname{div} \mathbf{u}$ in (5), we obtain

$$\frac{D}{Dt}\left(\frac{\omega}{\rho}\right) \equiv \left(\frac{\partial}{\partial t} + \mathbf{u} \cdot \nabla\right)\frac{\omega}{\rho} = \frac{\omega}{\rho} \cdot \nabla \mathbf{u}, \tag{6}$$

which expresses the rate of change of ω/ρ following a fluid particle in terms of its instantaneous value and the local velocity gradient.

The idea of a fluid particle and material line and the concept of moving with the fluid can be made precise as follows. A fluid particle or material point is a point lying on the trajectory $\mathbf{X}(t)$ satisfying

$$\frac{d\mathbf{X}}{dt} = \mathbf{u}(\mathbf{X}, t), \quad \mathbf{X}(0) = \mathbf{x}_0. \tag{7}$$

The description of the fluid motion contained in the set of path lines or trajectories $\mathbf{X}(t)$ is called a Lagrangian representation. The velocity of the fluid at time t and position $\mathbf{X}$ is $d\mathbf{X}/dt$, and is described as a function of t and the initial position $\mathbf{x}_0$, instead of in terms of the current position $\mathbf{X}$ as in the Eulerian representation. A material line is a curve $\mathbf{X}(t, \sigma)$ for constant t and varying σ, where σ is a Lagrangian particle specifying position along the original position of the line $\mathbf{x}_0(\sigma) = \mathbf{X}(t_0, \sigma)$. A streakline is the locus $0 < \tau < t$ of particles released from a fixed point $\mathbf{x}_0$ during the time interval t, and is obtained from the integration of

$$\frac{d\mathbf{X}}{dt}(t, \tau) = \mathbf{u}(\mathbf{X}, t), \quad \mathbf{X}(\tau, \tau) = \mathbf{x}_0. \tag{8}$$

In steady flow, streaklines and pathlines coincide, and are also streamlines (curves tangent to the instantaneous velocity).[12]

Let $\theta(\mathbf{x}, t)$ be a field variable. The value on a material line is denoted by

$$\theta_\sigma(t) = \theta(\mathbf{X}(t, \sigma), t). \tag{9}$$

The time derivative of θ moving with the fluid is

$$\frac{d\theta_\sigma}{dt} = \frac{\partial\theta}{\partial t} + \mathbf{u} \cdot \nabla\theta = \frac{D\theta}{Dt} \tag{10}$$

on using (7).[13] We can write (6) as

$$\frac{d}{dt}\left(\frac{\omega}{\rho}\right)_\sigma = \left(\frac{\omega}{\rho}\right) \cdot \nabla\mathbf{u}. \tag{11}$$

Equation (11) is an ordinary differential equation for $(\omega/\rho)_\sigma$, and from the standard theory of such equations it has a unique solution for some time interval if $\nabla\mathbf{u}$ is continuous. It follows from (1.5) that continuity of ω implies continuity of $\nabla\mathbf{u}$. Then $(\omega/\rho)_\sigma = 0$ for all time for which the solution exists if initially zero.[14] This is Helmholtz law I.

Consider now a line element $\delta\mathbf{s}_\sigma$ of a material line, which can be defined precisely as $\delta\mathbf{s}_\sigma = \epsilon\partial\mathbf{X}/\partial\sigma$ for some time-independent ϵ. Then

$$\frac{d}{dt}\delta\mathbf{s}_\sigma = \epsilon\frac{\partial}{\partial\sigma}\left(\frac{d\mathbf{X}}{dt}\right) = \epsilon\frac{\partial}{\partial\sigma}\mathbf{u}(\mathbf{X}, t) = \delta\mathbf{s}_\sigma \cdot \nabla\mathbf{u}. \tag{12}$$

Hence on a material line, we have from (11) and (12) that

$$\frac{d}{dt}\left(\delta\mathbf{s}_\sigma - \epsilon\frac{\omega}{\rho}\right) = \left(\delta\mathbf{s}_\sigma - \epsilon\frac{\omega}{\rho}\right) \cdot \nabla\mathbf{u}. \tag{13}$$

[12] In unsteady flow, great care must be exercised in the interpretation of streakline patterns from experiments employing flow visualization techniques. They may show apparent roll-up into vortices where no concentration of vorticity has occurred (Hama [1962]).

[13] $\theta(\mathbf{x}, t)$ is the Eulerian representation, and $\theta_\sigma(t)$ is the Lagrangian representation. To keep distinct the two representations, we shall use d/dt for time differentiation in the Lagrangian representation, even if it is strictly a partial derivative with σ or τ kept constant, and use D/Dt as shorthand for the derivative following the flow in the Eulerian representation.

[14] The existence of solutions of the Euler equations with smooth initial data for all time is currently an open question. See Kreiss and Lorenz [1989] for a recent text.

Suppose now that at the initial instant t_0, the material line coincided with a vortex line, so that $\delta\mathbf{s}_\sigma = \epsilon(\omega/\rho)_\sigma$ when $t = t_0$ for some $\epsilon(\sigma)$. Then again from the existence theory of solutions of ordinary differential equations, (13) has the unique solution

$$\delta\mathbf{s}_\sigma = \epsilon(\omega/\rho)_\sigma \tag{14}$$

while the solution exists and $\nabla\mathbf{u}$ is continuous. Hence the material line remains a vortex line and in this sense the vortex lines are said to move with the fluid.[15] This is Helmholtz law II.

Equation (11) can be written in terms of the rate of strain tensor e_{ij} given by (2.2),

$$\frac{d}{dt}\left(\frac{\omega_i}{\rho}\right)_\sigma = e_{ij}\frac{\omega_j}{\rho}, \tag{15}$$

since $\omega_j\,\Omega_{ij} = 0$. This shows that the rate of change of vorticity following a fluid particle is proportional to the local rate of strain tensor, and is not explicitly dependent on the local vorticity or rotation tensor. As discussed in §2, this does not mean that the vorticity vector does not rotate; part of the deformation by e_{ij} extends the vortex line and part rotates it (see also Lighthill [1963 §II.1.3]). (Separate consideration of rotation and extension may break up the vorticity field into non-solenoidal constituents.)

To obtain the third Helmholtz law, consider an infinitesimal vortex tube of cross-section δA_σ. Since the tube moves with the fluid, conservation of mass gives the differential expression

$$\rho_\sigma\,\delta s_\sigma\,\delta A_\sigma = \text{constant.} \tag{16}$$

Using (14) to substitute for δs_σ, we have

$$\omega_\sigma\,\delta A_\sigma = \text{constant.} \tag{17}$$

Hence the strength of an infinitesimal tube is constant and therefore also that of a tube of arbitrary size since it is composed of infinitesimal tubes.[16] Since $\delta A_\sigma \propto 1/\delta s_\sigma$ in an incompressible fluid, the result (17) implies that

[15] Lamb [1932 §146] deduces (14) by in effect integrating (13), giving Cauchy's equations (see §7). The present argument was given by Whitham [1963 §III.7]. See also Goldstein [1960 §4.2]. Helmholtz's argument was to assert that the vanishing of the left-hand side of (13) at some instant of time implied the vanishing at subsequent times. This fallacy was also contained in Lagrange's proof of the persistence of irrotational flow. A counter example occurs in the viscous diffusion of a vortex sheet. Further discussion and references can be found in these texts.

[16] An alternative argument which is based on Kelvin's circulation theorem and does not rely on infinitesimals is provided in §6.

vorticity is proportional to the length of the vortex line. A flow which stretches vortex lines therefore increases the vorticity.

The cases of planar two-dimensional flow and axisymmetric flow without swirl deserve special mention. In the first case, $w = 0$, and $\partial/\partial z = 0$. The streamlines and pathlines lie in planes perpendicular to the z-axis. Then

$$\boldsymbol{\omega} = (0, 0, \partial v/\partial x - \partial u/\partial y) = (0, 0, \omega), \tag{18}$$

and the vorticity is essentially a scalar, having only one non-zero component. Since the vorticity is perpendicular to the direction in which the flow changes, $\boldsymbol{\omega} \cdot \nabla \mathbf{u} = 0$ (the vortex lines are not stretched), and the Lagrangian form (11) of the Helmholtz equation becomes

$$\frac{d}{dt}\left(\frac{\omega}{\rho}\right) = 0. \tag{19}$$

That is, ω/ρ is conserved following a fluid particle. When $d\rho/dt = 0$, as happens when the fluid is incompressible or the density is constant, we have

$$\frac{\partial \omega}{\partial t} + \mathbf{u} \cdot \nabla \omega = 0, \tag{20}$$

and the vorticity is a conserved scalar.

In the case of axisymmetric flow without swirl, the streamlines lie in meridional planes through an axis and the vorticity is in the azimuthal direction. In cylindrical polar co-ordinates (r, θ, z), $\boldsymbol{\omega} = \omega_\theta \, \boldsymbol{\theta}$, $\mathbf{u} = u_r \mathbf{r} + u_z \mathbf{z}$ and (11) takes the form

$$\frac{d}{dt}\left(\frac{\omega_\theta}{\rho}\right) = \frac{\omega_\theta u_r}{\rho r}, \tag{21}$$

because $\boldsymbol{\omega} \cdot \nabla = (\omega_\theta/r) \, \partial/\partial\theta$ and $\partial\mathbf{u}/\partial\theta = u_r \boldsymbol{\theta}$. Since $dr/dt = u_r$, (21) can be written

$$\frac{d}{dt}\left(\frac{\omega_\theta}{\rho r}\right) = 0. \tag{22}$$

When the density is constant, this equation expresses the result that the vorticity is proportional to the length of the vortex line, which in axisymmetric flow is proportional to the distance r from the axis.

1.6   Kelvin's circulation theorem

The Helmholtz laws can also be approached through the conservation of circulation in an ideal barotropic fluid acted on by conservative forces.

Consider the circulation $\Gamma(t)$ around a closed material curve $C(t)$ lying entirely in the fluid. The curve need not be reducible. We write

$$\Gamma(t) = \oint_{C(t)} \mathbf{u} \cdot d\mathbf{s} = \oint \mathbf{u}_\sigma \cdot \frac{\partial \mathbf{s}_\sigma}{\partial \sigma} \, d\sigma, \tag{1}$$

where the last integral is between fixed (cyclic) values of σ. We now calculate the rate of change of this circulation as the curve moves with the fluid. We have

$$\frac{d\Gamma}{dt} = \oint \frac{d\mathbf{u}_\sigma}{dt} \cdot \frac{\partial \mathbf{s}_\sigma}{\partial \sigma} \, d\sigma + \oint \mathbf{u}_\sigma \cdot \frac{d}{dt} \left(\frac{\partial \mathbf{s}_\sigma}{\partial \sigma} \right) d\sigma. \tag{2}$$

Now $d\mathbf{u}_\sigma/dt = \partial \mathbf{u}/\partial t + \mathbf{u} \cdot \nabla \mathbf{u}$ is given by (5.2) and $d\{(\partial \mathbf{s}_\sigma/\partial \sigma) \, d\sigma\}/dt = d(d\mathbf{s}_\sigma/dt)$ is given by (5.12). Substituting, we have

$$\begin{aligned}
\frac{d\Gamma}{dt} &= \oint \left(-\frac{1}{\rho} \nabla p + \mathbf{F} \right) \cdot d\mathbf{s} + \oint (\mathbf{u} \cdot (d\mathbf{s} \cdot \nabla)\mathbf{u}) \\
&= \oint \left(-\frac{1}{\rho} \frac{\partial p}{\partial \sigma} + \frac{\partial}{\partial \sigma} \left(\tfrac{1}{2}\mathbf{u}^2 \right) \right) d\sigma + \oint \mathbf{F} \cdot d\mathbf{s}.
\end{aligned} \tag{3}$$

The contribution from the first integral in (3) vanishes when the fluid is barotropic, since p and $\mathbf{u}$ must be single-valued functions. If $\mathbf{F}$ is the gradient of a single valued potential V,

$$\oint \mathbf{F} \cdot d\mathbf{s} = \oint -\frac{\partial V}{\partial \sigma} \, d\sigma = 0. \tag{4}$$

Under these conditions, we have obtained Kelvin's circulation theorem that the circulation around a material circuit in an ideal fluid is an invariant of the motion. Note that it is assumed that the velocity $\mathbf{u}$ is continuous and that the circuit $C(t)$ is being deformed smoothly. Both conditions may be violated in flows of physical significance, leading to the creation of circulation (see §6.5).

To obtain the Helmholtz laws, we note that when the circuit is reducible, it follows from Stokes's theorem (3.2) that the flux of vorticity through the circuit is constant. Consider now the material surface which at some time t_0 is the surface bounding a vortex tube. The circulation around any circuit on the surface which does not go around the tube is initially zero. By the circulation theorem, it is always zero. Hence the normal to the surface is always perpendicular to the vorticity and the material surface remains a vortex tube during the motion. Further, the circulation around a circuit which goes around the tube remains constant, and hence the strength of the tube is constant. Since vortex lines are the intersection of vortex tubes, and vortex tubes are material surfaces, it follows that vortex lines are material lines. Moreover, if the vorticity in a simply connected region of fluid is initially zero, the flux of vorticity through any material surface is initially zero and remains zero, and hence the vorticity remains zero.

If the region occupied by the fluid is not singly connected, that is, there exist closed curves which cannot be shrunk continuously to a point, (3) shows that circulation may be created around a non-reducible circuit if the body force **F** is the gradient of a multivalued potential. This may happen, for example, in electrically conducting fluids set in motion by electromagnetic forces. However, the circulation around any two reconcileable circuits (i.e., which can be continuously deformed into one another without going out of the fluid) will be the same, since the difference between the two circulations can be expressed as the line integral of **F** around a closed reducible curve.

It may be noted that it is a corollary of Kelvin's circulation theorem that the vorticity distribution produced by an initial value problem for the incompressible Navier–Stokes equations, with an initial condition for which the vorticity is initially zero in a finite region, cannot be an analytical function of the time t unless the vorticity is everywhere zero. For if $\omega = 0$, then $\nabla^2 \mathbf{u} = -\operatorname{curl} \omega = 0$ when $\operatorname{div} \mathbf{u} = 0$. Hence for a circuit lying in this region, it follows that the viscous contribution to $d\Gamma/dt$ is zero, and that the higher time derivatives at the initial instant also vanish. Since analyticity then implies the vanishing of Γ and hence ω for at least a finite time, the vorticity cannot be analytic. An example of a solution of the Navier–Stokes equations in which the vorticity remains zero is provided by motion past a cylinder in a stream of velocity U when the wall of the cylinder is flexible and moves with speed $2U \sin \theta$ (with an obvious notation).

1.7   Cauchy's equations

The Helmholtz equation (5.6) or (5.11) can be integrated. We use the fact that a material element $\delta \mathbf{s}$ is related to its initial value $\delta \mathbf{s}_0$ at time t_0 through the Jacobian of the trajectory $\mathbf{X}(t)$ with respect to the original position $\mathbf{X}_0 = \mathbf{X}(t_0)$,

$$\delta \mathbf{s} = \frac{\partial(\mathbf{X})}{\partial(\mathbf{X}_0)} \, \delta \mathbf{s}_0. \tag{1}$$

Taking the material element to lie along a vortex line, it follows from (5.13) that the vorticity ω at $\mathbf{X}$ at time t is related to the vorticity ω_0 at

$\mathbf{X}_0$ at time t_0 by Cauchy's equations

$$\frac{\omega}{\rho} = \frac{\partial(\mathbf{X})}{\partial(\mathbf{X}_0)} \frac{\omega_0}{\rho_0}. \tag{2}$$

Direct verification of (2) is straightforward. Differentiating (2) with respect to time, we obtain, since $d\mathbf{X}/dt = \mathbf{u}$,

$$\frac{d}{dt}\left(\frac{\omega}{\rho}\right) = \frac{\partial(\mathbf{u})}{\partial(\mathbf{X}_0)} \frac{\omega_0}{\rho} = \frac{\partial(\mathbf{u})}{\partial(\mathbf{X})} \frac{\partial(\mathbf{X})}{\partial(\mathbf{X}_0)} \frac{\omega_0}{\rho_0} = \frac{\partial(\mathbf{u})}{\partial(\mathbf{X})} \frac{\omega}{\rho}, \tag{3}$$

which is (5.11).

An alternative derivation based on manipulations of the equations of motion is given by Lamb [1932 §146]. Goldstein [1960 §4.2] derives the result from Kelvin's circulation theorem.

1.8 Irrotational flow

When the vorticity is zero in any connected region, the flow is irrotational, and since curl $\mathbf{u} = 0$, there exists a scalar $\phi(x, y, z, t)$, called the velocity potential, such that

$$\mathbf{u} = \nabla\phi. \tag{1}$$

When the region is simply connected, the velocity potential is single valued since the circulation around any closed curve is zero, being equal to the flux of vorticity through a surface which lies entirely in the fluid and is bounded by the curve. But if the region is multiply connected, the circulation around irreducible circuits need not vanish and the velocity potential is then not single valued.

The velocity potential at a point P is given in terms of the velocity by the line integral

$$\phi(P) = \int_A^P \mathbf{u} \cdot d\mathbf{s}, \tag{2}$$

where the point A is arbitrary, reflecting the uncertainty of ϕ to the addition of an arbitrary function of time. Stokes's theorem and the vanishing of ω ensures that (2) is the same for all reconcileable paths from A to P; but the values of ϕ may differ by integral multiples of the circulation around non-reducible closed curves through P.[17]

---

[17] We shall always assume that we are dealing with regions in which the intuitive concepts of Euclidean geometry are satisfied. Reducible curves always bound open surfaces lying in the fluid, and the converse is the case for non-reducible curves.

The velocity potential in a multiply-connected region can be made single valued by the artifice of introducing barriers, or diaphragms, which are connected open surfaces with edges on the boundaries drawn across the 'holes' so that the region is simply connected. The velocity potential may then be discontinuous, that is, have a finite jump, across the barrier. But the velocity field is continuous and the value of the jump is therefore the same over the barrier. It is also constant in time, being the circulation around a closed curve which passes through the barrier once, if the conditions for Kelvin's circulation theorem are satisfied (see Lamb [1932 §§47–52], Batchelor [1967 §2.8]).

From the first Helmholtz law or the circulation theorem, it follows that a material volume of ideal barotropic fluid remains in irrotational motion if originally in such a state, provided that it is acted upon only by conservative forces. This will be the case for flow generated from rest by the motion of boundaries and accounts for the voluminous interest in irrotational motions. An additional reason is the fact that when the fluid is incompressible, $\operatorname{div} \mathbf{u} = 0$ and ϕ satisfies Laplace's equation

$$\nabla^2 \phi = 0, \tag{3}$$

for which an extensive analysis exists.

1.9　Bernoulli's equation

We suppose now that the fluid is ideal and the density ρ is constant. (To distinguish between cases with uniform or variable density, we shall put $\rho = 1$ when the density does not vary with position or time.) The equation of motion (5.2) can be written

$$\frac{\partial \mathbf{u}}{\partial t} = -\nabla \left(p + \tfrac{1}{2} \mathbf{u}^2 \right) + \mathbf{u} \times \boldsymbol{\omega} + \mathbf{F}. \tag{1}$$

Consider first steady motion when the external force per unit mass is conservative with a potential V. The streamlines are fixed and coincide with the pathlines. Taking the scalar product of (1) with $\mathbf{u}$, and writing $\mathbf{u} \cdot \nabla = q\, \partial/\partial s$, where $q = |\mathbf{u}|$ and s is distance along a streamline, we have

$$\frac{\partial}{\partial s} \left(p + \tfrac{1}{2} q^2 + V \right) = 0. \tag{2}$$

Bernoulli's equation follows

$$p + \tfrac{1}{2} q^2 + V = H, \tag{3}$$

where H is constant along each streamline but in general varies from streamline to streamline. The quantity H is called the Bernoulli constant. Substitution of (3) into (1) gives

$$\nabla H = \mathbf{u} \times \boldsymbol{\omega}, \tag{4}$$

which shows that H is constant throughout the region if and only if $\mathbf{u}$ and $\boldsymbol{\omega}$ are parallel, that is, streamlines and vortex lines coincide, or the flow is irrotational.

Flows with $\mathbf{u} \times \boldsymbol{\omega} = 0$, $\boldsymbol{\omega} \neq 0$, (i.e., $\mathbf{u} \parallel \boldsymbol{\omega}$) are called Beltrami flows (see Serrin [1959], Truesdell [1954]). If $\mathbf{u} \cdot \boldsymbol{\omega} = 0$, with $\boldsymbol{\omega} \neq 0$, the term 'complex lamellar' is used; special cases are two-dimensional planar flow and axisymmetric flow without swirl. The surfaces $H = $ constant are sometimes called Lamb surfaces.

In the engineering literature, p is often called the static pressure and $p + \frac{1}{2}q^2 = p_T$, say, is the total pressure. H/g is the total head, where g is the acceleration due to gravity.

Bernoulli's equation can be extended to unsteady flow if irrotational. In this case, (1) can be written

$$\nabla \frac{\partial \phi}{\partial t} = -\nabla \left(p + \tfrac{1}{2} q^2 + V \right), \tag{5}$$

and hence

$$\frac{\partial \phi}{\partial t} + p + \tfrac{1}{2} q^2 + V = F(t), \tag{6}$$

where $F(t)$ is an arbitrary function of time (usually specified by boundary conditions or convention, but may be an unknown depending upon the formulation). When the fluid is barotropic, p may be replaced by $\int dp/\rho$ in the Bernoulli equation (3) and its irrotational extension (6).

Another extension is possible to two-dimensional flow of incompressible fluid in regions of constant vorticity (Lamb [1932 §159a]). Let $\Psi(x,y,t)$ be the stream function, and let $\chi(x,y)$ be any function, independent of t, which satisfies $\nabla^2 \chi = -\omega$. Then $\Psi - \chi$ is harmonic and has a harmonic conjugate $\phi(x,y,t)$ such that $\phi + i(\Psi - \chi)$ is analytic. The velocity components are $(\phi_x + \chi_y, \phi_y - \chi_x)$. Then the Euler equations can be written in the region of constant vorticity

$$\nabla(\frac{\partial \phi}{\partial t} + \tfrac{1}{2} q^2 + \omega\Psi + p + V) = 0, \tag{7}$$

from which follows

$$\frac{\partial \phi}{\partial t} + p + \tfrac{1}{2} q^2 + \omega\Psi + V = F(t). \tag{8}$$

2

SINGULAR DISTRIBUTIONS OF VORTICITY

2.1 Vortex jumps

It has been supposed so far that the velocity and vorticity fields are continuous and have continuous derivatives. In a real viscous fluid, these assumptions are generally regarded as appropriate, and the velocity field is assumed analytic everywhere, except possibly at an initial instant or when the speed of boundaries changes in a non-analytic manner.[1] However, if we go to the the limit of vanishingly small viscosity and consider ideal fluids, to which the Helmholtz laws apply, the mathematics allows non-analytic behaviour and we cannot assert on physical grounds that only continuous fields should be considered. Velocity or vorticity fields with singularities or discontinuities are indeed of considerable importance. The singularities are, of course, not arbitrary and must be consistent with integral forms of the Euler equations or equivalently with the conservation of mass, momentum and energy. In particular, the dynamical constraint that pressure is continuous across a surface of discontinuity must be satisfied unless there are also singularities in the external force fields. We suppose in this chapter that the density is uniform and put it equal to unity unless explicitly stated otherwise. Also, external forces are supposed conservative unless non-conservative forces are explicitly introduced.

The simplest type of discontinuity is a simple jump in the tangential component of vorticity across a surface S,[2] at which the following conditions

---

[1] Consideration of the extent to which the assumptions of existence and smoothness have been proved is outside the scope of this work.

[2] Unless the contrary is explicitly stated, singularities will be confined to isolated smooth surfaces and lines, and the fields will be otherwise analytic with well-behaved 'smooth' limiting behaviour as the singularity is approached insofar as this is possible. Serrin [1959] refers to such discontinuities as singular surfaces of order one.

are satisfied between the limiting values, denoted by subscripts 1 and 2, on the two sides ($\mathbf{n}$ denotes the unit normal):

   (i) $\omega_1 \cdot \mathbf{n}_1 = \omega_2 \cdot \mathbf{n}_2$,
   (ii) $\mathbf{u}_1 = \mathbf{u}_2$,
   (iii) $p_1 = p_2$.

Condition (i) follows immediately from the kinematic requirement that $\operatorname{div} \omega = 0$. Condition (ii) is also kinematic. The vorticity is bounded across a simple jump, the strength of the jump being the difference in the limiting values of the vorticity on the two sides. The integral of the vorticity is continuous and hence the velocity field induced by a vorticity field with a simple jump is continuous. Condition (iii), which requires the absence of singular forces in the surface, such as surface tension, implies that the surface S_ω is a material surface and moves with the fluid.

To see this, note that the proof of the circulation theorem remains valid since it depends only on the continuity of pressure and velocity. If an infinitesimal material circuit were to cross the jump, its shape would change continuously because the velocity is continuous and the jump in vorticity would then produce a change in circulation, by Stokes's theorem, violating the circulation theorem. Therefore fluid cannot cross a vortex jump, which must be a material surface; that is,

$$\frac{DS_\omega}{Dt} = \frac{\partial S_\omega}{\partial t} + \mathbf{u} \cdot \nabla S_\omega = 0, \tag{1}$$

where $S_\omega(\mathbf{x}, t) = 0$ is the equation of the surface. Alternatively, substitute the local expansion

$$\omega = \sigma H(S_\omega) + \text{less singular terms}, \tag{2}$$

where σ is the magnitude of the vorticity jump and H is the Heaviside function, into the Helmholtz equation (1.5.5). This gives

$$\frac{\partial \omega}{\partial t} + (\mathbf{u} \cdot \nabla)\omega - (\omega \cdot \nabla)\mathbf{u} = H'\sigma \left(\frac{\partial S_\omega}{\partial t} + \mathbf{u} \cdot \nabla S_\omega \right) + \text{less singular terms}. \tag{3}$$

Setting the coefficients of $H'(S_\omega) = \delta(S_\omega)$ to zero gives (1).

Conversely, (1) and the continuity of velocity implies the continuity of pressure across the jump. This follows from (1.6.3) applied to a material circuit joining pairs of adjacent points, A, D and B, C lying on opposite sides of the surface. Since the velocity is continuous, and the surface is material, each pair stays together and the circulation along the open curve from A to B is equal to the circulation from D to C. Now

$$\frac{d}{dt} \int_A^B \mathbf{u} \cdot d\mathbf{s} = \left[\frac{1}{2} \mathbf{u}^2 - V - p \right]_A^B \tag{4}$$

from (1.6.3), and similarly with A, B replaced by D, C. Hence

$$p_B - p_A = p_C - p_D, \tag{5}$$

and the pressure is therefore continuous everywhere on the surface when made continuous at one place. Note that $(\boldsymbol{\omega} \cdot \nabla)\mathbf{u}$ is in general not continuous.

The boundary of a vortex is in general a vortex jump and is therefore also a material surface. Two examples of steady flows of ideal incompressible fluid with vortex jumps are now described.

The first is the uniform rectilinear vortex filament (Rankine vortex). We use cylindrical polar co-ordinates (r, θ, z). The z-component of vorticity inside the circle $r = a$ is equal to a constant ω_0, and the other components are zero. The flow is steady, with azimuthal velocity $(-\infty < z < \infty)$

$$u_\theta = \tfrac{1}{2}\omega_0 r, \quad r < a; \quad u_\theta = \tfrac{1}{2}\omega_0 a^2/r, \quad r > a. \tag{6}$$

The surface $r = a$ is a vortex jump. The strength of the filament is $\pi a^2 \omega_0$. the pressure obeys the equation

$$\partial p/\partial r = -u_\theta^2/r \tag{7}$$

and is

$$p = p_\infty - \omega_0^2(2a^2 - r^2)/8, \quad r < a; \quad p = p_\infty - \omega_0^2 a^4/8r^2, \quad r > a. \tag{8}$$

This solution can obviously be generalized to an arbitrary axially symmetric vorticity distribution $\omega(r)$, and a continuous axial velocity distribution $w(r)$ can also be added, giving a flow in which both the streamlines and the vortex lines are helices about the z-axis. For a general $\omega(r)$, the tangential velocity is given by[3]

$$u_\theta = \frac{1}{r} \int_0^r r\omega(r)\, dr, \tag{9}$$

which tends to $\tfrac{1}{2}r\omega(0)$ as $r \to 0$. The circulation $\Gamma(r)$ around a circle of radius r is given by

$$\Gamma(r) = 2\pi r u_\theta = 2\pi \int_0^r r\omega(r)\, dr. \tag{10}$$

---

[3] This velocity is given by the integral (1.1.5), even though condition (vi) of §1.1 is violated. But the same velocity holds if the vortex is bounded by planes perpendicular to the z-axis, and now (1.1.5) is incorrect because condition (v) is not satisfied.

The only restriction on $\omega(r)$ is that the strength of the filament, $\Gamma_\infty = \Gamma(\infty)$, is finite. The value of du_θ/dr at $r = 0$ is the value of ω on the axis. Two radii can conveniently be defined for the general distribution. The outer or vortex radius, r_o, is the value of r at which $\Gamma = \Gamma_\infty$ (or $\Gamma = 0.99\Gamma_\infty$, say, if the vorticity extends out to infinity and the vortex jump at r_o is replaced by a smooth tail). The inner or core radius r_1 is the value at which the tangential velocity attains its maximum v, say. When the distribution of vorticity is continuous inside the vortex, the tangent at r_1, where $\Gamma = \Gamma_1 = 2\pi r_1 v$, to the curve of $\Gamma(r)$ vs r passes through the origin. For the uniform vortex, $r_1 = r_o$. For 'real' vortices, the experimental measurements (e.g., Baker et al. [1974]) indicate that ω may be sharply peaked and consequently $r_1 \ll r_o$. The rectilinear vortex filament, either uniform or with arbitrary distribution of vorticity and axial velocity, is one of the basic building blocks for vortex models of real flows.

A second example is Hill's [1894][4] spherical vortex in which the vorticity is confined to the interior of a uniformly translating sphere of radius a, say. The vortex lines are circles about an axis passing through the centre of the sphere, and the streamlines lie in meridional planes. Outside the sphere, the flow is irrotational. We again use cylindrical polar co-ordinates (r, θ, z) with origin coinciding instantaneously with the centre of the sphere and consider the distribution of vorticity in which $\boldsymbol{\omega} = (0, \omega_\theta, 0)$, where for some constant A

$$\omega_\theta = Ar, \quad r^2 + z^2 < a^2; \qquad \omega_\theta = 0, \quad r^2 + z^2 > a^2. \tag{11}$$

The velocity field can be found from integration of (1.1.5), but it is more convenient to proceed via the calculation of the vector potential or equivalently, in the case of axisymmetric flow, the Stokes stream function. This is the scalar $\Psi(r, z, t)$ such that

$$u_z = (1/r)\partial\Psi/\partial r, \quad u_r = -(1/r)\partial\Psi/\partial z. \tag{12}$$

The streamlines are the curves $\Psi = \text{const}$. The vector potential is

$$\mathbf{A} = (0, \Psi/r, 0). \tag{13}$$

The equation for Ψ is

$$\omega_\theta = \partial u_r/\partial z - \partial u_z/\partial r$$
$$= -(1/r)(\partial^2\Psi/\partial z^2 + \partial^2\Psi/\partial r^2 - \partial\Psi/r\partial r)$$
$$= Ar \quad \text{for } r^2 + z^2 < a^2, \tag{14}$$

[4] Villat [1930] incorrectly claims that generalization to an ellipsoidal vortex is possible.

and zero otherwise. The surface $(r^2 + z^2)^{1/2} = a$ is a vortex jump, and continuity of velocity requires that Ψ and $\partial\Psi/\partial z$ be continuous there. By inspection,[5] a solution of (14) is

$$\Psi = \begin{cases} \dfrac{-A}{10}(r^4 + r^2z^2 + \tfrac{5}{3}r^2a^2), & r^2 + z^2 < a^2, \\[2ex] \dfrac{Ar^2a^5}{15(r^2 + z^2)^{3/2}}, & r^2 + z^2 > a^2. \end{cases} \tag{15}$$

It is easily verified that the irrotational motion for $r^2 + z^2 > a^2$ is also described by the velocity potential

$$\phi = -Aza^5/15(r^2 + z^2)^{3/2}, \tag{16}$$

which is that due to a rigid sphere moving through ideal fluid with velocity in the z-direction

$$U = 2Aa^2/15. \tag{17}$$

The velocity field obtained from (15) is for the instant when the vorticity is given by (11). The vorticity then evolves according to the Helmholtz equation (1.5.5). The remarkable feature of (11) is that this distribution of vorticity propagates with speed U without change of shape. Since $\partial\omega/\partial t = 0$ and

$$(\boldsymbol{\omega} \cdot \nabla)\mathbf{u} = (\omega_\theta/r)\,\partial(u_r\hat{\mathbf{r}} + u_z\hat{\mathbf{z}})/\partial\theta = \omega_\theta(u_r/r)\boldsymbol{\theta}, \tag{18}$$

$$(\mathbf{u} \cdot \nabla)\boldsymbol{\omega} = (u_r\,\partial/\partial r + u_z\,\partial/\partial z)\omega_\theta\boldsymbol{\theta} = u_r(\partial\omega_\theta/\partial r)\boldsymbol{\theta}, \tag{19}$$

it follows immediately that the Helmholtz equation is satisfied inside the sphere and is trivially satisfied outside. Even more simply, we note that $D(\omega_\theta/r)/Dt = 0$. It then suffices to demonstrate that the surface of the sphere is a material surface, which is immediate from (16) and (17).

The pressure can be found by using a frame of reference moving with the vortex in which the motion is steady and applying Bernoulli's equation. Outside the sphere, the Bernoulli constant H is a global constant and can be taken to be zero. Then the pressure is

$$p = p_\infty - \frac{2A^2a^7}{225}\left(\frac{r^2 - 2z^2}{(r^2 + z^2)^{3/2}} + \frac{a^3(r^2 + 4z^2)}{r(r^2 + z^2)^4}\right), \quad r^2 + z^2 < a^2. \tag{20}$$

Inside the sphere, the variation of H can be found from integrating (1.9.4), from which follows

$$H = A^2r^2(z^2 + r^2)/10. \tag{21}$$

---

[5] The results can be deduced by changing into spherical polar co-ordinates and separating variables.

The pressure is then found from (1.9.3) with $V = 0$ and

$$q^2 = \frac{A^2}{25} \left(z^2 r^2 + z^2 (z^2 + r^2)^2 + \frac{a^2 z^2}{9} \right). \tag{22}$$

There is no need to verify that the pressure is continuous across the surface of the sphere as this is automatically ensured by the continuity of velocity. For further details, see Hill [1894].[6]

The velocity field of the Hill spherical vortex satisfies the Navier–Stokes equations inside and outside the vortex, provided an additional pressure $2A\mu z$ is added inside the sphere, where μ is the viscosity, since $\nabla^2 \mathbf{u} = 2a\hat{\mathbf{z}}$. The kinematic condition of continuous velocity is satisfied across the boundary. However, the dynamical condition of continuous stress is not satisfied, since the normal and tangential stresses differ by terms of order $\mu U/a^2$, so the entire flow field is not an exact steady solution of the Navier–Stokes equations.

Moffatt and Moore [1978] have investigated the stability of the Hill's spherical vortex to axisymmetric disturbances. They show that the disturbance at the rear stagnation point can grow algebraically in time without bound. The vortex is thus linearly unstable in an L^∞ norm.[7]

2.2 Vortex sheets

Surfaces on which the vorticity is infinite are kinematically possible and do not necessarily violate dynamical principles. They give surfaces across which the tangential component of the velocity field is discontinuous, called vortex sheets. They can be introduced formally as surfaces of discontinuity or imagined to arise from limiting processes in which the vorticity and vortex lines are confined to a thin, surface-like region of thickness ϵ, and the limit is taken with $\epsilon \to 0$ and the magnitude of the vorticity, ω, going to infinity in such a way that $\epsilon\omega \to \kappa$, where κ is finite

---

[6] Synge and Lin [1943] employed a linear superposition of Hill's spherical vortices to model homogeneous turbulence. Moffatt [1969] generalized the flow field to include swirl about the axis, i.e., non-zero $u_\theta(r, z)$; see also Hicks [1899]. Moffatt [1986] describes a general method based on the analogy between the steady Euler equations and the equations of magnetostatic equilibrium in a perfectly conducting fluid to construct general classes of three-dimensional vortices which propagate without change of shape.

[7] Conservation of energy and momentum considerations can be employed to demonstrate the non-linear stability to axisymmetric disturbances in an L^1 and L^2 norm. This approach to global stability was pioneered by Kelvin [1875]. See also Benjamin [1976].

and is in general a function of position over the surface. Since vorticity is a vector, there is a direction associated with the strength. We can express the vorticity in a vortex sheet in the form

$$\omega = \kappa \, \delta(n),$$ (1)

where $\kappa \cdot \mathbf{n} = 0$, $\kappa = |\kappa|$, n is distance along the sheet normal $\mathbf{n}$ and $\delta(n)$ is the Dirac delta function. Sheet vortex lines are defined as curves parallel to κ on the sheet. The solenoidal property of vorticity gives

$$\operatorname{div} \kappa = 0,$$ (2)

where the divergence operator is two-dimensional on the surface. It is a consequence of (2) that sheet vortex lines are either closed or end on boundaries.[8]

It has been assumed implicitly that the vortex sheet separates regions of fluid in irrotational motion, so that the vortex lines are confined to the sheet. This restriction is not strictly necessary, and vortex sheets can be defined when there is smooth vorticity adjacent to the sheet. In this case, the vortex lines in the fluid could be parallel to the sheet, or they could intersect it. For example, a vortex sheet could be placed in a fluid in solid body rotation which is threaded uniformly by vortex lines, perpendicular to the axis of rotation. The dynamics of vortex sheets intersected by vortex lines form an interesting class of problems. Non-trivial changes are called for as (1) needs modifying and (2) fails. It is replaced by

$$\operatorname{div} \kappa = [\omega \cdot \mathbf{n}]_2^1,$$ (3)

where $[\ ]_2^1$ denotes the jump between side 1 and side 2, that is, the value on side 1 minus the value on side 2. See Figure 2.2-1 for a sketch of the orientation. Sheet vortex lines could start and end in the sheet. In fact, it is sometimes useful to consider cases in which the vortex sheets are bounded or edged by line vortices. Little work seems to have been done on the properties of vortex sheets in rotational flow, except by Kiya and Arie [1979], who have studied the Kelvin–Helmholtz instability of a vortex sheet in a uniform shear.

The velocity $\mathbf{u}_v$ induced by the vortex sheet follows from the integral expression (1.1.5). The substitution of (1) into this formula converts the volume integral into a surface integral over the sheet S_v. We obtain

$$\mathbf{u}_v(\mathbf{x}, t) = \frac{1}{4\pi} \int_{S_v} \kappa(\mathbf{S}, t) \times (\mathbf{x} - \mathbf{S}) / |\mathbf{x} - \mathbf{S}|^3 \, dS,$$ (4)

---

[8] The ability of vortex lines to twist around each other in three dimensions and produce space-filling curves of infinite length does not exist in two dimensions.

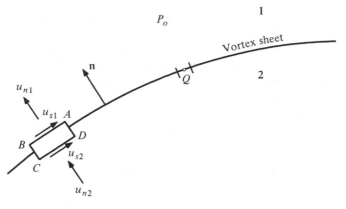

2.2-1 Sketch of co-ordinates for a vortex sheet.

where $\mathbf{S}$ is position vector on the sheet and dS is the element of surface area. The velocity defined by (4) is finite when the field point $\mathbf{x}$ is off the sheet. We now consider how it behaves as the point P with position vector $\mathbf{x}$ approaches a point Q on the sheet. We write (4) in the form

$$\mathbf{u}_v(P) = \frac{1}{4\pi}\, \kappa(Q) \times \int_{S_v} \mathbf{r}_{PS}/r_{PS}^3 \, dS$$

$$+ \frac{1}{4\pi} \int_{S_v} (\kappa(S) - \kappa(Q)) \times \mathbf{r}_{PS}/r_{PS}^3 \, dS. \tag{5}$$

By standard processes of analysis,[9] it can be shown that the second term on the right-hand side of (5) has a unique limit as $P \to Q$. The limit of the first term is finite but discontinuous, the value depending upon the side from which the sheet is approached:

$$\lim_{P_1 \to Q} \int_{S_v} \frac{\mathbf{r}_{PS}}{r_{PS}^3}\, dS - \lim_{P_2 \to Q} \int_{S_v} \frac{\mathbf{r}_{PS}}{r_{PS}^3}\, dS = 4\pi \mathbf{n}(Q), \tag{6}$$

where $\mathbf{n}$ is from side 2 into side 1. These results can be put together into the form

$$\lim_{P \to Q} \mathbf{u}_v(P) = \pm \tfrac{1}{2} \kappa(Q) \times \mathbf{n}(Q) + \mathbf{q}_v(Q), \tag{7}$$

where

$$\mathbf{q}_v(Q) = \frac{1}{4\pi} \oint \kappa(S) \times \frac{\mathbf{x}(Q) - \mathbf{x}(S)}{|\mathbf{x}(Q) - \mathbf{x}(S)|^3}\, dS \tag{8}$$

[9] Proofs of the statements made here are reasonably straightforward. One procedure is to prove the results for an infinite plane sheet and then show that the difference between a plane sheet and a curved sheet is a bounded integral.

is a principal value integral analogous to the Cauchy principal value, with the vorticity in an infinitesimal circle centred on Q excluded from the integral. The sign in (7) is positive if P approaches from the side into which $\mathbf{n}$ points. $\mathbf{q}_v$ is also the arithmetic mean of the two limits.

Thus the velocity produced by a vortex sheet is finite but has a simple jump in the tangential component on the sheet, the normal component being continuous. The jump $[\mathbf{u}]_2^1$ in the tangential velocity is

$$[\mathbf{u}]_2^1 = \kappa \times \mathbf{n}. \tag{9}$$

These results can also be established simply from considerations of an infinitesimal pill-box and rectangle, given that the velocities exist and are finite. See Figure 2.2-1 for an explanation of the notation. Conservation of mass through the sides of the pill-box implies that the normal velocities are continuous, that is, $u_{n_1} = u_{n_2}$. The circulation around the infinitesimal rectangle ABCD is $AB \times (u_{s_2} - u_{s_1})$, and this is equal to the flux of vorticity $\kappa AB \sin\theta$ threading the circuit, where θ is the angle between the plane of the circuit and the direction of the vorticity. It follows that

$$[\mathbf{u}]_2^1 = \kappa \times \mathbf{n}, \quad \text{or} \quad \kappa = \mathbf{n} \times [\mathbf{u}]_2^1. \tag{10}$$

The complete determination of the velocity field still requires the evaluation of the integral (8) and the addition of the contribution from other distributions of vorticity and boundary effects.

If the vortex sheet is not introduced as the singular distribution of vorticity (1), but as a surface discontinuity of velocity, then the strength is given by (10). In a real viscous fluid, vortex sheets must diffuse and broaden under the action of viscosity and can exist for at most an initial instant, but for fluids of small viscosity they provide a valuable idealization.

An alternative representation of the velocity induced by the sheet can be given in terms of an equivalent surface distribution of dipoles. Suppose the sheet separates regions of fluid in irrotational motion, and let ϕ_1 and ϕ_2 denote the velocity potentials on opposite sides of the surface at a point Q on the sheet. Then

$$\phi_2 - \phi_1 = \int^Q [\mathbf{u}]_1^2 \cdot d\mathbf{s} = \int^Q \mathbf{n} \times \kappa \cdot d\mathbf{s} \tag{11}$$

is independent of the path from some origin on the sheet to Q and hence single valued, provided all paths on the sheet are reducible. If this is not the case – for example, the sheet may be the surface of a torus – then barriers are to be introduced into the region to ensure that the paths are reducible, and these barriers (across which the velocity potential is discontinuous) must now be considered part of the sheet. The velocity potential at a point off the sheet is given by

$$\phi(\mathbf{x}, t) = -\frac{1}{4\pi} \int_{S_v} [\phi(\mathbf{x}')]_1^2 \, \mathbf{n}' \cdot \frac{\mathbf{x} - \mathbf{x}'}{|\mathbf{x} - \mathbf{x}'|^3} \, dS', \tag{12}$$

which is the potential due to a surface distribution of dipoles of strength density $\mathbf{n}(\phi_2 - \phi_1)/4\pi$. It can be shown that ϕ as defined by (12) has the jump $\phi_2 - \phi_1$ across the sheet and that $\partial\phi/\partial n$ is continuous, and hence the velocity field given by the gradient of (12) is that due to the vortex sheet. It can be shown directly that substituting (11) into (12), introducing barriers as appropriate and taking the gradient gives the velocity field (5). See also Lamb [1932 §58 §151].

So far the discussion has been entirely kinematic. We now turn to the dynamics of vortex sheets and their motion. Since vortex lines move with the fluid when the flow is inviscid and under the influence of conservative external forces, we expect that under the same circumstances vortex sheets move with the fluid and are composed of the same fluid particles. Thus, in particular, the velocity of the sheet normal to itself is equal to the velocity of the fluid. The result is true, but the argument is inadequate as it assumes that the proof of the Helmholtz laws commutes with the limiting process which gives the vortex sheet. Also, the proof of Kelvin's circulation theorem, which provides an alternative approach to the Helmholtz laws, depends upon the velocity being continuous and is incomplete in the presence of vortex sheets.

The existence of velocity jumps suggests that we employ an integrated form of the Euler equation (1.5.2). Integrating this equation over a fixed volume V, we obtain

$$\frac{\partial}{\partial t} \int_V \mathbf{u} \, dV = -\int_{\partial V} [p \, d\mathbf{S} + \mathbf{u} (\mathbf{u} \cdot d\mathbf{S})] + \int_V \mathbf{F} \, dV. \tag{13}$$

Take as the volume the infinitesimal pill-box of Figure 2.2-1 and suppose that the sheet moves with velocity $\mathbf{v}$. The left-hand side of (13) contributes $(\mathbf{u}_2 - \mathbf{u}_1) \mathbf{v} \cdot \mathbf{n}$ per unit area of sheet surface. The first term on the right-hand side contributes $(p_2 - p_1) \mathbf{n} + (\mathbf{u}_2 - \mathbf{u}_1) \mathbf{u}_1 \cdot \mathbf{n}$. Contributions from the edge and volume of the pill-box are negligible unless the force density in the sheet is singular. Neglecting this possibility for the present, we have, on equating components of (13) parallel to $\mathbf{n}$ and $\mathbf{u}_2 - \mathbf{u}_1$, respectively,

$$p_2 = p_1, \quad \mathbf{v} \cdot \mathbf{n} = \mathbf{u}_1 \cdot \mathbf{n} = \mathbf{u}_2 \cdot \mathbf{n}, \tag{14}$$

showing that consistency with the dynamics expressed by the Euler equations requires that the pressure is continuous across the sheet and the sheet

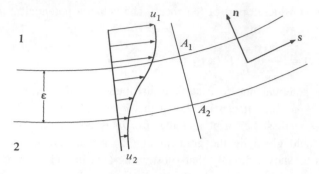

2.2-2 Local intrinsic co-ordinates for a vortex sheet.

moves with the fluid. Hence if $S_v(\mathbf{x}, t) = 0$ is the equation of the sheet,

$$\partial S_v/\partial t + \mathbf{u}_1 \cdot \nabla S_v = \partial S_v/\partial t + \mathbf{u}_2 \cdot \nabla S_v = 0. \tag{15}$$

Note that $\mathbf{u}_1 - \mathbf{u}_2$ and ∇S_v are perpendicular.

We can give a meaning to motion of the vortex lines in the sheet. Define a fluid particle on the sheet as a point on the sheet moving with velocity $\mathbf{q}_v + \nabla\Phi = \mathbf{q}$, where $\mathbf{q}_v$ is given by (8) and $\nabla\Phi$ is the external irrotational velocity field, and $q = |\mathbf{q}|$ is the average of the velocities on the two sides. We shall now show that sheet vortex lines move with these fluid particles. Consider first the integrated strength of the sheet, or circulation function, or relative potential jump, defined for two points A and B of the sheet by

$$\Gamma_{AB} = \int_B^A [\mathbf{u}]_1^2 \cdot d\mathbf{s}. \tag{16}$$

(Γ_{AB} may be multivalued if the sheet contains irreducible curves.) Now Γ_{AB} is also the flux of sheet vorticity across the line joining A and B, and is zero if they lie on the same sheet vortex line. The stated result therefore follows if we can show that

$$d\Gamma_{AB}/dt = 0, \tag{17}$$

when A and B move with their local value of $\mathbf{q}$.

We give two arguments, the first intuitive and the second formal. Note that (17) is equivalent to saying that there is no flux of vorticity relative to a point moving with velocity $\mathbf{q}$. Suppose that the sheet has a small thickness ϵ. Consider a section A_1A_2 of the sheet by a plane normal to the vorticity. Use intrinsic co-ordinates (s, n) as shown in Figure 2.2-2 in a frame of reference moving with the sheet. The tangential component of velocity u_s can be approximated by

$$u_s = u_1, \quad n > \tfrac{1}{2}\epsilon; \quad u_s = u_2, \quad n < -\tfrac{1}{2}\epsilon;$$
$$u_s = q + (u_1 - u_2)n/\epsilon, \quad -\tfrac{1}{2}\epsilon < n < \tfrac{1}{2}\epsilon, \tag{18}$$

where $q = \tfrac{1}{2}(u_1 + u_2)$. The vorticity (out of the page) is $\omega = -\partial u_s/\partial n$, and the strength of the sheet is $\kappa = \int \omega \, dn = u_2 - u_1$. The flux of vorticity across $A_1 A_2$ is

$$\int \omega u_s \, dn = -\tfrac{1}{2}(u_1^2 - u_2^2) = \kappa q. \tag{19}$$

Hence there is no vorticity flux relative to a point moving with speed q.

The formal argument proceeds along the lines of that for Kelvin's circulation theorem, §1.6. Define d/dt to be differentiation following a fluid particle on the sheet, d_1/dt to be differentiation following a particle on side 1 and similarly for d_2/dt. We have

$$\frac{d}{dt} \int_B^A \mathbf{u}_2 \cdot d\mathbf{s} = \int_B^A (d_2\mathbf{u}_2/dt + (\mathbf{q} - \mathbf{u}_2) \cdot \nabla \mathbf{u}_2) \cdot d\mathbf{s} + \int_B^A \mathbf{u}_2 \cdot (d\mathbf{s} \cdot \nabla \mathbf{q}). \tag{20}$$

Replacing $d_2\mathbf{u}_2/dt$ by $-\nabla p - \nabla V$, and $\mathbf{u}_2$ by $\mathbf{q} + \tfrac{1}{2}[\mathbf{u}]_1^2$, and carrying out some integration, we obtain

$$\frac{d}{dt} \int_B^A \mathbf{u}_2 \cdot d\mathbf{s} = p_B - p_A + V_B - V_A + \tfrac{1}{2}\mathbf{q}^2\big|_B^A + (1/8)(\mathbf{u}_1 - \mathbf{u}_2)^2\big|_B^A \tag{21}$$

$$+ \int_B^A \tfrac{1}{2}[(\mathbf{u}_2 - \mathbf{u}_1) \cdot (d\mathbf{s} \cdot \nabla)\mathbf{q} - d\mathbf{s} \cdot ((\mathbf{u}_2 - \mathbf{u}_1) \cdot \nabla)\mathbf{q}].$$

A similar equation obtained from interchanging 1 and 2 holds on side 1. Expanding the integrand in the last term in (21), we recognise it as

$$\tfrac{1}{2}(\mathbf{u}_2 - \mathbf{u}_1) \times d\mathbf{s} \cdot \operatorname{curl} \mathbf{q}. \tag{22}$$

Now $\operatorname{curl} \mathbf{q} = \tfrac{1}{2}(\operatorname{curl} \mathbf{u}_1 + \operatorname{curl} \mathbf{u}_2) = 0$ because the flow is irrotational on each side. (Actually (22) would also vanish even if the vortex sheet separated fluid in rotational motion provided the external average vorticity were tangent to the sheet, since (22) is then the triple scalar product of three coplanar vectors.) The result (17) follows immediately from the continuity of pressure.[10]

In principle, equations (8) and (17), together with the value of the external velocity, provide sufficient information to follow the development in

[10] The argument requires explicitly that the densities are the same on the two sides and that there is no discontinuity in the force potential V on the two sides. These conditions imply that no vorticity is created if the sheet is given a small but finite thickness.

time of a vortex sheet in homogeneous fluid in the presence of conservative forces. The sheet strength is related to Γ, defined by (16), relative to an origin on the sheet by

$$\kappa = \nabla\Gamma \times \mathbf{n}, \quad \nabla\Gamma = \mathbf{n} \times \kappa, \tag{23}$$

the gradient operator being two-dimensional in the sheet. Thus the two equations provide the information to update in time the position and strength of the sheet given its position and strength at any instant. However, it remains an open question if such problems are well posed, that is, possess smooth solutions for more than a finite time (see §8.3).

It is of interest, particularly for aerodynamical applications, to study the shapes of vortex sheets in steady motion. Since streamlines and pathlines then coincide, the vortex sheet is a stream surface for fluid on both sides, and each of $\mathbf{u}_1$ and $\mathbf{u}_2$ are parallel to the sheet. From (17), $\mathbf{q} \cdot \nabla\Gamma$ is constant on sheet streamlines, that is, curves parallel to $\mathbf{q}$. The constant is not necessarily zero, because (17) expresses the difference in $\mathbf{q} \cdot \nabla\Gamma$ at two points on the streamline. Using (23), we deduce that on sheet streamlines

$$\mathbf{q} \cdot \nabla\Gamma = \mathbf{q} \cdot \mathbf{n} \times \kappa = -c, \tag{24}$$

where $c = $ constant. Now $\mathbf{q} \times \kappa$ is parallel to $\mathbf{n}$. Hence

$$\mathbf{q} \times \kappa = c\,\mathbf{n}. \tag{25}$$

Since $p_1 = p_2$ and $V_1 = V_2$, we have for the difference in the Bernoulli constants

$$H_1 - H_2 = \tfrac{1}{2}(\mathbf{u}_1^2 - \mathbf{u}_2^2) = \mathbf{q} \times \kappa \cdot \mathbf{n} = c. \tag{26}$$

The Bernoulli constants are uniform over the sheet when the flow outside is irrotational and c is in fact constant over the sheet. If the Bernoulli constants are the same, then $c = 0$, and the sheet streamlines coincide with the sheet vortex lines; that is, $\mathbf{q}$ and κ are parallel. This is the usual case.

We emphasise that the results given above are for the case of ideal uniform fluids and no singularities in the external flow fields. Note that the material surface separating two regions each containing ideal uniform fluids with conservative external force fields, but such that the density or force potential are discontinuous across the surface, will become a vortex sheet, as follows immediately from (21). (When the density is not equal to unity, terms involving the pressure are divided by the density.) Note also that discontinuities in the force potentials are required, and hence discontinuities in the normal components of external force fields will not create vorticity in the sheet; it is jumps in tangential components which produce a vortex sheet. The jumps in the normal component can be absorbed into the pressure. A singular force (i.e., a force with a finite magnitude per unit area) localised in, but normal to, the sheet gives rise to a pressure jump across the sheet but does not change the conclusion that the sheet moves with the fluid; this is immediate from the argument going from (13) to (14). However, (17) now fails, and vortex sheet lines no longer move with the average velocity $\mathbf{q}$. A singular force in, and parallel to, the sheet will cause the sheet to move relative to the fluid. Such singular forces could be produced, for example, by Lorentz forces in perfectly conducting fluid where the vortex sheet is also a current sheet, or by having a high-speed jet with finite momentum flux buried in the vortex sheet. The latter occurs in the jet flap for wings (see, e.g., Thwaites [1960]). At the surface between two immiscible fluids, surface tension gives rise to a pressure jump across the interface equivalent to a singular normal force. The theory of water waves or Rayleigh–Taylor instability can be regarded as a special case of the dynamics of vortex sheets, the motion being due to the vortex sheet created at the surface by the density discontinuity (Zarodny and Greenberg [1973].) The generation of sheet strength follows, for all cases when the sheet moves with the fluid, from (20). There are alternative formulations based on Bjerknes's theorem (Lamb [1932 §166a]).[11]

The simplest example of a vortex sheet is the uniform plane sheet, separating uniform regions of constant but different velocity. If the velocity has components $(-\frac{1}{2}U, 0, 0)$ for $z > 0$ and $(\frac{1}{2}U, 0, 0)$ for $z < 0$, the sheet has strength $U\mathbf{j}$ in the $z = 0$ plane. Another example is the cylindrical vortex sheet of circular cross-section, radius a, with the sheet vortex lines along the generators. The velocity field is zero inside the cylinder and azimuthal with value $\kappa a/r$ outside, where κ is the strength of the sheet and r is the distance from the axis of the cylinder. A uniform velocity inside the cylinder may be imposed on this flow, giving a cylindrical vortex sheet with helical vortex lines and sheet streamlines. It is easy to verify that (25) is satisfied. A case of great practical importance is the trailing vortex system behind a wing (see §8.4).

2.3 Line vortices

A singular distribution of considerable utility is that in which infinite vorticity is concentrated on a line in space, such that the circulation

---

[11] For a review and a careful discussion of numerical approximations, see Moore [1984].

around a closed circuit threaded once by the line is finite. This distribution is called a line vortex, and its strength, Γ, is the circulation around it, which is independent of the particular circuit when the line vortex threads it once in the same sense. Line vortices (not to be confused with vortex lines) are the result of a limiting process in which a vortex filament of finite strength is contracted to a curve, the strength being kept constant. Formally, the vorticity can be expressed as

$$\omega = \Gamma \, \delta(n) \, \delta(b) \, \mathbf{s}, \tag{1}$$

where $\mathbf{s}$ is a unit vector (the tangent vector) along the curve, and n, b denote co-ordinates in the normal and binormal direction.

A space curve is given parametrically by the vector equation $\mathbf{x} = \mathbf{R}(\sigma)$, where σ is a parameter specifying position along the curve. When $\sigma = s$, where s is distance along the curve, we have the Frenet–Serret formulae

$$d\mathbf{R}/ds = \mathbf{s}, \quad d\mathbf{s}/ds = \mathbf{n}/\rho, \quad d\mathbf{n}/ds = -\mathbf{s}/\rho + \tau\mathbf{b}, \quad d\mathbf{b}/ds = -\tau\mathbf{n}, \tag{2}$$

with the triad (s, n, b) forming a right-handed set of mutually perpendicular axes. The radius of curvature is denoted by ρ (the context will distinguish it from the density), and τ is the torsion of the curve and measures the rotation rate of the tangent plane. Substituting (1) into (1.1.5) and using $d\mathbf{x}' = ds \, dn \, db$, we obtain the Biot–Savart law (so called in view of the analogy with the magnetic field induced by a line current)

$$\mathbf{u}_{\mathrm{v}}(\mathbf{x}) = \frac{\Gamma}{4\pi} \oint \frac{\mathbf{s} \times (\mathbf{x} - \mathbf{R}(s))}{|\mathbf{x} - \mathbf{R}(s)|^3} \, ds = \frac{\Gamma}{4\pi} \operatorname{curl} \oint \frac{ds}{|\mathbf{x} - \mathbf{R}(s)|} \tag{3}$$

for the velocity $\mathbf{u}_{\mathrm{v}}$ induced by a line vortex of strength Γ along the closed curve $\mathbf{R}(s)$. The last member of (3) shows that the vector potential of a line vortex has a particularly simple form.

Because a vortex tube is either closed or ends on boundaries, the same must be true for a line vortex. However, the integrals in (3) must always be along a closed curve, for otherwise the velocity field $\mathbf{u}_{\mathrm{v}}(\mathbf{x})$ will not be solenoidal. A vortex tube ending on a wall violates condition (v) of §1.1, and the formula cannot be applied to a vortex spanning the gap between boundaries unless the vortex is closed in some way (see §2.4). But with this proviso, we can interpret (3) as saying that each element ds of the line induced a velocity $\Gamma \sin \chi \, ds/4\pi r^2$ at a point P distance r away, χ being the angle between the element and the radius vector to P, and the direction of the velocity being perpendicular to these directions.

By an application of Stokes's theorem, (3) can be converted into an integral over a surface S bounded by the line vortex, provided the curve is reducible. Since

$$\oint d\mathbf{s} \times \mathbf{a} = \int [(\nabla \mathbf{a}) \cdot d\mathbf{S} - \operatorname{div} \mathbf{a} \, d\mathbf{S}],$$

$$\operatorname*{div}_{R} \frac{\mathbf{x} - \mathbf{r}}{|\mathbf{x} - \mathbf{R}|^3} = 0, \tag{4}$$

$$\left(\frac{\partial}{\partial \mathbf{x}} + \frac{\partial}{\partial \mathbf{R}} \right) \frac{(\mathbf{x} - \mathbf{R})}{|\mathbf{x} - \mathbf{R}|^3} = 0,$$

we can write (3) as (putting $\mathbf{a} = (\mathbf{x} - \mathbf{R})/|\mathbf{x} - \mathbf{R}|^3$)

$$\mathbf{u}_v = -\Gamma/4\pi \, \nabla\Omega, \quad \Omega(\mathbf{x}) = \int_S (\mathbf{x} - \mathbf{R})/|\mathbf{x} - \mathbf{R}|^3 \cdot d\mathbf{S}, \tag{5}$$

where $\mathbf{R}$ now denotes the position vector of a point on the surface S. The quantity Ω is the solid angle[12] subtended at $\mathbf{x}$ by the line vortex, and (5) shows that the velocity potential of the flow induced by the line vortex is simply $-\Gamma\Omega/4\pi$. The region occupied by the fluid outside the vortex is not simply connected, and the velocity potential is not single valued; Ω increases by 4π in describing a circuit going once around the vortex. It can be made single valued by introducing a barrier consisting of a surface bounded by the vortex, across which the velocity potential jumps by amount $-\Gamma$.

A dipole of strength $\mathbf{m}$ at $\mathbf{R}$ produces a velocity potential $-\mathbf{m} \cdot (\mathbf{x} - \mathbf{R})/|\mathbf{x} - \mathbf{R}|^3$ at $\mathbf{x}$. Thus the line vortex is kinematically equivalent to a surface distribution of dipoles with uniform density $\Gamma/4\pi$, the axes of the dipoles being aligned along the normal to the barrier. An infinitesimal line vortex of area $\delta\mathbf{S}$ and strength Γ is equivalent to an infinitesimal dipole of strength $\Gamma \, \delta\mathbf{S}$.[13]

The velocity induction law (1.1.5), and hence the Biot–Savart law (3), can be applied to an infinitely long line vortex, going to infinity, because the curve can be imagined closed by a large semi-circle of radius R_∞, and it is immediate from (3) that the contribution from this portion is $O(R_\infty^{-1})$ and hence zero.

The special case of the infinitely long straight line vortex, the rectilinear line vortex, is of extreme importance. The streamlines of the induced velocity are circles with axis on the vortex, and the azimuthal speed is

---

[12] The solid angle subtended by an area at a point is defined as the area of the unit sphere centred at the point intersected by the cone joining the point to the edges of the area.

[13] This is an alternative way of seeing the equivalence of a vortex sheet and a non-uniform surface distribution of dipoles, as the vorticity in a sheet can be discretized into a superposition of line vortices lying, for instance, along an infinitesimal mesh formed by the sheet vortex lines and orthogonal curves.

$\Gamma/2\pi r$, where r is distance from the vortex. The velocity is unbounded on the vortex itself, but it turns out that there is no dynamical inconsistency for two-dimensional flows (in planes perpendicular to the axis of the vortex) and the vortex moves with a finite velocity (see (10)–(19)), which is usually that of the fluid with the infinity subtracted.

The curved line vortex is a useful kinematic concept, but it does not exist as a sensible dynamical limit. The reason is that the speed of a curved vortex filament becomes infinite as its cross-section tends to zero if the strength is kept constant. This singularity is associated with the singularity of the Biot–Savart integral (3) as $\mathbf{x}$ approaches a point on the vortex. Let this point be the origin and the approach be perpendicular to the curve, and align the x, y, z axes parallel to n, b, s at the origin. We have

$$\mathbf{R} = s(\partial \mathbf{R}/\partial s)_o + \tfrac{1}{2} s^2 (\partial^2 \mathbf{R}/\partial s^2)_o + \tfrac{1}{6} s^3 (\partial^3 \mathbf{R}/\partial s^3)_o + \cdots . \tag{6}$$

The derivatives are expressible via the Frenet–Serret formulae in terms of $\mathbf{n}, \mathbf{b}, \mathbf{s}$ and the derivatives of ρ and τ at the origin. Thus

$$\mathbf{R} = s\,\mathbf{s}_o + \tfrac{1}{2}\,(s^2/\rho_o)\,\mathbf{n}_o + O(s^3/\rho^2). \tag{7}$$

After some elementary algebra, it is found that the integrand in the first integral in (3) behaves like

$$\frac{x\,\mathbf{b}_o - y\,\mathbf{n}_o}{(r^2 + s^2)^{3/2}} \left[1 + \frac{3xs^2}{2(r^2 + s^2)\rho_o} \right] + \frac{y\,s\,\mathbf{s}_o + \tfrac{1}{2} s^2\,\mathbf{b}_o}{(r^2 + s^2)^{3/2}\rho_o} + O(\rho_o^{-2}), \tag{8}$$

where $r = \sqrt{x^2 + y^2}$ denotes the shortest distance from the line vortex. To determine the singular behaviour of the Biot–Savart integral as $r \to 0$, substitute (8) into (3), integrate in the range $-\epsilon < s < \epsilon$, change variable from s to $\sigma = s/r$ and evaluate the integral in the limit $\epsilon \to 0$, $\epsilon/r \to \infty$. It is found that the limit is[14]

$$\mathbf{u}_{vo} \sim \frac{\Gamma}{2\pi} \left(\frac{x}{r^2} \mathbf{b}_o - \frac{y}{r^2} \mathbf{n}_o \right) - \frac{\Gamma \log r}{4\pi \rho_o} \mathbf{b}_o$$

$$- \frac{\Gamma}{4\pi \rho_o} \left(\frac{y^2}{r^2} \mathbf{b}_o + \frac{xy}{r^2} \mathbf{n}_o \right)$$

$$+ \frac{\Gamma \log \epsilon}{4\pi \rho_o} \mathbf{b}_o + \lim_{\epsilon \to 0} \frac{\Gamma}{4\pi} \oint_{|s| > \epsilon} \frac{\mathbf{s} \times (\mathbf{R}(s_o) - \mathbf{R}(s))}{|\mathbf{R}(s_o) - \mathbf{R}(s)|^3} \, ds$$

$$+ O\left(\frac{\Gamma r}{\rho_o^2} \right). \tag{9}$$

---

[14] The first two and last terms are given by Batchelor [1967 §7.1]. The third term was obtained by Moore and Saffman [1972].

The first term in the expression is the circulation about the tangent rectilinear line vortex. The second term is an unbounded contribution, due to the curvature, in the direction of the binormal. The third term is bounded but not continuous at $r = 0$. The fourth term cancels the singularity as $\epsilon \to 0$ from the last integral. To see this, note that the value at the origin of this contribution is $(\Gamma/4\pi) \int_{|s|>\epsilon} \mathbf{s} \times \mathbf{r}/R^3 \, ds$, and the substitution of (7) shows that the integrand of this integral is singular like $\mathbf{b}_o/(2\rho_o s)$ and hence the integral is asymptotic to $-(\Gamma/4\pi\rho_o) \mathbf{b}_o \log \epsilon$, which cancels the fourth term in (9). When the line vortex is a circle of radius ρ_o, it can be shown from detailed evaluation of the integrals in terms of elliptic functions (Lamb [1932 §161]) that (9) gives the velocity induced by the circular line vortex with the given error when ϵ is put equal to $8\rho_o$ in the fourth term and the integral is dropped.

The logarithmically infinite, non-circulatory, velocity along the binormal as $r \to 0$ inhibits the curved line vortex (of zero cross-section) from being a useful dynamical model. However, this restriction does not apply to the rectilinear line vortex, for which $\rho_o \to \infty$, that moves with the finite speed obtained by adding the external velocity produced by boundaries and other vortices to the self-induced velocity with the infinite circulatory part subtracted. This result is usually stated as an immediate consequence of the Helmholtz laws. However, since the vorticity distribution is singular, it is appropriate to give a direct argument based on momentum conservation. It is also of interest to examine the changes produced by external singular non-conservative forces acting on the vortex core.

Consider a rectilinear line vortex parallel to the z-axis moving with velocity components (U_v, V_v). Take moving polar co-ordinates centred on the vortex. The velocity potential in an infinitesimal fixed circle of radius ϵ whose centre instantaneously coincides with the vortex is

$$\phi = \Gamma\theta/2\pi + Ur \cos\theta + Vr \sin\theta + O(\epsilon^2), \tag{10}$$

where (U, V) is the finite part of the velocity field at the vortex. From the pressure equation for unsteady irrotational flow (1.9.6), we have for the pressure on the perimeter of this small circle,

$$\begin{aligned} p &= -\tfrac{1}{2}(\operatorname{grad}\phi)^2 - \partial\phi/\partial t \\ &\sim -\Gamma^2/8\pi^2\epsilon^2 + \Gamma/2\pi\epsilon[(U - U_v)\sin\theta - (V - V_v)\cos\theta] \end{aligned} \tag{11}$$

since $\partial\theta/\partial t = (U_v \sin\theta - V_v \cos\theta)/r$ at a point fixed in space. The force exerted on the fluid inside the circle by the pressure over the boundary is to leading order

$$\oint_{r=\epsilon} p\,\mathbf{n}\,ds = -\tfrac{1}{2}\Gamma[(U - U_\mathrm{v})\mathbf{j} - (V - V_\mathrm{v})\mathbf{i}], \qquad (12)$$

where $\mathbf{n} = (-\cos\theta, -\sin\theta)$ is the inward pointing normal. The rate of change of fluid momentum inside the fixed circle is non-zero because the vortex moves relative to the circle and there is a flux of momentum across its boundary. The rate of increase of momentum due to the relative motion is

$$\oint_{r=\epsilon} \frac{\Gamma}{2\pi\epsilon}(\cos\theta\,\mathbf{j} - \sin\theta\,\mathbf{i})(-U_\mathrm{v}\cos\theta - V_\mathrm{v}\sin\theta)\,ds. \qquad (13)$$

The momentum flux across the circumference of the circle is

$$\oint_{r=\epsilon} \frac{\Gamma}{2\pi\epsilon}(\cos\theta\,\mathbf{j} - \sin\theta\,\mathbf{i})(U\cos\theta + V\sin\theta). \qquad (14)$$

The sum of the two is a net momentum gain at a rate

$$\tfrac{1}{2}\Gamma(U - U_\mathrm{v})\mathbf{j} - \tfrac{1}{2}\Gamma(V - V_\mathrm{v})\mathbf{i}, \qquad (15)$$

which must be supplied by the pressure field acting over the circumference. Equating (15) to the pressure force (12), we have

$$U_\mathrm{v} = U, \qquad V_\mathrm{v} = V. \qquad (16)$$

Thus a rectilinear line vortex moves with the fluid and is a consistent dynamical concept. We can say that the rectilinear line vortex

$$\omega = \delta(\mathbf{x} - \mathbf{X}(t))\,\mathbf{k}, \qquad d\mathbf{X}(t)/dt = \nabla\Phi, \qquad (17)$$

where Φ is the velocity potential induced by other sources and boundaries, is a 'quasi-weak' solution of the two-dimensional Euler equations in the sense that vorticity, momentum and angular momentum are conserved,[15] and it is a weak solution in the usual sense if the singular spatial distribution is supposed to have circular symmetry (Saffman and Meiron [1986]). The approximation of replacing a two-dimensional continuous distribution of vorticity, which may have vortex jump and vortex sheet singularities, by an assembly of rectilinear line vortices, replaces one Euler flow by another Euler flow. How close the second one can be made to the first is an interesting question, since (see §7.3) the motion of line vortices can often be reduced to the integration of a set of ordinary differential equations of Hamiltonian form.[16]

[15] The finite part of the energy is also conserved, see §7.3.
[16] Estimates of the convergence of the difference between a smooth field of vorticity and a line vortex representation have been given by Goodman, Hou and Lowengrub

If an external force **F**, per unit length of vortex, is applied to the rectilinear line vortex in the plane of the motion, the balance of momentum gives

$$U_v = U - F_y/\Gamma, \quad V_v = V + F_x/\Gamma, \quad \mathbf{F} = \Gamma(U - U_v)\mathbf{j} - \Gamma(V - V_v)\mathbf{i},$$

$$(18)$$

which can be expressed compactly as

$$\mathbf{U}_v = \mathbf{U} + \mathbf{k} \times \mathbf{F}/\Gamma. \tag{19}$$

Thus rectilinear line vortices will move relative to the fluid if subjected to an external force. (For an alternative argument using vortex force, see §3.1.)

It can be anticipated from the expression (9) for the velocity induced by a curved line vortex that the velocity of a curved vortex filament becomes infinite, like the logarithm of its cross-section radius as this quantity goes to zero. This is also to be expected from Kelvin's [1867b] formula for the speed V of a vortex filament in the shape of a circle of radius R, with (to leading order) circular cross-section of radius a and uniform vorticity, given in equation (10.2.1).

Chapters 10 and 11 contain further discussion of vortex rings and curved vortex filaments.

2.4 Image vorticity

Suppose the values of a vector field **a** are given on a smooth closed surface S, subject to the condition that

$$\int_S \mathbf{a} \cdot \mathbf{n}\, dS = 0. \tag{1}$$

Then the vector field can be extended into the interior of S in such a way that it is solenoidal inside S, so that

$$\operatorname{div} \mathbf{a} = 0, \tag{2}$$

and the components of **a** are continuous across S. The extension is not unique and can be done in an infinite number of ways. Geometrically, it suffices to join the field lines at the surface by any set of non-intersecting

[1990]. The accuracy with which a vorticity field is approximated by a set of in general overlapping, non-deforming circular vortex patches is not a problem of vortex dynamics, since they do not constitute an Euler flow, but is a question of numerical analysis.

smooth curves parallel to **a** on S, the magnitude of the field inside being inversely proportional to the cross-section of the field tubes. Equation (1) ensures that as many tubes enter as leave. If the field lines kink at the surface S, the normal component of **a** will be continuous but the tangential components will have a discontinuity; this will be called a partial extension.

The vorticity field ω must satisfy (1) for any closed surface S. This follows from an application of Stokes's theorem to a circuit dividing the surface, since contributions from the two subdivisions of the surface make equal and opposite contributions to the line integral. The vorticity can therefore be extended into the interior of any body surrounded by or bounding the fluid. With the understanding that the vorticity field is to be extended if $\omega \cdot \mathbf{n} \neq 0$ on bounding surfaces, the formula (1.1.5) for the velocity as an integral of the vorticity is valid (cf. Batchelor [1967 §2.4]). The condition (v) of §1.1 can in this way be bypassed.[17] The arbitrary function $\nabla \Phi$ in the expression for the induced velocity will depend on the particular method of extension. It is often convenient to confine the extension to a singular surface, which is a vortex sheet, lying just outside the surface of the body. Then by this device, $\omega \cdot \mathbf{n} = 0$ on S. The Biot–Savart law (3.3) can be applied to line vortices which end on surfaces if the line vortex is closed by extension through the body. Condition (1) ensures that the number of lines entering equals the number leaving, the lines being weighted, of course, according to their strengths.

The incompressible velocity field can be extended into a body of constant volume (which need not be rigid or at rest or in steady motion) since (1) is then satisfied with $\mathbf{a} = \mathbf{u}$. Denote an extension of velocity into the body by $\mathbf{u}_B$; then $\omega_B = \operatorname{curl} \mathbf{u}_B$ is a partial extension of ω, since the continuity of velocity implies that the normal components of ω and ω_B are equal. The surface is then a vortex jump. A partial extension of velocity would lead to a vortex sheet on the surface; this is convenient for some applications.

When $\omega \cdot \mathbf{n} = 0$ on the body, the velocity **u** outside the body can be expressed in the form $\mathbf{u}_v + \nabla \Phi$, where the first term is produced by the vorticity outside the body and is defined inside by the same formula (1.1.5). As discussed in §I.1, the irrotational component is chosen to satisfy the condition on the normal velocity on the surface or some equivalent requirement. The extension of **u** into the body of assumed constant volume gives a velocity field $\mathbf{u}_I$ such that $\mathbf{u}_v + \mathbf{u}_I$ is continuous inside and on the body. The body can therefore be taken away and replaced by the vorticity

---

[17] This point was made by Kelvin [1868 §60o].

field

$$\omega_I = \text{curl } \mathbf{u}_I, \tag{3}$$

since curl $\mathbf{u}_v = 0$ inside the body. We refer to the distribution ω_I as the image vorticity. The original velocity field is the same as that given by the integral formula (1.1.5) with the image vorticity (for all the bodies) included in the integrand.[18]

Note that the image vorticity is not unique, but the velocity field induced outside the body by the image vorticity must be independent of the particular method of construction because it is an irrotational solenoidal field taking given values on a surface. Note also that the image vorticity is not a property of the body alone, but depends upon the other bodies that are present insofar as Φ depends upon the entire geometry.

It is sometimes convenient to use an irrotational partial extension of the contribution to the velocity field constructed as follows. Define $\nabla\Phi_B$ inside the body by

$$\nabla^2\Phi_B = 0, \quad \partial\Phi_B/\partial n = \partial\Phi/\partial n \quad \text{on } S, \tag{4}$$

and the requirement that Φ_B is analytic inside the body. The potential Φ_B is thereby uniquely defined. In general, the analytic continuation of Φ into the body is not the same as Φ_B, and the tangential components of $\nabla\Phi$ and $\nabla\Phi_B$ are discontinuous, the surface being equivalent to a vortex sheet of strength

$$\kappa = \mathbf{n} \times (\nabla\Phi - \nabla\Phi_B) \tag{5}$$

or an equivalent surface distribution of dipoles of strength $\nabla\Phi - \nabla\Phi_B$. This gives the representation of image vorticity as a vortex sheet.

In general, it is not possible to describe image vorticity by closed-form expressions. There are, however, a number of simple cases in which it can be written down by inspection.

An image system for a rectilinear line vortex in the z-direction parallel to a rigid plane $y = 0$ is an equal and opposite vortex at the mirror image. The velocity induced at the vortex by its image is $\Gamma/4\pi h$ parallel to the wall, where Γ is the strength of the vortex and h is the distance from the wall. By the considerations of §2.3, the vortex moves parallel to the wall with this speed in the absence of external forces on the core. In this case, $\Phi = -(\Gamma/2\pi)\,\tilde{\theta}$, where $\tilde{\theta}$ is the angle between the line joining the field point with the image vortex and the x-axis; Φ_B is $-(\Gamma/2\pi)\,\theta$ where

---

[18] If the volume is not constant, sources and sinks must be introduced.

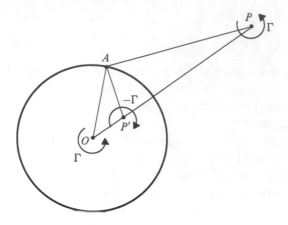

2.4-1 Image system for a point vortex outside a cylinder.

θ is the angle for the actual vortex. A vortex sheet on $y = 0$ of strength $-\Gamma h/\pi(h^2 + x^2)^{3/2}$ is then an alternative image vorticity system. It is easily verified that the two systems give the same velocity in $y > 0$.

A rectilinear line vortex outside a rigid circular cylinder has the image system of an equal and opposite vortex at the inverse point together with an equal vortex at the centre, when the circulation around the surface of the cylinder is zero. This follows from the fact that $\sin(PAO)/PA - \sin(P'AO)/P'A = 0$ (see Figure 2.4-1), where $OP.OP' = a^2$ defines the inverse point, and hence the velocity at A is tangential to the cylinder. The vortex moves clockwise around the cylinder in a circle with constant velocity

$$\Gamma/2\pi(r - a^2/r) - \Gamma/2\pi r = \Gamma a^2/2\pi r(r^2 - a^2). \tag{6}$$

Note that the vortex moves around the cylinder, although the circulation around the cylinder is zero. This is another example of the misleading information that can be produced by visualisation of unsteady flows.

At moderate Reynolds number there exist two standing, oppositely rotating vortices behind a cylinder in a uniform stream. An inviscid model of this flow consists of two equal and opposite rectilinear (free) line vortices standing symmetrically behind the cylinder. The image system consists of two equal and opposite (bound) vortices at the inverse points, and there is a dipole at the centre of the cylinder (or equivalently an infinitesimal vortex pair) due to the uniform stream. Existence of a steady solution reduces to finding roots of the algebraic equation posed by the condition that the velocity induced at each free vortex by the other free vortex and the two images and the uniform stream and its image should vanish. This problem was investigated by Föppl [1913] (see Lamb [1932 §155]) who demonstrated loci of possible vortex positions. The loci are the curves (see §7.2)

$$r^2 - a^2 = 2ry,\tag{7}$$

and the vortex strengths are

$$4\pi Uy(1 - a^4/r^4),\tag{8}$$

where U is the velocity of the free stream taken in the x−direction and a is the radius of the cylinder. The model has some interesting features. It can be shown that the arrangement is unstable to antisymmetrical two-dimensional disturbances.[19] Also the drag on the cylinder vanishes, as can be shown either by direct calculation of the pressure on the cylinder or more easily by noting that at large distances the flow is asymptoted by a two-dimensional dipole superposed on a uniform stream, and the standard argument for D'Alembert's paradox proves that the momentum flux at infinity is zero and hence there is no drag.

The image system of a curved line vortex in a rigid plane is just the mirror reflection of the line vortex with reversed circulation. The image system for a sphere was worked out by Lighthill [1956] and is quite complicated. Consider an element $d\mathbf{s}$ of a line vortex of circulation Γ at a point distance $r > a$ from the centre of a sphere of radius a. The image of the element is the inverse of the element with a circulation $-\Gamma r/a$, together with a line vortex stretching radially from the inverse point to the centre of the sphere with circulation $\Gamma ds_r/a$, where ds_r is the radial component of $d\mathbf{s}$ (see Figure 2.4-2). To verify this statement, note first that according to the Biot–Savart law (3.3) the element outside the sphere produces a normal velocity

$$\Gamma \, ds_t \sin\theta \sin\phi/4\pi(a^2 - 2ar\cos\theta + r^2)^{3/2}$$

at a point on the surface with spherical polar co-ordinates (a, θ, ϕ), with the axis of the co-ordinates passing through the vortex and the plane $\phi = 0$ containing the element (ds_t denotes the transverse component of $d\mathbf{s}$). Since this expression is unchanged when r is replaced by a^2/r, ds_t by $(a^2/r^2) \, ds_t$ and Γ by $(r/a)\Gamma$, it follows that the element and the inverse make equal

---

[19] I. Soibelman has pointed out (private communication) that the algebraic formulae given by Föppl for the eigenfrequencies are in error, but the qualitative considerations are correct.

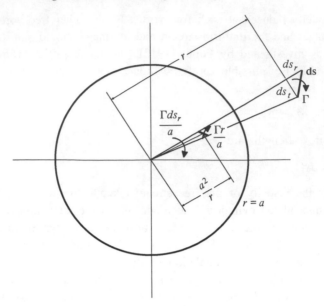

2.4-2 Image of a vortex element in a sphere.

and opposite contributions to the normal velocity. However, the inverse does not preserve the solenoidal property of the vorticity field; this is accomplished by having a line vortex of circulation $\Gamma\,ds_r/a$ stretching from the inverse point to the centre of the sphere along the radius to account for the change in circulation at the ends of the image element. This line vortex leaves the surface of the sphere a streamline. Lighthill [1956] gives the lengthy algebraic expressions for the velocities induced by the three elements at an arbitrary point, but note that he defines the strength of a vortex element as its circulation multiplied by its length, in contrast to present practice, which takes circulation and strength to be synonymous.

Thus the image system of a line vortex in a sphere is a line vortex of variable strength along the inverse of the curve, together with a conical vortex sheet joining the centre of the sphere to the inverse curve. The strength of the vortex sheet at its edge with the inverse vortex is $\Gamma\,(r^2/a^3)\cot\xi$, where ξ is the angle made by the tangent to the vortex at r with the radial direction. The vortex sheet is absent when the line vortex lies in a spherical surface concentric with the fixed sphere, as would be the case for a circular line vortex with its axis passing through the centre of a rigid sphere. In this case, the image system is easily derived by replacing the vortex by a uniform distribution of dipoles over the concentric surface and using the result (Lamb [1932 §96]) that the image of a dipole is one of opposite sign placed at the inverse point with strength reduced by a factor a^3/r^3. The density of dipoles is, however, increased by a factor R^4/a^4, owing to the reduced area of the surface bounded by the inverse curve, giving the result that the strength of the image vortex is greater by a factor r/a (Larmor [1889]).

The image for a line vortex interior to a sphere is obtained in a similar manner and consists of a line vortex on the inverse curve with a conical vortex sheet extending out to infinity.

Dhanak [1981] has calculated the interaction of a vortex filament with a rigid sphere.

3

VORTEX MOMENTUM

3.1 Vortex force and bound vorticity

The term $\mathbf{u} \times \boldsymbol{\omega}$ in the Euler equation of motion (1.9.1) has been interpreted as an equivalent body force called the vortex force.[1] The idea originated with Prandtl [1918] and was used extensively by Karman and Burgers [1934] in their account of flow over wings. It seems most useful when combined with the concept of bound vorticity in steady motions.[2] Integrating (1.9.1) through a fixed volume V, we have

$$\frac{\partial}{\partial t} \int \mathbf{u} \, dV = - \int p_T \mathbf{n} \, dS + \int (\mathbf{u} \times \boldsymbol{\omega} + \mathbf{F}) \, dV, \qquad (1)$$

where $p_T = p + \frac{1}{2} \mathbf{u}^2$ is called the total head. (We use the convention that $\mathbf{n}$ is directed out of the fluid.) The left-hand side is the rate of change of momentum of the fluid inside V (not in general the same as the rate of change of the momentum of the material volume of fluid instantaneously occupying the volume V). It vanishes when the flow is steady, which we now suppose to be the case. Then (1) expresses the equilibrium of the total pressure force acting over the surface of V together with the external force and the vortex force. If no external forces act outside V, and the Bernoulli constant is uniform over the boundary (this is so when it is a stream surface), the surface integral vanishes since $\int \mathbf{n} \, dS = 0$ over a closed surface. Equilibrium then requires that the external force needed to maintain a steady flow is determined by the vortex force. Alternatively,

---

[1] The requirement of constant density seems to be essential for the momentum invariants discussed in this chapter to be conserved quantities. We shall therefore put $\rho = 1$, unless variations of density are explicitly considered.

[2] It has, however, been used effectively by Lighthill [1962] in a study of momentum transfer in a perturbed shear flow.

we can say that if in a steady flow $\mathbf{u} \times \omega$ is not the gradient of a single-valued scalar, which can be absorbed into the pressure, then an external non-conservative body force must be applied to maintain equilibrium.

As was shown in §2.4, a body B moving relative to a fluid can be replaced kinematically (in an infinite number of ways) by a distribution of image vorticity. In steady motion, this distribution of vorticity is fixed relative to the body and is referred to as bound vorticity. It does not in general satisfy the Helmholtz laws, because to maintain equilibrium it must be imagined that an external body force is applied to the bound vorticity in order to balance the vortex force. Vorticity satisfying the Helmholtz laws (i.e., outside the body when external forces are conservative) is referred to as free vorticity. (In steady flow with globally constant Bernoulli constant H, free vorticity is parallel to the streamlines, as shown in §I.9.) Suppose now that the total force exerted on the body by the fluid is $\mathbf{D}$. Then the total external force applied to the fluid is $-\mathbf{D}$. Then applying (1) to a volume V which includes the bound vorticity and on whose boundary the total head is constant, we have

$$\mathbf{D} = \int \mathbf{u} \times \omega \, dV. \tag{2}$$

It follows that the total vortex force is independent of the particular image system that is employed. This also follows directly from the identity

$$\int_B \mathbf{u} \times \omega \, dV = \int_B \left(\nabla \tfrac{1}{2} \mathbf{u}^2 - \mathbf{u} \cdot \nabla \mathbf{u} \right) dV$$
$$= \int_S \left(\tfrac{1}{2} \mathbf{u}^2 \, \mathbf{n} - \mathbf{u}(\mathbf{u} \cdot \mathbf{n}) \right) dS. \tag{3}$$

The right-hand side is determined by the velocity on the body and is independent of the way in which the velocity and vorticity are extended into the body to give the image system.

As an example, consider irrotational flow past a circular cylinder of radius a with circulation Γ. The cylinder axis lies along the z-axis and the motion is two-dimensional in the xy-plane. We require that the velocity potential satisfies Laplace's equation, $\partial\phi/\partial r = 0$ on $r = a$, $\phi \sim -Ur\cos\theta$ as $r \to \infty$, and the increase in ϕ in going once counter-clockwise around the cylinder is Γ. The unique solution (obtained, for example, by separation of variables) is

$$\phi = -Ur\cos\theta - Ua^2/r \cos\theta + \Gamma\theta/2\pi. \tag{4}$$

We now take the body away and replace it by a distribution of bound vorticity. The vortex lines will be in the z-direction, that is, $\omega = (0, 0, \omega)$.

The vorticity distribution will be chosen to give an extension of the velocity field. One of an infinite number of possibilities is

$$\omega = \Gamma/\pi a^2 + 8Uy/a^2, \qquad r < a, \tag{5}$$

which is consistent with the solenoidal velocity field inside the cylinder

$$u = -U(x^2 + 3y^2)/a^2 + U - \Gamma y/2\pi a^2,$$
$$v = 2Uxy/a^2 + \Gamma x/2\pi a^2. \tag{6}$$

Substituting in (2) and evaluating the total force, we find that the component parallel to the free stream vanishes, and the lift (i.e., the force perpendicular to the free stream) is

$$L = -\int_{r<a} u\,\omega\,dx\,dy = U\Gamma. \tag{7}$$

This is the famous Kutta–Joukowski lift per unit span on a wing with circulation Γ moving with speed U. There are, of course, simpler ways of deriving this result, which also show that for given circulation the lift is independent of the shape of the body.

The concept of vortex force can also be used to give the result for the velocity of a rectilinear line vortex subject to an external force. In a frame moving with the vortex, the vortex force is $(\mathbf{U} - \mathbf{U_v}) \times \boldsymbol{\Gamma}$ per unit length of vortex, where $\boldsymbol{\Gamma} = \Gamma\mathbf{k}$. Equilibrium requires

$$\mathbf{F} + (\mathbf{U} - \mathbf{U_v}) \times \boldsymbol{\Gamma} = 0, \tag{8}$$

which is equivalent to (2.3.18).

3.2 **Hydrodynamic impulse**

We consider first unbounded three-dimensional flows at rest at infinity in which the vorticity vanishes outside some finite region.[3] The case of bounded flows or flows with bodies is considered in Chapter 4. If $\omega = 0$ everywhere, then $\mathbf{u} = 0$. At large distances, the flow is irrotational; that is, $\mathbf{u} = \nabla\phi$, where ϕ has an expansion

$$\phi = \sum_0^\infty a_n S_n(\theta, \lambda)/r^{n+1}. \tag{1}$$

---

[3] The results remain true and the modifications to the discussion are obvious if the vorticity is exponentially small at infinity.

The S_n are surface harmonics, (r, θ, λ) are spherical polar co-ordinates and the a_n are coefficients, depending on the distribution of vorticity. In the absence of sources and sinks, $a_0 = 0$, and the leading term of (1) can be written

$$\phi = -\mathbf{M} \cdot \mathbf{x}/r^3. \tag{2}$$

Since S_1 is a linear combination of $x/r, y/r, z/r$, we can find $\mathbf{M}$ from the asymptotic expansion of (1.1.5). Alternatively, we note that a dipole of strength $\mathbf{M}$ at the origin produces a vector potential $\mathbf{M} \times \mathbf{x}/r^3$ and expand the vector potential of the vorticity distribution given by (1.1.18). We have $(r = |\mathbf{x}|)$

$$1/|\mathbf{x} - \mathbf{x}'| = 1/r + \mathbf{x} \cdot \mathbf{x}'/r^3 + O(r^{-3}), \tag{3}$$

and hence

$$\mathbf{A}(\mathbf{x}) \sim (1/4\pi r) \int \boldsymbol{\omega}(\mathbf{x}')\, d\mathbf{x}' + (1/4\pi r^3) \int \mathbf{x} \cdot \mathbf{x}'\, \boldsymbol{\omega}(\mathbf{x}')\, d\mathbf{x}' + O(r^{-3}). \tag{4}$$

The leading term in (4) vanishes, since[4]

$$\int \omega_i(\mathbf{x})\, d\mathbf{x} = \int \partial(x_i \omega_j)/\partial x_j\, d\mathbf{x} = \int_{S_\infty} x_i \boldsymbol{\omega} \cdot \mathbf{n}\, dS = 0. \tag{5}$$

For the second term in (4), a trick is employed. First note that

$$(\mathbf{x} \cdot \mathbf{x}')\boldsymbol{\omega}' = \mathbf{x} \times (\boldsymbol{\omega}' \times \mathbf{x}') + (\boldsymbol{\omega}' \cdot \mathbf{x})\mathbf{x}',$$

and second that

$$\int (\boldsymbol{\omega}' \cdot \mathbf{x})\mathbf{x}'\, d\mathbf{x}' + \int (\mathbf{x}' \cdot \mathbf{x})\boldsymbol{\omega}'\, d\mathbf{x}' = 0$$

for the sum of the integrands is

$$x_j\, (x_i' \omega_j' + x_j' \omega_i') = x_j \partial(x_i' x_j' \omega_k')/\partial x_k'$$

whose integral with respect to $\mathbf{x}'$ vanishes. Hence,

$$\mathbf{A}(\mathbf{x}) \sim (1/8\pi r^3)\mathbf{x} \times \int \boldsymbol{\omega}' \times \mathbf{x}'\, d\mathbf{x}'. \tag{6}$$

Thus

$$\mathbf{M} = \mathbf{I}/4\pi, \tag{7}$$

[4] This result that the volume integral of a finite three-dimensional vorticity field is zero implies that the centroid of a vorticity field needs to be defined in a non-conventional way (see §9). Note that the volume integral vanishes over any volume bounded by a surface on which $\boldsymbol{\omega} \cdot \mathbf{n} = 0$.

where[5]

$$\mathbf{I} = \tfrac{1}{2} \int \mathbf{x} \times \boldsymbol{\omega} \, d\mathbf{x}. \tag{8}$$

Note that the order of magnitude of the velocity field at infinity is $O(|\mathbf{I}|/r^3)$.

The quantity $\mathbf{I}$ is called the hydrodynamic impulse (per unit density of the flow). It plays the part of momentum because for unbounded flows, as will now be proved,

$$d\mathbf{I}/dt = \int \mathbf{F} \, d\mathbf{x}, \tag{9}$$

where $\mathbf{F}$ denotes the non-conservative external body forces (per unit density). The force distribution must be confined to a finite region (or be exponentially small at infinity), for otherwise vorticity is generated at infinity. In the absence of non-conservative forces, the impulse is an invariant of the motion, even though the flow field is in general unsteady. The argument for (9) goes as follows. First differentiate (8) with respect to time and substitute for $d\boldsymbol{\omega}/dt$ from the incompressible form of (1.5.5) to give

$$\begin{aligned} d\mathbf{I}/dt &= \tfrac{1}{2} \int \mathbf{x} \times \partial \boldsymbol{\omega}/\partial t \, d\mathbf{x} \\ &= \tfrac{1}{2} \int \mathbf{x} \times \operatorname{curl} \mathbf{F} \, d\mathbf{x} + \tfrac{1}{2} \int \mathbf{x} \times \operatorname{curl}(\mathbf{u} \times \boldsymbol{\omega}) \, d\mathbf{x}. \end{aligned} \tag{10}$$

The following vector identity is readily established. (Note that this identity is not valid in two dimensions.)

$$\int \mathbf{x} \times \operatorname{curl} \mathbf{a} \, d\mathbf{x} = 2 \int \mathbf{a} \, d\mathbf{x} + \int \mathbf{x} \times (\mathbf{n} \times \mathbf{a}) \, dS. \tag{11}$$

Applying this to (10) with $\mathbf{a} = \mathbf{F}$ and $\mathbf{a} = \mathbf{u} \times \boldsymbol{\omega}$, together with the condition that $\mathbf{F}$ and $\boldsymbol{\omega}$ vanish at infinity so that the surface integrals vanish, we obtain

$$d\mathbf{I}/dt = \int \mathbf{F} \, d\mathbf{x} + \int \mathbf{u} \times \boldsymbol{\omega} \, d\mathbf{x}. \tag{12}$$

The second term is the integral of the vortex force and is convertible into a surface integral of the velocity by (1.3), which has an integrand $O(r^{-6})$ at infinity and therefore vanishes. The result follows.

Since $\int \nabla \Omega \, d\mathbf{x} = \int \Omega \mathbf{n} \, dS = 0$ if Ω is constant on S, we can add any conservative force field with a uniform potential at infinity to the non-conservative field $\mathbf{F}$, which is always arbitrary to this extent.

---

[5] Note that it has just been shown that $\int x_i \omega_j \, d\mathbf{x}$ is an antisymmetrical tensor.

The real momentum is not well defined in an infinite region, since in general $\int \mathbf{u} \, dx$ is only conditionally and not absolutely convergent. From the identity (11) with $\mathbf{a} = \mathbf{u}$, we obtain

$$\int \mathbf{u} \, dx = \tfrac{1}{2} \int \mathbf{x} \times \omega \, dx + \tfrac{1}{2} \int [\mathbf{u}(\mathbf{n} \cdot \mathbf{x}) - \mathbf{n}(\mathbf{u} \cdot \mathbf{x})] \, dS. \tag{13}$$

This shows that the fluid momentum (per unit density) in a volume V is the hydrodynamic impulse plus a surface integral. The latter is bounded as the surface goes to infinity, since the integrand is in general $O(r^{-2})$, but its value depends upon the shape of the surface. If V is a sphere of radius R, then in the surface integral $\mathbf{x} = R\mathbf{n}$ and $\mathbf{u} \sim (3(\mathbf{M} \cdot \mathbf{n})\mathbf{n} - \mathbf{M})/R^3$. Substitution and evaluation give a value $-\mathbf{I}/3$ for the surface integral. Thus (13) gives

$$\int_V \mathbf{u} \, dx = \tfrac{2}{3} \, \mathbf{I}, \tag{14}$$

when V is a sphere of infinite radius.

Of course, a fluid of constant density bounded by rigid walls at rest has zero momentum, because its centre of mass is at rest. Equivalently, this follows from the identity

$$\int_V \mathbf{u} \, dx = \int_{\partial V} \mathbf{x}(\mathbf{u} \cdot \mathbf{n}) \, dS, \tag{15}$$

where $\mathbf{u} \cdot \mathbf{n} = 0$ on walls at rest. The identity (15) can also be used to derive (14) for unbounded flow.

The value of the impulse $\mathbf{I}$ given by (8) is independent of the choice of origin of $\mathbf{x}$ because of (5).

3.3 Impulsive generation from rest

Another interpretation of hydrodynamic impulse is provided by considering the impulsive forces necessary to generate instantaneously the motion from rest (Kelvin [1868]). Suppose that the fluid is at rest and impulsive forces $\mathbf{F}$ are instantaneously applied at time t_0 to produce the velocity field $\mathbf{u}(\mathbf{x}, t_0)$. The impulsive field can be written

$$\mathbf{F}(\mathbf{x}, t) = \mathbf{f}(\mathbf{x}) \, \delta(t - t_0). \tag{1}$$

The velocity and pressure fields will have the expansions

$$\mathbf{u}(\mathbf{x}, t) = \mathbf{u}(\mathbf{x}) \, H(t - t_0) + \mathbf{u}^1(\mathbf{x}) \, H^1(t - t_0) + \cdots \tag{2}$$

$$p(\mathbf{x}, t) = P(\mathbf{x}) \, \delta(t - t_0) + p^1(\mathbf{x}) \, H(t - t_0) + \cdots, \tag{3}$$

where δ denotes the Dirac delta function, H is the Heaviside step function ($H' = \delta$), H^1 the indefinite integral of the Heaviside function, and so on. Substituting into the Euler equation (1.5.2), and equating the leading terms, we obtain

$$\mathbf{u}(\mathbf{x}) = -\nabla P + \mathbf{f}. \tag{4}$$

The quantity P is called the impulsive pressure. It appears because of the assumption that the fluid is incompressible.

Taking the divergence of (4), we obtain

$$\nabla^2 P = \operatorname{div} \mathbf{f}. \tag{5}$$

When the fluid is unbounded and $\mathbf{f}$ is of finite extent, this serves to define P apart from an arbitrary constant. The initial velocity field is then given by (4).

If the velocity field is given, then the impulsive force is first determined by taking the curl of (4), which gives

$$\operatorname{curl} \mathbf{f} = \boldsymbol{\omega}(\mathbf{x}, t). \tag{6}$$

This equation determines $\mathbf{f}$ to the extent of the gradient of a single-valued potential which can be absorbed into P. In simply connected regions where $\boldsymbol{\omega} = 0$, we can take $\mathbf{f} = 0$, and (4) shows that $P = -\phi$, where ϕ is the velocity potential which exists in irrotational regions. In multiconnected regions with $\boldsymbol{\omega} = 0$, it is still possible to assume $\mathbf{f} = 0$, but it may be necessary to introduce barriers across which the impulsive pressure jumps if ϕ is not single valued. Such a barrier with a pressure jump $[P]$ is equivalent to a singular distribution of impulsive force $\mathbf{f} = [P]\mathbf{n}\,\delta(n)$, where $\mathbf{n}$ is the normal to the barrier.

The total impulse applied to the fluid is (using (2.11))

$$\int \mathbf{f}\,d\mathbf{x} = \tfrac{1}{2} \int \mathbf{x} \times \operatorname{curl} \mathbf{f}\,d\mathbf{x} = \tfrac{1}{2} \int \mathbf{x} \times \boldsymbol{\omega}\,d\mathbf{x} = \mathbf{I}. \tag{7}$$

The invariance of the hydrodynamic impulse in unbounded flow is equivalent to the constancy of the total impulsive force required to generate the motion from rest, or equivalently to bring the flow instantaneously to rest.

The difference between the impulse and the momentum in an infinite region of given shape can be related to the momentum flux across the infinite surface due to the pressure forces. For an arbitrary fixed volume, we have (combining (1.1) and (1.3), or directly from the Euler equations)

$$\frac{\partial}{\partial t} \int \mathbf{u}\,d\mathbf{x} = -\int [p\,\mathbf{n} + \mathbf{u}(\mathbf{u} \cdot \mathbf{n})]\,dS + \int \mathbf{F}\,d\mathbf{x}. \tag{8}$$

At infinity, the contribution of the velocity to the surface integral is zero, and

$$p \sim -\frac{\partial \phi}{\partial t} \sim \frac{1}{4\pi r^3} \mathbf{x} \cdot \frac{d\mathbf{I}}{dt} \tag{9}$$

from Bernoulli's theorem and (2.2) and (2.7). For a spherical surface, the integral of the pressure force given by (9) over the surface is $-\frac{1}{3} d\mathbf{I}/dt$.

If the fluid is confined within rigid walls at infinity, on which $\mathbf{u} \cdot \mathbf{n} = 0$, the asymptotic formulae do not hold on the walls and the fluid momentum is zero. In this case, (8) shows that the impulsive force communicated to the fluid is transmitted instantaneously to the walls by pressure forces, where it will be balanced by external forces needed to keep the container at rest, in accordance with Newton's third law.

3.4 Effect of compressibility

The apparent paradoxical nature of some of the results connected with momentum is due to the assumption of incompressible fluid, in which pressure forces are transmitted with infinite speed. The introduction of a small compressibility eliminates the divergence of the momentum integral. To illustrate this point, we consider the motion generated by the application of a concentrated impulsive force in a slightly compressible fluid. The problem is made tractable by linearising in the velocity and disturbances to the pressure and density. The treatment can be justified as giving the dominant part of the flow field at large distances (see Landau and Lifshitz [1959 §73]).

The equations of motion are

$$\rho_0 \, \partial \mathbf{u}/\partial t = -\nabla p + \rho_0 \mathbf{I} \, \delta(\mathbf{x}) \, \delta(t), \tag{1}$$

$$\partial \rho/\partial t = -\rho_0 \, \text{div} \, \mathbf{u}, \tag{2}$$

$$p - p_0 = c^2 \, (\rho - \rho_0), \tag{3}$$

where p_0, ρ_0 are the undisturbed values of pressure and density, $\delta(\mathbf{x}) = \delta(x) \, \delta(y) \, \delta(z)$ is the three-dimensional delta function, c is the velocity of sound and $\rho_0 \mathbf{I}$ is the hydrodynamic impulse. The initial conditions are $\mathbf{u} = 0$, $p = p_0$, $\rho = \rho_0$ for $t < 0$. It follows from (1) that the vorticity is concentrated at the origin,

$$\omega = \nabla \delta(\mathbf{x}) \times \mathbf{I} \, H(t), \tag{4}$$

so that $\mathbf{u} = \nabla\phi$ for $\mathbf{x} \neq 0$. It is easily shown (using $\int \mathbf{x} \cdot \nabla\delta(\mathbf{x})\,d\mathbf{x} = -3$, $\int (\mathbf{x} \cdot \mathbf{I})\nabla\delta(\mathbf{x})\,d\mathbf{x} = -\mathbf{I}$) that

$$\frac{1}{2}\int \mathbf{x} \times \omega\,d\mathbf{x} = \mathbf{I}. \tag{5}$$

The elimination of ρ and $\mathbf{u}$ leads to the equation

$$(1/c^2)\,\partial^2 p/\partial t^2 - \nabla^2 p = -\mathbf{I} \cdot \nabla\,\delta(\mathbf{x})\,\delta(t). \tag{6}$$

Now it can be shown that the solution of

$$(1/c^2)\,\partial^2 \chi/\partial t^2 - \nabla^2 \chi = \delta(\mathbf{x})\,H(t) \tag{7}$$

is[6]

$$\chi = (1/4\pi r)\,H(t - r/c). \tag{8}$$

Hence, differentiating (8) with respect to t and taking the scalar product of the spatial gradient with $\mathbf{I}$, we obtain

$$\begin{aligned} p &= p_0 - (\mathbf{I}/4\pi) \cdot \nabla[\delta(t - r/c)/r] \\ &= p_0 + \frac{\mathbf{I} \cdot \mathbf{x}}{4\pi r^3}\,\delta(t - r/c) + \frac{\mathbf{I} \cdot \mathbf{x}}{4\pi c r^2}\,\delta'(t - r/c). \end{aligned} \tag{9}$$

Since $\partial\phi/\partial t = -(p - p_0)/\rho_0$, the velocity potential is

$$\phi = \frac{\mathbf{I}}{4\pi\rho_0} \cdot \nabla\frac{H(t - r/c)}{r}. \tag{10}$$

For $r > ct$, $p = p_0$ and $\phi = 0$. Inside the sphere $r = ct$, the velocity is that due to a dipole of strength $\mathbf{I}/4\pi$, with momentum $2\rho_0\mathbf{I}/3$. This follows from (2.13) and (4). The pressure, given by (9), is constant inside the sphere because the linearization neglects terms quadratic in the velocity. The 'lost' momentum is carried by the spherical front of radius ct, as the amount transmitted is

$$-\int dt \int (p - p_0)\mathbf{n}\,dS = -\tfrac{1}{3}\rho_0\mathbf{I}, \tag{11}$$

on substituting (9) and integrating first over the surface of the sphere and then with respect to t.

---

[6] For example, take the Laplace transform with respect to t and use the result that $\exp(-sr/c)/4\pi rs$ solves $(s^2/c^2)\chi - \nabla^2\chi = \delta(\mathbf{x})/s$.

3.5 Angular impulse

The moment of the impulsive force system which generates the motion from rest is called the angular impulse. Denoting it by $\mathbf{A}$, we have

$$\mathbf{A} = \int \mathbf{x} \times \mathbf{f}\,dV = -\tfrac{1}{2} \int r^2\,\mathrm{curl}\,\mathbf{f}\,dV - \tfrac{1}{2} \int r^2\,(\mathbf{n} \times \mathbf{f})\,dS. \tag{1}$$

Substituting for $\mathrm{curl}\,\mathbf{f}$ from (3.6) and noting that the surface integral vanishes for a finite force distribution, we have

$$\mathbf{A} = -\tfrac{1}{2} \int r^2 \boldsymbol{\omega}\,dV. \tag{2}$$

The angular impulse is also an invariant of the motion, in the sense that its rate of change in unbounded flow is the moment of the external non-conservative forces. It is easy to show that

$$\frac{d\mathbf{A}}{dt} = -\tfrac{1}{2} \int r^2 \frac{d\boldsymbol{\omega}}{dt}\,dV$$

$$= \int \mathbf{x} \times \mathbf{F}\,dV \tag{3}$$

$$+ \tfrac{1}{2} \int [r^2\,(\mathbf{n} \times \mathbf{F}) + r^2\,\mathbf{n} \times (\mathbf{u} \times \boldsymbol{\omega}) - \tfrac{1}{2}\mathbf{u}^2\,(\mathbf{x} \times \mathbf{n}) + \mathbf{x} \times \mathbf{u}\,(\mathbf{u} \cdot \mathbf{n})]\,dS.$$

Since the surface integrals vanish over a surface at infinity,

$$\frac{d\mathbf{A}}{dt} = \int \mathbf{x} \times \mathbf{F}\,dV. \tag{4}$$

The expression (2) was given by Lamb[1932 §152]. An alternative form,

$$\mathbf{A} = \tfrac{1}{3} \int \mathbf{x} \times (\mathbf{x} \times \boldsymbol{\omega})\,dV, \tag{5}$$

is given by Batchelor [1967 §7.2]. It can be shown that

$$\tfrac{1}{3} \int \mathbf{x} \times (\mathbf{x} \times \boldsymbol{\omega})\,dV + \tfrac{1}{2} \int r^2 \boldsymbol{\omega}\,dV = -\tfrac{1}{6} \int r^2\,\mathbf{x}\,(\boldsymbol{\omega} \cdot \mathbf{n})\,dS, \tag{6}$$

so the two definitions are equivalent when $\boldsymbol{\omega} \cdot \mathbf{n} = 0$ on the boundary.

The angular momentum of the fluid is in general a divergent integral. From the identity expressed in (1) with $\mathbf{f}$ replaced by $\mathbf{u}$,

$$\int \mathbf{x} \times \mathbf{u}\,dV = \mathbf{A} - \tfrac{1}{2} \int r^2\,(\mathbf{n} \times \mathbf{u})\,dS. \tag{7}$$

However, the surface integral vanishes if S is a sphere containing all the vorticity, since then $r^2 = \mathrm{constant}$ and

$$\int \mathbf{n} \times \mathbf{u}\,dS = \int \boldsymbol{\omega}\,dV = \int \mathbf{x}\,(\boldsymbol{\omega} \cdot \mathbf{n})\,dS = 0. \tag{8}$$

In this case the angular impulse is the angular momentum of a spherical region of fluid containing all the vorticity. This is to be expected since the pressure forces exert no torque on a sphere.

3.6 Effect of viscosity

In unbounded fluid, the invariance of $\mathbf{I}$ and $\mathbf{A}$ remains true in the presence of viscosity. The Navier–Stokes momentum equation for a fluid of uniform density is

$$\frac{d\mathbf{u}}{dt} = -\nabla p + \mathbf{F} + \nu \nabla^2 \mathbf{u}. \tag{1}$$

Thus the action of viscosity is equivalent to an extra body force $\nu \nabla^2 \mathbf{u}$. Since

$$\int \nabla^2 \mathbf{u}\, dV = -\int (\mathbf{n} \cdot \nabla)\mathbf{u}\, dS \tag{2}$$

and

$$\int \mathbf{x} \times \nabla^2 \mathbf{u}\, dV = -\int \omega\, dV - \int \mathbf{x} \times (\mathbf{n} \cdot \nabla)\mathbf{u}\, dS, \tag{3}$$

where the right-hand sides vanish when the integral is over all space since $\mathbf{u} = O(r^{-3})$, it follows that viscous forces have no effect on the rate of change of hydrodynamic impulse and angular impulse.

Note, however, that $\int \omega\, dV$ is not necessarily zero in two-dimensional flow, and hence in this case $\mathbf{A}$ may not be invariant in the presence of viscosity (see §10).

3.7 Impulse of isolated vortices

In the discussion so far, it has been supposed that the fluid was unbounded and the invariance of the impulse of the entire flow was demonstrated. Now we suppose that we have an isolated vortex, that is, a closed vorticity distribution of finite extent surrounded by irrotational fluid in an external velocity field produced by other vortices, or equivalently the motion of bodies. Let V_{v} denote the volume occupied by the vortex and S_{v} its surface. By hypothesis,

$$\omega \cdot \mathbf{n} = 0 \quad \text{on } S_{\mathrm{v}}, \tag{1}$$

and hence[7]

$$\int \omega \, dV_v = - \int \mathbf{x} \, (\omega \cdot \mathbf{n}) \, dS_v = 0. \tag{2}$$

If non-conservative forces are acting, the volume occupied by vorticity may not be material. In this case, it is understood that V_v is a material volume containing the finite region in which the non-conservative forces act. Since

$$\frac{d}{dt} \int \omega \, dV_v = \int \frac{D\omega}{Dt} \, dV_v = \int (\omega \cdot \mathbf{n}) \mathbf{u} \, dS_v + \int \mathbf{n} \times \mathbf{F} \, dS_v, \tag{3}$$

it is sufficient for consistency that $\mathbf{F}$ either vanishes on, or is normal to, the bounding surface so that

$$\mathbf{n} \times \mathbf{F} = 0 \quad \text{on } S_v. \tag{4}$$

(A surface across which $\mathbf{n} \times \mathbf{F}$ is discontinuous becomes a vortex sheet; see §2.2.)

We define the impulse $\mathbf{I}_v$ and angular impulse $\mathbf{A}_v$ of the vortex by the expressions used for unbounded flow, that is,

$$\mathbf{I}_v = \tfrac{1}{2} \int \mathbf{x} \times \omega \, dV_v, \tag{5}$$

$$\mathbf{A}_v = -\tfrac{1}{2} \int \mathbf{x}^2 \, \omega \, dV_v. \tag{6}$$

The velocity in the vortex is $\mathbf{u} = \mathbf{u}_v + \mathbf{u}_e$, where $\mathbf{u}_v$ is given by the integral (1.1.5) over V_v, and $\mathbf{u}_e (= \nabla \Phi)$ is due to other vortices or the motion of bodies. The external velocity $\mathbf{u}_e$ contains the velocity induced by the image vorticity produced by the vortex.

We can interpret $\mathbf{I}_v$ and $\mathbf{A}_v$ again in terms of the impulse and moment of impulse of the impulsive force system $\mathbf{f}$ inside V_v required to generate the vorticity instantaneously. The relations between the impulsive forces and the velocity and vorticity fields are given by (3.4) and (3.6). From the identity (2.11), with $\mathbf{a} = \mathbf{f}$, we obtain immediately

$$\mathbf{I}_v = \int \mathbf{f} \, dV_v, \tag{7}$$

provided (4) holds for the impulsive force on the boundary S_v. Under the same condition, it follows from (5.1) that

$$\mathbf{A}_v = \int \mathbf{x} \times \mathbf{f} \, dV_v. \tag{8}$$

---

[7] It is supposed that the motion is three-dimensional. A separate discussion is required for two-dimensional motion where (2) need not be true.

A comment is in order here concerning the impulsive pressure P. When the impulsive forces are given, P is determined by the Poisson equation (3.5). In the presence of boundaries, however, it is necessary to give a boundary condition. The appropriate condition is a Neumann one on the normal derivative of P given by the requirement that the normal component of equation (3.4) is satisfied on the boundary; that is,

$$\partial P/\partial n = (\mathbf{f} \cdot \mathbf{n} - \mathbf{U} \cdot \mathbf{n}), \tag{9}$$

where $\mathbf{U}$ is the velocity of the boundary and $\mathbf{u} \cdot \mathbf{n} = \mathbf{U} \cdot \mathbf{n}$ must be satisfied. The tangential components of the velocity field will in general not equal the tangential velocity of the boundary, and hence a vortex sheet is usually produced at walls.

The impulses of an isolated vortex are in general not invariant in the presence of external velocities. We have

$$\begin{aligned}
\frac{d\mathbf{I}_v}{dt} &= \tfrac{1}{2} \int \left(\mathbf{x} \times \frac{D\omega}{Dt} + \mathbf{u} \times \omega \right) dV_v \\
&= \int (\mathbf{F} + \tfrac{1}{2}\mathbf{x} \times (\omega \cdot \nabla)\mathbf{u} + \tfrac{1}{2}\mathbf{u} \times \omega) \, dV_v,
\end{aligned} \tag{10}$$

where we have used (4) and the identity (2.11) for the contribution from F. Now

$$\int \mathbf{x} \times (\omega \cdot \nabla)\mathbf{u} \, dV_v = \int \mathbf{u} \times \omega \, dV_v - \int (\mathbf{u} \times \mathbf{x})\omega \cdot \mathbf{n} \, dS_v. \tag{11}$$

Hence,

$$\frac{d\mathbf{I}_v}{dt} = \int \mathbf{F} \, dV_v + \int \mathbf{u} \times \omega \, dV_v - \tfrac{1}{2} \int (\mathbf{u} \times \omega)\omega \cdot \mathbf{n} \, dS_v. \tag{12}$$

It follows from (1.3), the fact that $\omega = \operatorname{curl} \mathbf{u}_v$ inside S_v and that the surface integral of $\mathbf{u}_v$ in (1.3) may be taken over the surface at infinity and therefore vanishes, that

$$\int \mathbf{u}_v \times \omega \, dV_v = 0. \tag{13}$$

Equation (13) says that the total vortex force exerted by a vortex on itself (excluding effect of images) is zero. Hence, combining (12) and (1), we have

$$\frac{d\mathbf{I}_v}{dt} = \int \mathbf{F} \, dV + \int \mathbf{u}_e \times \omega \, dV_v. \tag{14}$$

The second integral in (14) can be interpreted as the vortex force due to the external velocity. Suppose that we imagine the vortex to be confined in a vessel with rigid walls and made steady by applying an external force to the vortex in order to balance the vortex force. From Newton's third law, the external force is equal to the force exerted by the fluid on the walls in the steady motion. Hence, if the vortices are free to move inside the rigid vessel, the rate of change of impulse is the force exerted by the walls of the vessel on the fluid in the instantaneously equivalent steady motion. This also follows from the identity (1.3), which gives, using (12),

$$\int \mathbf{u}_e \times \boldsymbol{\omega} \, dV_v = \int \mathbf{u} \times \boldsymbol{\omega} \, dV_v = \int (\tfrac{1}{2} \mathbf{u}^2 \mathbf{n} - \mathbf{u}(\mathbf{u} \cdot \mathbf{n})) \, dS_v, \tag{15}$$

and Bernouilli's equation (1.9.3), combined with the observation that H is constant and $\mathbf{u} \cdot \mathbf{n} = 0$ on the walls. Of course, the vessel can be hypothetical; any stream surface containing the vortex would be a possible containing surface. As a simple application of this principle, enunciated by Betz [1932], it follows that for a vortex bounded by a plane wall, the components of impulse parallel to the wall stay constant.

Similarly,

$$\frac{d\mathbf{A}_v}{dt} = -\tfrac{1}{2} \int \left(r^2 \frac{D\boldsymbol{\omega}}{Dt} + 2(\mathbf{u} \cdot \mathbf{x})\boldsymbol{\omega} \right) dV_v$$

$$= \int \mathbf{x} \times \mathbf{F} \, dV_v - \int (\tfrac{1}{2} r^2 (\boldsymbol{\omega} \cdot \nabla) \mathbf{u} + (\mathbf{u} \cdot \mathbf{x})\boldsymbol{\omega}) \, dV_v, \tag{16}$$

where we have used (5.1) and (4) for the contribution from $\mathbf{F}$. Now

$$\int \tfrac{1}{2} r^2 (\boldsymbol{\omega} \cdot \nabla \mathbf{u} + (\boldsymbol{\omega} \cdot \mathbf{x})\mathbf{u}) \, dV_v = - \int \tfrac{1}{2} r^2 (\boldsymbol{\omega} \cdot \mathbf{n}) \mathbf{u} \, dS_v = 0. \tag{17}$$

Hence the second term on the right-hand side of (16) is

$$\int \mathbf{x} \times (\mathbf{u} \times \boldsymbol{\omega}) \, dV_v = \int ((\mathbf{n} \times \mathbf{x}) \tfrac{1}{2} \mathbf{u}^2 + (\mathbf{x} \times \mathbf{u})\mathbf{u} \cdot \mathbf{n}) \, dS_v, \tag{18}$$

and the right-hand side vanishes when $\mathbf{u} = \mathbf{u}_v$ and the surface is then moved to infinity. Hence,

$$\frac{d\mathbf{A}_v}{dt} = \int \mathbf{x} \times \mathbf{F} \, dV_v + \int \mathbf{x} \times (\mathbf{u}_e \times \boldsymbol{\omega}) \, dV_v, \tag{19}$$

where the right-hand side can be interpreted as the moment of the external forces and vortex force due to the external velocity. The latter is equivalently the couple applied by the (hypothetical) containing vessel in the instantaneously equivalent steady motion.

3.8 Impulse of a line vortex

The expressions for impulse are finite when the singular distribution of a line vortex is substituted for ω. A line vortex of strength Γ lying on the curve $\mathbf{R}(s)$ has impulse

$$\mathbf{I} = \tfrac{1}{2}\Gamma \oint \mathbf{R} \times d\mathbf{s} \tag{1}$$

and angular impulse

$$\mathbf{A} = -\tfrac{1}{2}\Gamma \oint R^2 \, d\mathbf{s}. \tag{2}$$

The fluid outside the vortex is in irrotational motion with a multivalued velocity potential ϕ, which increases by Γ around a circuit threaded once by the line vortex. It can be made single valued by introducing a barrier edged by the vortex, across which the potential jumps. A possible impulsive force that would generate the vortex is an impulsive pressure $-\phi$, which is equivalent to a pressure difference Γ acting over the barrier S. Thus

$$\mathbf{I} = \Gamma \int d\mathbf{S}, \tag{3}$$

$$\mathbf{A} = \Gamma \int \mathbf{x} \times d\mathbf{S}, \tag{4}$$

where the direction of $d\mathbf{S}$ is that of the circulation taken in a right-handed sense. The equivalence of (1) and (2) with (3) and (4), respectively, is a consequence of the Stokes theorem extensions,

$$\oint \psi \, d\mathbf{s} = \int d\mathbf{S} \times \nabla\psi, \qquad \oint \mathbf{a} \times d\mathbf{s} = \int d\mathbf{S} \operatorname{div} \mathbf{a} - (\nabla \mathbf{a}) \cdot d\mathbf{S}.$$

The use of (3) may lead to apparently paradoxical results. Consider the example of a vortex ring of radius R in a tube of radius R_0, the axes being parallel as shown in Figure 3.8-1. The impulse is $\pi\Gamma R^2$ along the axis, either directly from (1) or from (3) applied on the barrier BC across which the potential ϕ has a jump of $-\Gamma$ from left to right. An alternative way to make the potential and impulsive pressure single valued is to introduce an annular barrier AB, CD connecting the vortex to the walls. Across this barrier, the potential, now denoted by Φ, has jumps of Γ from left to right, giving an impulse of $\pi\Gamma(R^2 - R_0^2)$. The apparent paradox is resolved by noting that Φ does not vanish at infinity. From its value of $-\tfrac{1}{2}\Gamma$ at infinity on the left and $\tfrac{1}{2}\Gamma$ at infinity on the right comes a contribution $\pi\Gamma R_0^2$.

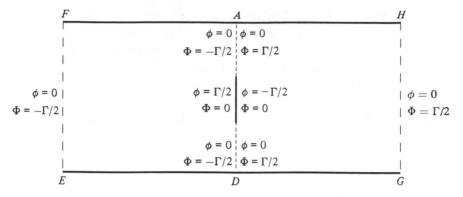

3.8-1 Velocity potential for impulsive motion in a channel.

3.9 Vortex centroid

It follows from (2.5) that

$$\int \omega \, dV_v = 0, \tag{1}$$

since $\omega \cdot \mathbf{n} = 0$ on the surface of an isolated vortex. Thus the centroid for a single vortex cannot be defined in the conventional manner as the first moment of the vorticity divided by the total vorticity. An alternative approach employed by Saffman [1970] in a study of viscous vortex rings was to define the centroid by the expression

$$\mathbf{X}_v = \tfrac{1}{2} \int \frac{\mathbf{x} \times \omega \cdot \mathbf{I}}{I^2} \, \mathbf{x} \, dV_v, \tag{2}$$

where $I = |\mathbf{I}|$ and the impulse $\mathbf{I}$ is defined by (7.5). The velocity $\mathbf{U}_v$ of the vortex can then be defined simply as

$$\mathbf{U}_v = d\mathbf{X}_v/dt. \tag{3}$$

These definitions localise the centroid in the region of large vorticity, put the centroid at the geometrical centroid of a symmetrical vortex such as a vortex ring or Hill spherical vortex and move the centroid with the correct velocity when the distribution translates unchanged. For an axisymmetric vortex with azimuthal vorticity $\omega_\theta(r, z)$, using cylindrical polar co-ordinates, the centroid is on the axis and the z-co-ordinate coincides with the weighted first moment of ω_θ; that is,

$$Z_v = \int 2\pi r z \omega_\theta \, dr \, dz \Big/ \int\int 2\pi r \omega_\theta \, dr \, dz. \tag{4}$$

One drawback of the definition is that in two dimensions this centroid is not related in a simple way to the conventional one which can then be employed because the average vorticity is not necessarily zero. An alternative approach could be to take the centroid as a point on the Poinsot axis of the impulsive forces which generate the motion. This is the line parallel to $\mathbf{I}_v$ and passing through $\mathbf{R} = \mathbf{I}_v \times \mathbf{A}_v/I_v^2$ such that the resultant of the impulsive forces is along the line together with a parallel couple $\mathbf{A}_v - \mathbf{R} \times \mathbf{I}_v$.

3.10 Impulse in two dimensions

A flow is said to be two-dimensional when all flow quantities are independent of a Cartesian co-ordinate, usually denoted by z or x_3. Then the equations of motion for the components of velocity (u, v) in the transverse direction and for the z-component of vorticity, ω_z, uncouple from the velocity, w, along the z-axis and the transverse components of vorticity. For motion in the plane, we may therefore assume without loss of generality that $w = 0$ and that the vorticity is perpendicular to the plane of motion. The facts that $\boldsymbol{\omega}$ is now essentially a conserved scalar (there is no amplification by vortex line stretching), and that one scalar (the stream function) determines both u and v, make problems more tractable, both analytically and numerically. Two-dimensional flows have attracted most of the theoretical attention devoted to vortex motion.

Much of the preceding discussion applies, with obvious modifications, to vortices in two dimensions, but some changes are needed because the extent of the vorticity field is now unbounded in the three-dimensional sense.

We consider the expression, analogous to (1.1.5), for the velocity components in the plane induced by the vorticity in the z-direction. It can be obtained by carrying out the integration with respect to z (the so-called method of descent). Denoting by $\mathbf{r}$ the two-dimensional vector (x, y) and by $\mathbf{R}$ the three-dimensional vector (x,y,z), we have

$$\int (\mathbf{R} - \mathbf{R}')/|\mathbf{R} - \mathbf{R}'|^3 \, dz' = 2(\mathbf{r} - \mathbf{r}')/|\mathbf{r} - \mathbf{r}'|^2.$$

Hence from (1.1.5) when $\boldsymbol{\omega} = \omega\mathbf{k}$,

$$\mathbf{u}_v(\mathbf{r}) = (1/2\pi) \int \omega(\mathbf{r}')\, \mathbf{k} \times (\mathbf{r} - \mathbf{r}')/|\mathbf{r} - \mathbf{r}'|^2 \, dS' \tag{1}$$

is the formula in two dimensions ($dS' = dx'\,dy'$). Condition (vi) of §1.1 is violated, but it is easy to verify, by methods similar to those employed in three dimensions, that div $\mathbf{u}_v = 0$ and curl $\mathbf{u}_v = \omega\mathbf{k}$ if ω is of finite extent (or exponentially small at infinity) in the plane. Condition (v) of §1.1 is irrelevant, as the vorticity is automatically tangent to the boundaries.

Alternatively, we introduce the (Lagrange) stream function $\psi(x, y, t)$, where the vector potential of the motion in the plane is $\psi\mathbf{k}$ and

$$u = \psi_y, \quad v = -\psi_x, \tag{2}$$

the existence of ψ being guaranteed in incompressible fluid by the equation of continuity

$$u_x + v_y = 0. \tag{3}$$

The curves $\psi = $ const. are the (instantaneous) streamlines of the flow. Evaluating curl $\mathbf{u}$, we find

$$\omega = -\nabla^2\psi, \tag{4}$$

where ∇^2 is now the two-dimensional Laplacian. A solution of the Poisson equation (4) is

$$\psi(\mathbf{r}) = -(1/2\pi) \int \omega' \log|\mathbf{r} - \mathbf{r}'|\,dS', \tag{5}$$

from which follows, on differentiating under the integral sign (a procedure which can be justified), that

$$u = -(1/2\pi) \int \omega'\,(y - y')/|\mathbf{r} - \mathbf{r}'|^2\,dS',$$

$$v = (1/2\pi) \int \omega'\,(x - x')/|\mathbf{r} - \mathbf{r}'|^2\,dS', \tag{6}$$

which is (1).

For uniform fluid, the Helmholtz equation takes the simple form

$$d\omega/dt = \partial\omega/\partial t + u\,\partial\omega/\partial x + v\,\partial\omega/\partial y = \mathbf{k} \cdot \text{curl}\,\mathbf{F}, \tag{7}$$

and the statement that vortex lines move with the fluid when $\mathbf{F} = 0$ says that vorticity is conserved following a fluid particle.

For two-dimensional motion, a finite region in the plane with non-zero vorticity surrounded by irrotational fluid will be referred to as a vortex. An example is the uniform rectilinear vortex filament discussed in §2.3. If the vorticity is concentrated along a curve, we have a vortex sheet. When the vorticity is concentrated into a point, that is,

$$\omega = \Gamma\delta(x - x_o)\,\delta(y - y_o) = \Gamma\delta(\mathbf{r} - \mathbf{r}_o), \tag{8}$$

we have the limit of a rectilinear line vortex, conveniently called a point vortex of strength Γ for motion in a plane, with stream function

$$\psi(\mathbf{r}) = -\Gamma/2\pi \log |\mathbf{r} - \mathbf{r}_o|. \tag{9}$$

As discussed in §2.3, the point vortex is a meaningful dynamical concept.

In unbounded fluid, $\mathbf{u}_v$ given by (1) is the unique velocity produced by a vorticity field of finite extent. To determine the asymptotic behaviour of the velocity at infinity, we note that

$$\log |\mathbf{r} - \mathbf{r}'| = \log r - \mathbf{r} \cdot \mathbf{r}'/r^2 + O(1/r^2), \tag{10}$$

where $r = |\mathbf{r}|$. Substitution into (5) gives

$$\psi \sim -\Gamma_\infty/2\pi \log r + \mathbf{M} \times \mathbf{r} \cdot \mathbf{k}/r^2 + O(1/r^2), \tag{11}$$

where

$$\Gamma_\infty = \int \omega \, dS, \quad \mathbf{M} = \frac{1}{2\pi} \int \omega \mathbf{r} \times \mathbf{k} \, dS. \tag{12}$$

Equation (11) shows that the flow at infinity is a circulation of strength Γ_∞ plus a two-dimensional dipole of strength $\mathbf{M}$, as the latter gives rise to a velocity potential and stream function given by, respectively,

$$\phi = -\mathbf{M} \cdot \mathbf{r}/r^2, \qquad \psi = \mathbf{M} \times \mathbf{r} \cdot \mathbf{k}/r^2. \tag{13}$$

Although the total vorticity, or equivalently the circulation at infinity, need not vanish, it is an invariant of the motion. For

$$d\Gamma/dt = \int \partial\omega/\partial t \, dS = \int (\mathbf{k} \cdot \text{curl} \, \mathbf{F} - \mathbf{u} \cdot \nabla\omega) \, dS$$
$$= \oint_\infty \mathbf{F} \cdot d\mathbf{s} + \oint_\infty (\mathbf{u} \cdot \mathbf{n})\omega \, ds = 0, \tag{14}$$

when the external force and vorticity vanish at infinity.

The impulse and other quantities characteristic of the flow as a whole, such as the energy, are now to be understood in terms of per unit length in the z-direction, and their dimensions are therefore different from the corresponding quantities in three dimensions. The impulse is now taken to be the invariant satisfying (2.9) and is found to be

$$\mathbf{I} = \int \omega \mathbf{r} \times \mathbf{k} \, dS. \tag{15}$$

Notice the absence of the factor $\frac{1}{2}$ which is present in the three-dimensional expression (2.8). This result follows straightforwardly from direct manipulation of the two-dimensional Euler equations. (The missing factor can in fact be traced to the vortex lines not being closed in two-dimensional flow. If a unit length of flow is considered, and the vortex lines are closed over the end planes $z = 0$ and $z = 1$, it can be shown that the contribution to the three-dimensional integral (2.8) from the end planes accounts exactly for the missing factor in the two-dimensional formula.)

In terms of the impulsive force $\mathbf{f}$ required to generate the motion impulsively from rest, we use the identity in two dimensions

$$\int \mathbf{r} \times \operatorname{curl} \mathbf{a}\, dS = \int \mathbf{a}\, dS - \oint \mathbf{r} \times (\mathbf{n} \times \mathbf{a})\, ds. \tag{16}$$

(Notice the factor-2 difference from the analogous three-dimensional identity (2.11).) Substitute $\mathbf{a} = \mathbf{f}$, note that $\omega \mathbf{k} = \operatorname{curl} \mathbf{f}$, and obtain

$$\int \mathbf{f}\, dS = \int \omega \mathbf{r} \times \mathbf{k}\, dS = \mathbf{I}. \tag{17}$$

It is also assumed here that $\Gamma_\infty = 0$, for otherwise the impulsive force is not of finite extent and the contour integral may not vanish. If $\Gamma_\infty \neq 0$, then $\mathbf{f} \sim (1/2\pi r)\theta$ at ∞, since $\oint \mathbf{f} \cdot d\mathbf{s} = \int w\, dS = \Gamma_\infty$.

The definition (15) may still be applied when $\Gamma_\infty \neq 0$, except that the impulse will depend on the choice of origin. However, it follows from (14) that the contribution depending on the choice of origin is independent of time and so the invariance is unaffected. The case $\Gamma_\infty \neq 0$ is physically somewhat unrealistic, since the momentum and kinetic energy of such flows are unbounded, and for applications the existence of other vortices of opposite strength should be kept in mind. The identity (5.1) for the angular impulse has the same form in two dimensions as in three, and hence we can define the scalar angular momentum by the z-component of the same expression, that is,

$$\mathbf{A} = \int \mathbf{r} \times \operatorname{curl} \mathbf{f}\, dS = \frac{1}{2} \int r^2 \omega\, dS. \tag{18}$$

For an isolated single vortex in an external velocity field, $\mathbf{u}_e$, with impulse $\mathbf{I}_v$ and angular impulse $\mathbf{A}_v$ occupying an area S_v, the results of §7 still hold, namely,

$$d\mathbf{I}_v/dt = \int \mathbf{F}\, dS_v + \int \omega\, \mathbf{u}_e \times \mathbf{k}\, dS_v, \tag{19}$$

$$d\mathbf{A}_v/dt = \int \mathbf{r} \times \mathbf{F}\, dS + \int \omega\, \mathbf{r} \times (\mathbf{u} \times \mathbf{k})\, dS. \tag{20}$$

The proofs proceed as for three dimensions; the contour integrals at infinity all vanish when the vorticity field is of finite extent and the velocity is $O(1/r)$ at infinity. By the use of (7.15) and (7.18), which apply in two dimensions, the vortex force contributions to $d\mathbf{I}_v/dt$ and $d\mathbf{A}_v/dt$ can be expressed as integrals of the velocity over surrounding contours or equivalently in terms of the force and couple on a containing wall in the instantaneously kinematically identical steady motion. The total vorticity Γ in the vortex is constant from (14).

For the case of a single vortex moving steadily without change of shape with velocity $\mathbf{U}$ in the absence of any external velocity field, it follows from (19) on working in a co-ordinate system moving with the vortex that a total external force $\Gamma \mathbf{k} \times \mathbf{U}$ is required to maintain the motion (see §2.3 and §1.)

Since in two dimensions, the total vorticity $\Gamma = \int \omega \, dS_v$ need not vanish, we can in this case define the centroid of vorticity by the conventional formula

$$\overline{\mathbf{R}}_v = \int \omega \, \mathbf{r} \, dS_v/\Gamma = \mathbf{k} \times \mathbf{I}_v/\Gamma, \qquad \mathbf{I}_v = \Gamma \, \overline{\mathbf{R}}_v \times \mathbf{k}. \tag{21}$$

This definition is not, however, simply related to (9.2) used for vortices in three dimensions, which gives

$$\mathbf{R}_v = (\mathbf{i} \, \overline{Y}_v + \mathbf{j} \overline{X}_v) \int \omega \, xy \, dS_v/(\Gamma \overline{\mathbf{R}}_v^{-2}). \tag{22}$$

The two-dimensional centroid (21) has the property that the impulse measured relative to it is zero. The angular impulse relative to the centroid is a minimum and defines a radius of gyration R_g by

$$\Gamma R_g^2 = \int |\mathbf{r} - \overline{\mathbf{R}}_v|^2 \, \omega \, dS_v. \tag{23}$$

Since

$$A_v = -\tfrac{1}{2} \Gamma \, (\overline{X}_v^2 + \overline{Y}_v^2) - \tfrac{1}{2} \Gamma R_g^2, \tag{24}$$

and $d\overline{X}_v/dt$ and $d\overline{Y}_v/dt$ follow from (19), equation (20) can be written in the form

$$\Gamma \, dR_g^2/dt = -2\mathbf{k} \cdot \int (\mathbf{r} - \overline{\mathbf{R}}_v) \times \mathbf{F} \, dS_v + 2 \int \omega \, (\mathbf{r} - \overline{\mathbf{R}}_v) \cdot \mathbf{u}_e \, dS_v. \tag{25}$$

Note that R_g is constant if the external forces and the vorticity weighted external velocity have no moment about the centroid.[8]

It follows (Betz [1932]) from (19) and (21) that the velocity $\overline{\mathbf{U}}_v$ of the two-dimensional centroid in the absence of external forces is the vorticity

---

[8] This is the basis of the Betz approximation for the roll-up of vortex sheets (see §8.4).

weighted centroid of the vortex velocity; that is,

$$\overline{\mathbf{U}_v} = \int \omega \, \mathbf{u}_e \, dS/\Gamma. \tag{26}$$

This formula is particularly useful when ω is constant.

The effects of viscosity in two-dimensional flow are slightly different. In unbounded flow, the total impulse $\mathbf{I}$ remains invariant, because on putting $\mathbf{F} = \nu\nabla^2\mathbf{u}$, it can be seen that the boundary integral in (6.2) vanishes when the velocity has the form given asymptotically by (11). On the other hand, the total angular impulse $\mathbf{A}$ is not invariant. The two terms on the right-hand side of (6.3) make equal contributions $-\Gamma_\infty\mathbf{k}$, so that the angular impulse of the entire distribution obeys the equation

$$dA/dt = -2\nu\Gamma_\infty. \tag{27}$$

Hence,

$$R_g^2 = 4\nu t + \text{const.}, \tag{28}$$

when R_g is defined. The constant decrease in angular impulse is due to a finite transfer of angular momentum by viscous stresses across the circle at infinity in two dimensions.

The relation between momentum and impulse in two dimensions follows from the identity (16) with $\mathbf{a} = \mathbf{u}$,

$$\int \mathbf{u} \, dS = \int \omega \mathbf{r} \times \mathbf{k} \, dS + \oint \mathbf{r} \times (\mathbf{n} \times \mathbf{u}) \, ds. \tag{29}$$

If the area is taken to be a large circle of radius R, on which

$$\mathbf{u} \sim (-\mathbf{I}R^2 + 2(\mathbf{I} \cdot \mathbf{r})\mathbf{r})/2\pi R^4, \tag{30}$$

that is, $\Gamma_\infty = 0$, we obtain the result (cf (2.14)),

$$\lim_{R \to \infty} \int_{r<R} \mathbf{u} \, dS = \tfrac{1}{2}\mathbf{I}. \tag{31}$$

Thus in two dimensions half the applied impulse is communicated to the momentum of unbounded fluid and the other half is transmitted to infinity.

3.11 Kinetic energy of vortices

It is known that the invariance of the hydrodynamic impulse in unbounded fluid in the absence of non-conservative forces corresponds to the invariance of the Euler equations to spatial displacement. Likewise, the invariance of the Euler equations to time and reflection is associated with two further invariants, the energy and the helicity.

The kinetic energy of the fluid in a volume V is

$$T = \int \tfrac{1}{2} \rho \mathbf{u}^2 \; dV. \tag{1}$$

If the fluid is of constant density, the following identity holds in three dimensions

$$\frac{1}{2} \int \mathbf{u}^2 \, dV = \int \mathbf{u} \cdot \mathbf{x} \times \boldsymbol{\omega} \, dV + \int [(\mathbf{u} \cdot \mathbf{x})(\mathbf{n} \cdot \mathbf{u}) - \tfrac{1}{2} \mathbf{u}^2 (\mathbf{n} \cdot \mathbf{x})] \, dS. \tag{2}$$

The surface integral vanishes in unbounded fluid over the surface at ∞, where $|\mathbf{u}| = O(1/r^3)$, and we have the formula (Lamb [1932 §153])

$$T = \int \mathbf{u} \cdot \mathbf{x} \times \boldsymbol{\omega} \, dV, \tag{3}$$

giving the kinetic energy as an integral over the vortices, independent of the origin since $\int \mathbf{u} \times \boldsymbol{\omega} \, dV = 0$ from (1.3). Note that this formula does not hold in two dimensions, where the right-hand side of (2) can be shown to vanish identically.

Introducing the vector potential $\mathbf{A}$, we have

$$\int \mathbf{u}^2 \, dV = \int \mathbf{A} \cdot \boldsymbol{\omega} \, dV + \int \mathbf{u} \times \mathbf{A} \, dS. \tag{4}$$

The surface integral vanishes for unbounded flow, giving

$$T = \tfrac{1}{2} \int \mathbf{A} \cdot \boldsymbol{\omega} \, dV \tag{5}$$

$$= (1/8\pi) \iint \boldsymbol{\omega}(\mathbf{x}) \cdot \boldsymbol{\omega}(\mathbf{x}') \, |\mathbf{x} - \mathbf{x}'|^{-1} \, dV \, dV'. \tag{6}$$

The expressions (4) and (5) also hold in two dimensions, provided $\Gamma_\infty = 0$, and in this case using the relation between the vector potential and the stream function, the kinetic energy (per unit length) is

$$T = \tfrac{1}{2} \int \omega \psi \, dS. \tag{7}$$

For axisymmetric motion, $\boldsymbol{\omega} = \omega_\theta \boldsymbol{\theta}$, $\mathbf{A} = \psi/r \, \boldsymbol{\theta}$. Then

$$T = \tfrac{1}{2} \int \omega_\theta \psi/r \, dV = \pi \iint \omega \psi \, dr \, dz \tag{8}$$

gives the kinetic energy of an axisymmetric distribution of vorticity.

The kinetic energy is invariant if the fluid is unbounded or confined by walls at rest, the external forces are conservative with a single-valued potential Ω and the viscosity is zero. Including the viscous term in the equation of motion, we have for incompressible fluid

$$\frac{d}{dt} \int \tfrac{1}{2} \mathbf{u}^2 \, dV = \int \mathbf{u} \cdot D\mathbf{u}/Dt \, dV$$

$$= \int [-\mathbf{u} \cdot \nabla p + \mathbf{u} \cdot \mathbf{F} + \nu \mathbf{u} \cdot \nabla^2 \mathbf{u}] \, dV \qquad (9)$$

$$= \int [p + \Omega](\mathbf{u} \cdot \mathbf{n}) \, dS + \nu \int \mathbf{u} \cdot \nabla^2 \mathbf{u} \, dV.$$

The surface integral vanishes if $\mathbf{u} \cdot \mathbf{n} = 0$ or the surface is at infinity. The viscous term can be transformed into the equivalent forms

$$\nu \int \mathbf{u} \cdot \nabla^2 \mathbf{u} \, dV = -\nu \int \tfrac{1}{2} (\partial u_i/\partial x_j + \partial u_j/\partial x_i)^2 \, dV$$

$$- \nu \int u_i n_j (\partial u_i/\partial x_j + \partial u_j/\partial x_i) \, dS \qquad (10)$$

$$= -\nu \int \omega^2 \, dV - \nu \int \mathbf{n} \cdot \mathbf{u} \times \omega \, dS. \qquad (11)$$

The second term on the right-hand side of (10) is the rate of working over the surface of the surface stresses. The integrand of the first term can then be interpreted as the energy dissipation rate per unit volume. If the surface integral in (11) vanishes, the total rate of energy dissipation can be computed as an integral over the vortices. (It should be kept in mind that in general viscosity will generate vorticity at walls, and the created vorticity should be included in (11).)

3.12 Helicity

The invariant associated with invariance under reflection is the helicity, defined by (Moffatt [1969])

$$J = \int \mathbf{u} \cdot \omega \, dV_v. \qquad (1)$$

The helicity of an isolated vortex acted upon by conservative forces with a single-valued potential in inviscid incompressible fluid is constant. Let V_v be a material volume containing the vortex, so that $\omega \cdot \mathbf{n} = 0$ on its surface S_v, and suppose that the fluid is incompressible, then

$$dJ/dt = \int (D\mathbf{u}/Dt \cdot \omega + \mathbf{u} \cdot D\omega/Dt) \, dV_v \qquad (2)$$

$$= \int \omega \cdot \mathbf{n} \left(\tfrac{1}{2} \mathbf{u}^2 - p - \Omega \right) dS_\mathrm{v} + 2 \int \mathbf{F} \cdot \omega \, dV_\mathrm{v} + \int \mathbf{n} \times \mathbf{F} \cdot \mathbf{u} \, dS_\mathrm{v},$$

where $\mathbf{F}$ is the non-conservative part of the external force field and Ω is the potential of the conservative part. Hence it follows that J is constant if $\mathbf{F} = 0$.[9]

Moffatt gives an interpretation of the helicity of a flow consisting of closed vortex filaments with axial flow in terms of the knottedness of the configuration (see also Kelvin [1868]). Suppose that at some initial instant the spiral vortex lines in each filament are closed, and moreover that each vortex line can be continuously deformed into the axis of the filament, defined in some suitable way. Then the circulation K around a circuit which consists of a vortex line will be the same for each vortex line in any one filament. Suppose the value is $K(i)$ for the ith filament, whose strength is Γ_i. The helicity of the filament is $J_i = \Gamma_i K_i$ (not summed) and is invariant by Kelvin's circulation theorem. Now the value of K_i is the flux of vorticity through a surface spanning the circuit. If the filament is not knotted – that is, can be shrunk continuously into a point – the contribution to K_i from the filament itself is zero, and its value is the number of times (counted algebraically) the surface is intersected by the other filaments weighted by the filament strength. That is,

$$K_i = \sum_j \alpha_{ij} \Gamma_j, \tag{3}$$

where α_{ij} is the winding number of filaments i and j and is a positive or negative integer (depending on orientation), or zero. If the ith filament is knotted, it makes a contribution α_{ii} to its own helicity, where α_{ii} is the self-winding number. Conservation of helicity is thus equivalent to the conservation of winding number under continuous deformations together with the conservation of circulation.

Since the velocity field due to one filament can be approximated by a line vortex along its axis, the Biot–Savart law gives an expression for the circulation α_{ij} along the ith filament due to the velocity induced by the jth. From Stokes's theorem the velocity flux through a circuit is the line integral of the vector potential around the circuit. Using the Biot–Savart law for the velocity induced by the ith line vortex and replacing $\omega_j \, dV$ by $\Gamma_j \, d\mathbf{s}_j$, we obtain the analytical expression for the winding number (due to

---

[9] Moffatt [1969] points out that the assumption of uniform density is not necessary and that J is invariant if the fluid is barotropic.

Gauss)

$$\alpha_{ij} = (1/4\pi) \oint_{C_i} \oint_{C_j} d\mathbf{s}_i \times d\mathbf{s}_j \cdot (\mathbf{R}_i(s_i) - \mathbf{R}_j(s_j))/|\mathbf{R}_i - \mathbf{R}_j|^3, \qquad (4)$$

where C_i is the axis and $\mathbf{R}_i(s_i)$ is the equation of the ith filament. (For further discussion of helicity invariants and topological structure, see Moffatt [1990], and references cited therein.)

For motion generated impulsively from rest (cf §3), the helicity is

$$J = \int \mathbf{f} \cdot \operatorname{curl} \mathbf{f} \, dV_{\mathrm{V}}. \qquad (5)$$

3.13 Axisymmetric motion with swirl

To obtain an example of a vortex with helicity, we consider first the conditions for axisymmetric incompressible flow with a swirl velocity about the axis of symmetry. We use cylindrical polar co-ordinates (r, θ, z), with velocity components (u_r, u_θ, u_z) which are functions only of r, z and t, that is, $\partial/\partial\theta = 0$. The equation of continuity is satisfied by the existence of the Stokes stream function Ψ, in terms of which the velocity and vorticity components are

$$\mathbf{u} = (-\partial\Psi/r\partial z, \; u_\theta, \; \partial\Psi/r\partial r), \qquad (1)$$

$$\boldsymbol{\omega} = (-\partial u_\theta/\partial z, \; -1/r \, D^2\Psi, \; \partial u_\theta/r\partial r), \qquad (2)$$

where

$$D^2 = \partial^2/\partial r^2 - \partial/r\partial r + \partial^2/\partial z^2. \qquad (3)$$

The vortex lines are not closed about the r axis due to the swirl.

The azimuthal component of the Helmholtz equation (1.5.6) reduces to

$$D(\omega_\theta/r)/Dt = r^{-2} \, \partial u_\theta^2/\partial z, \qquad (4)$$

which is (1.5.22) when the swirl u_θ is zero. The meridional components reduce to

$$D(ru_\theta)/Dt = 0. \qquad (5)$$

This equation is, of course, just Kelvin's circulation theorem for circuits which are circles about the axis of symmetry. The two equations (4) and (5) describe the evolution of the vorticity field for axisymmetric motion with swirl.

Suppose that the vorticity propagates without change of shape and hence that the motion is steady in some moving frame relative to which Ψ is independent of t and the solution of (5) is

$$ru_\theta = C(\Psi), \tag{6}$$

where C is an arbitrary function. Since $d\Psi = -ru_r\,dz + ru_z\,dr$, the right-hand side of (4) can be written

$$r^{-2}\partial(C^2/r^2)/\partial z = -u_r r^{-3}\,dC^2/d\Psi = \tfrac{1}{2}D(r^{-2}\,dC^2/d\Psi)/Dt, \tag{7}$$

where the last step follows from $D\Psi/Dt = 0$ and $Dr/Dt = u_r$. Then (4) can be integrated to give

$$\omega_\theta/r = -dH(\Psi)/d\Psi + (C/r^2)\,dC/d\Psi, \tag{8}$$

where H is an arbitrary function of Ψ. In fact, H is the Bernoulli constant, as can be seen from evaluating (1.9.4), since by definition $\nabla H = (ru_z\,dH/d\Psi,\ 0,\ -ru_r\,dH/d\Psi)$ and from (1), (2) and (6) it follows that $\mathbf{u} \times \boldsymbol{\omega} = (C/r^2\,\partial C/\partial r - u_z\omega_\theta,\ 0,\ C/r^2\,\partial C/\partial z + u_r\omega_\theta)$. These two expressions are equal when ω_θ is given by (8).

Substituting for ω_θ into the θ-component of (2), we obtain the (in general) non-linear equation for the stream function

$$D^2\Psi = r^2\,dH/d\Psi - C\,dC/d\Psi \tag{9}$$

in steady axisymmetric motion (Bragg and Hawthorne [1950]). The Hill spherical vortex discussed in §2.1 corresponds to $H = -A\Psi$, $C = 0$, inside a sphere of radius a propagating relative to the fluid at infinity with speed $2Aa^2/15$, and zero outside.

Moffatt [1969] (see also Hicks [1899]) has pointed out that (9) can be solved when

$$H = -A\Psi, \qquad C = \pm\alpha\Psi, \tag{10}$$

inside a sphere of radius a, and H is constant and C is zero outside the sphere. Equation (9) is then linear and can be solved by separation of variables in spherical polar co-ordinates $(R,\ \Theta,\ \phi)$. A solution is

$$\Psi = R^2 \sin^2\Theta[-A/\alpha^2 + c(a/R)^{3/2}J_{3/2}(\alpha R)], \tag{11}$$

where c is a constant. There are other solutions, but (11) is of interest because it can be matched to the steady irrotational flow outside the sphere with velocity $(0,0,-U)$ at infinity and stream function

$$\Psi = -\tfrac{1}{2}U(R^2 - a^3/R)\sin\Theta, \tag{12}$$

which clearly satisfies $D^2\Psi = 0$. It is necessary to have all velocity components continuous in order for the surface to be a vortex jump, the pressure being then automatically continuous, and it is readily verified that the gradients of (11) and (12) are equal on $R = a$ if

$$U = 2/15\, Aa^2[5J_{5/2}(\alpha a)/\alpha a J_{3/2}(\alpha a)], \tag{13}$$

$$c = A/\alpha^2\, J_{3/2}(\alpha a). \tag{14}$$

Owing to the oscillatory properties of the Bessel functions, the velocity can be backwards or forwards. The vortex is at rest when $J_{5/2}(\alpha a) = 0$; this occurs first when $\alpha a \sim 5.6$.

Since $H = H(\Psi)$, both the streamlines and vortex lines lie on the surfaces $H = $ const. These consist of a family of nested tori. For an enumerable set of values of α, the vortex lines lying on any one torus are closed, each line forming a torus knot. The helicity, impulse, angular impulse and kinetic energy of the swirling vortex can be calculated by substituting the expressions for Ψ and u_θ into the appropriate formulae. For example, the helicity is

$$J = \pm 16/3\, \pi a^2 c^2 f(\alpha a), \tag{15}$$

where

$$f(x) = \tfrac{1}{2}x^2 \left\{x(J_{3/2}(x))^2 - xJ_{1/2}(x)J_{5/2}(x) - 2J_{3/2}(x)J_{5/2}(x)\right\}. \tag{16}$$

As $x \to 0$, $f(x) \sim \pi x^6/1560$.

4

MOTION WITH SURFACES

4.1 Virtual momentum of a moving body

Consider a finite body B, surface S_B, moving through an inviscid incompressible fluid of uniform density. Suppose that the external forces acting on the fluid are conservative with a potential Ω, which is assumed to be single valued in the event the body is not simply connected. Suppose now that the fluid surrounding the body is in irrotational motion with a single-valued velocity potential ϕ.

The virtual momentum of the body is defined to be

$$\mathbf{I}_B = \int \phi\, \mathbf{n}\, dS_B, \tag{1}$$

with the convention that the normal is always out of the fluid into the body. Since $-\phi$ is the impulsive pressure required to generate the fluid motion impulsively from rest (§3.3), $\mathbf{I}_B$ is the impulsive force that needs to be applied to the body to set it in motion against the inertia of the fluid. It is also usual (e.g., Lamb [1932 chap. VI]) to call $\mathbf{I}_B$ the impulse of the fluid. However, it is convenient to distinguish between $\mathbf{I}_B$ and the hydrodynamic impulse of the image vorticity equivalent to the body, which was introduced in §2.4. The two quantities are not equal, although the relation between them is simple. To avoid confusion, we shall use the more descriptive term 'virtual', or 'apparent', momentum for the quantity defined by (1). As discussed in §3.2, the fluid momentum $\int \mathbf{u}\, dV$ is not well defined if the fluid is infinite. If the fluid is bounded externally by rigid stationary walls, or is in a parallel-sided channel or tube, the fluid momentum is the negative of that which would have been possessed by the fluid displaced by the body.

The force exerted by the fluid on the body can be separated into two parts. First there is the hydrostatic buoyancy force due to external conservative forces, namely

$$\int \Omega \mathbf{n}\, dS_B = -\int \nabla \Omega\, dV_B. \tag{2}$$

Second, there is the drag $\mathbf{D}$, which is the force on the body due to the motion, when the hydrostatic pressure from the external conservative forces has been subtracted,

$$\mathbf{D} = \int (p + \Omega)\mathbf{n}\, dS_B. \tag{3}$$

The significance of $\mathbf{I}_B$ is that in unbounded fluid when no other bodies are present,

$$\mathbf{D} = -d\,\mathbf{I}_B/dt, \tag{4}$$

and under these conditions the action of the fluid on the body is as if the fluid were absent and the body had additional momentum $\mathbf{I}_B$. If $\mathbf{f}_B$ is the external force applied to the body (including hydrostatic forces, if present) and $M\mathbf{U}$ is the momentum of the body, the expression of Newton's second law is

$$d\,(M\mathbf{U})/dt = \mathbf{D} + \mathbf{f}_B. \tag{5}$$

Hence when (4) holds,

$$d(M\mathbf{U} + \mathbf{I}_B)/dt = \mathbf{f}_B. \tag{6}$$

To prove (4), note that if S_B' denotes the position of the surface at time $t + \delta t$,

$$\int \phi(S', t)\mathbf{n}'\, dS_B' - \int \psi(S, t)\mathbf{n}\, dS_B = \int \nabla \psi\,(dV_B' - dV_B)$$

$$= \delta t \int (\mathbf{u} \cdot \mathbf{n})\nabla \phi\, dS_B.$$

Hence

$$\frac{d}{dt}\int \phi \mathbf{n}\, dS_B = \int \partial \phi / \partial t\, \mathbf{n}\, dS_B + \int (\mathbf{u} \cdot \mathbf{n})\nabla \phi\, dS_B. \tag{7}$$

It is essential for this identity that the velocity potential is single valued.[1] Now, $\partial \phi / \partial t$ is given by the unsteady Bernoulli equation (1.9.6). Substituting into (7) and using (3), and noting that $\int \mathbf{n}\, dS = 0$ over a closed body,

---

[1] If ϕ is not single valued, a δ-function contribution to $\partial \phi / \partial t$ arises from motion of the cut.

we obtain

$$d\mathbf{I}_B/dt = -\mathbf{D} + \int (\tfrac{1}{2}\mathbf{u}^2\,\mathbf{n} - (\mathbf{u}\cdot\mathbf{n})\mathbf{u})\,dS_B$$

$$= -\mathbf{D} + \int \mathbf{u}\times\omega\,dV_B, \tag{8}$$

where **u** in the integral through the body is the extension of the fluid velocity into the body and ω is the associated bound vorticity. The surface and volume in (8) can be replaced by any surface or volume surrounding the body provided there is no vorticity between the body and the new surface. For unbounded fluid in the absence of other bodies, we can use the sphere at infinity over which the surface integral vanishes since $|\mathbf{u}|$ is at most $O(r^{-2})$, giving (4).

There is no requirement in the above argument that the body be rigid or even of constant volume.[2] Equation (1) defines the virtual momentum for a flexible body of arbitrarily varying shape and volume. The restriction that the potential be single valued is also easily relaxed (at least in three dimensions). It suffices to include barriers or diaphragms to make the fluid region simply connected and the velocity potential single valued, and to include these barriers in the surface S_B of integration. Moreover, the body need not be connected but can consist of several separate bodies.

The identity (7) and the deduction (4) hold equally well in two dimensions, provided there is no net circulation about the body or group of bodies. In this case, **u** is $O(r^{-2})$ at infinity, and the contour integral vanishes over the circle at infinity. The modification required when there is circulation about the body is not simple and is considered in §4.4. In two dimensions, the virtual momentum is, of course, to be interpreted as per unit length in the direction normal to the plane of the flow.

4.2   Virtual momentum and impulse

Since the body can be completely flexible, it need not be a body but can equally well be a mass of fluid moving rotationally, that is, a vortex or group of vortices bounded by a material surface S_B, at least when the volume of the body is constant. The virtual momentum must therefore be connected with the impulse of the vorticity inside the body. Consider in three dimensions an arbitrary volume of fluid V_B, enclosed by a surface S_B

---

[2] But if the volume changes, image sources and sinks are required in addition to the image vorticity.

(which includes appropriate barriers), such that $\omega \times \mathbf{n} = 0$ on S_B, $\omega = 0$ outside S_B, and a single-valued velocity potential ϕ exists on and outside S_B.

We require the following lemma:

$$\tfrac{1}{2} \int \mathbf{x} \times (\mathbf{n} \times \mathbf{u}) \, dS_B = - \int \phi \mathbf{n} \, dS_B. \tag{1}$$

To prove (1), note that the integral on the left-hand side involves only the tangential component of $\mathbf{u}$. We define ϕ inside the body to be the analytic harmonic function which takes the value ϕ on S_B.[3] Then $\mathbf{n} \times \mathbf{u} = \mathbf{n} \times \nabla \phi$ on S. Application of the vector identity (3.2.11) to the left-hand side, with $\mathbf{a}$ replaced by $\nabla \phi$, and the divergence theorem to the right-hand side shows that each side is equal to $- \int \nabla \phi \, dV_B$.

It now follows from (3.2.13) that

$$\int \mathbf{u} \, dV_B = \tfrac{1}{2} \int \mathbf{x} \times \omega \, dV_B - \int \phi \mathbf{n} \, dS_B. \tag{2}$$

Hence, we have the relation between the hydrodynamic impulse $\mathbf{I}_\mathrm{v}$ of the vorticity, the virtual impulse $\mathbf{I}_B$ of the body and the actual momentum $\mathbf{P}_F = \int \mathbf{u} \, dV_B$ of the fluid in the region occupied by the body when the body is replaced by fluid,

$$\mathbf{I}_\mathrm{v} = \mathbf{P}_F + \mathbf{I}_B. \tag{3}$$

Since $\mathbf{P}_F$ is equal to $\int \mathbf{u} \cdot \mathbf{n} \, dS_B$, it is independent of the particular way the fluid velocity is extended. Equation (3) gives the simple relation between the hydrodynamic impulse and the virtual momentum of a body.

We can rederive the law (3.7.14) for the rate of change of hydrodynamic impulse from the equation of motion for the body. Note first that if $\mathbf{F}$ denotes the force density required to maintain the image vorticity,

$$d\mathbf{P}_F/dt = \mathbf{D} + \int \mathbf{F} \, dV_B, \tag{4}$$

since the drag $\mathbf{D}$, which is the integral of the pressure over the surface of the body, is the same when the body is replaced by fluid. Then from (3), (4) and (1.8), we obtain

$$\frac{d\mathbf{I}_\mathrm{v}}{dt} = \int \mathbf{F} \, dV_B + \int \mathbf{u}_\mathrm{e} \times \omega \, dV_B, \tag{3.7.14}$$

since $\mathbf{u}$ may be replaced by $\mathbf{u}_\mathrm{e}$ in the integral of the vortex force.

---

[3] It is not the analytic continuation into the body of the velocity potential outside S_B, which in general has a singularity inside S_B.

These results hold also in two dimensions, provided there is no net circulation about the body; the absence of the factor $\frac{1}{2}$ in the definition of impulse is compensated for by the absence of the factor 2 in (3.10.16) compared with (3.2.11), and the above argument works in detail with the $\frac{1}{2}$ dropped from the left-hand side of (1) and the first term on the right-hand side of (2).

4.3 **Virtual angular momentum**

The virtual angular momentum of the body relative to some origin is defined to be

$$\mathbf{A}_B = \int \phi \mathbf{x} \times \mathbf{n} \, dS_B. \tag{1}$$

The dynamic torque $\mathbf{T}$ on the body due to the motion is

$$\mathbf{T} = \int p \mathbf{x} \times \mathbf{n} \, dS_B. \tag{2}$$

By arguments similar to those just employed, but algebraically slightly more complicated, we obtain

$$d\mathbf{A}_B/dt = -\mathbf{T} + \int_S (\tfrac{1}{2}(\mathbf{n} \times \mathbf{x})\mathbf{u}^2 + (\mathbf{x} \times \mathbf{u})\mathbf{u} \cdot \mathbf{n}) \, dS, \tag{3}$$

where S is any surface surrounding the body such that there is no vorticity between it and the body. Further,

$$\mathbf{A}_v = \mathbf{A}_B + \int \mathbf{x} \times \mathbf{u} \, dV_B \tag{4}$$

relates the virtual angular momentum to the angular impulse and actual angular momentum of the virtual fluid displaced by the body.

The identity analogous to (2.1) used in the derivation of (4) from the identity expressed by 3.5.1, and proved in the same way, is

$$\tfrac{1}{2} \int \mathbf{x}^2 (\mathbf{n} \times \mathbf{u}) \, dS = \int \phi \mathbf{x} \times \mathbf{n} \, dS. \tag{5}$$

The integral in (4) does not appear to have a simple expression in terms of the angular momentum of the rigid body. However, its rate of change can be expressed as a surface integral of the pressure and the torque of external forces and is therefore independent of the particular way $\mathbf{u}$ is extended into the body.

The equation of motion of the body is

$$dJ/dt = T + g, \tag{6}$$

where J is the angular momentum of the body and g is the torque about the origin applied by external forces. Equation (6), with J replaced by the angular momentum of the displaced fluid and g replaced by $\int x \times F \, dV_B$, combined with (3) and (4), is equivalent to (3.7.19).

These formulae apply as they stand in two dimensions, provided there is no net circulation about the body or group of bodies.

4.4 Two-dimensional motion with circulation

The concept of the virtual momentum of a body in two-dimensional flow requires modification if there is a circulation Γ about the body. We now write $\phi = \phi_0 + \phi_\Gamma$, where ϕ_0 is the velocity potential in the absence of circulation and ϕ_Γ is the potential if the body is stationary and alone in unbounded fluid at rest at infinity but with circulation Γ. Then

$$n \cdot \nabla \phi_\Gamma = 0 \tag{1}$$

on S_B, and $[\phi_\Gamma]_C = \Gamma$ for a circuit surrounding the body once in the counter-clockwise direction. The bound vorticity ω_Γ corresponding to ϕ_Γ produces a velocity field u_Γ (which includes the extension of $\nabla \phi_\Gamma$ into the body) with a two-dimensional centroid $\overline{R}_\Gamma$ given by (3.10.21). It follows from (3.10.29), (3.2.15) and (1) that the centroid is independent of the way u_Γ is extended into the body as it is defined by the value of u on the body.

The virtual momentum of the body (per unit span) is now defined to be

$$I_B = \oint \phi_0 \, n \, ds_B + \Gamma \, \overline{R}_\Gamma \times k, \tag{2}$$

because this satisfies (1.8), and the boundary integral still vanishes when the fluid is unbounded, even if $\Gamma \neq 0$.

To see this, we start with (3.10.29), written as

$$P_F = I_v + \oint r \times (n \times u) \, ds_B. \tag{3}$$

Then from (2.4),

$$D + \int F \, dS_B = I_v + \oint r \times (n \times u) ds_B. \tag{4}$$

Differentiating (3) with respect to time and using (3.7.14), we have

$$\frac{d}{dt} \oint \mathbf{r} \times (\mathbf{u} \times \mathbf{n}) \, ds_B = -\mathbf{D} + \oint (\tfrac{1}{2} u^2 \mathbf{n} - \mathbf{u}(\mathbf{u} \cdot \mathbf{n})) \, ds, \tag{5}$$

where the contour integral on the right-hand side is around any circuit such that there is no vorticity between it and the body.

Now write $\mathbf{u} = \mathbf{u}_0 + \mathbf{u}_\Gamma$. We have (see (2.1))

$$\oint \mathbf{r} \times (\mathbf{u}_0 \times \mathbf{n}) \, ds_B = \oint \phi_0 \mathbf{n} \, ds_B. \tag{6}$$

From (3), since $\int \mathbf{u}_\Gamma \, dS_B = 0$, because it is a flow around a fixed body,

$$\oint \mathbf{r} \times (\mathbf{u}_\Gamma \times \mathbf{n}) \, ds = \Gamma \overline{\mathbf{R}}_\Gamma \times \mathbf{k}. \tag{7}$$

Thus (5) gives

$$d\mathbf{I}_B / dt = -\mathbf{D} + \oint (\tfrac{1}{2} u^2 \mathbf{n} - \mathbf{u}(\mathbf{u} \cdot \mathbf{n})) \, ds, \tag{8}$$

when $\mathbf{I}_B$ is given by (2). Note that (2.3) still holds.

The virtual momentum $\mathbf{I}_B$ with circulation is not Galilean invariant, and hence momentum is not conserved. To preserve a steady motion, it is necessary to apply an external force. Suppose a body moves steadily without change of shape with velocity $\mathbf{U}$ in unbounded fluid; then,

$$d\mathbf{I}_B / dt = \Gamma \mathbf{U} \times \mathbf{k} = -\mathbf{D}. \tag{9}$$

The drag force of magnitude $U\Gamma$ is again the Kutta lift.

Similarly, the virtual angular momentum is defined to be

$$\mathbf{A}_B = \oint \phi_0 (\mathbf{r} \times \mathbf{n}) \, ds_B + \tfrac{1}{2} \oint r^2 (\mathbf{n} \times \mathbf{u}_\Gamma) \, ds_B. \tag{10}$$

It can be shown from (3.4) and (3.10.20), together with (3.5.1), that (3.3) is satisfied. Lamb [1932 §134a] presents an alternative treatment of the force on a cylindrical body with circulation.

5

SOME APPLICATIONS

5.1 Virtual mass

Suppose a rigid body is moving with velocity $\mathbf{U}$ without rotation through an unbounded inviscid incompressible uniform fluid free of vorticity (outside the body). The velocity potential is the solution of $\nabla^2 \phi = 0$, satisfying

$$\partial \phi / \partial n = \mathbf{U} \cdot \mathbf{n} \quad \text{on } S_B \tag{1}$$

and vanishing at infinity. If the body is simply connected, ϕ is determined uniquely by (1). If not, ϕ is undetermined to the extent of cyclic motions fixed by constant jumps Γ_α across the barriers which would make the body simply connected. Because of the linearity of the equations,

$$\phi = \Phi^j U_j + \Gamma_\alpha \chi_\alpha, \tag{2}$$

where $\partial \Phi^j / \partial n = n_j$ on S_B and Φ^j, χ_α are functions only of the shape of the body and position relative to the body. It follows from substitution into (4.1.1) that the virtual momentum of the body can be expressed as

$$\mathbf{I}_B = \mathbf{I}_B^j U_j + \Gamma_\alpha \mathbf{I}^\alpha, \tag{3}$$

where

$$\mathbf{I}_B^j = \int \Phi^j \mathbf{n} \, dS_B, \quad \mathbf{I}^\alpha = \int \chi_\alpha \mathbf{n} \, dS_B. \tag{4}$$

The Cartesian components M_{ij} of $\mathbf{I}_B^j$ constitute the virtual mass tensor. The tensor is symmetrical, since by construction $n_i = \partial \Phi^i / \partial n$ and by Green's theorem applied to the region between the body and infinity

$$\int \left(\Phi^j \partial \Phi^i / \partial n - \Phi^i \partial \Phi^j / \partial n \right) \, dS_B = 0.$$

It can also be shown to be positive definite, most easily from consideration of the kinetic energy of the fluid.

The momentum of the body is $M_B\mathbf{U}$, where M_B is the mass of the body. The cyclic constants Γ_α are invariant because of Kelvin's theorem. The equation of motion (4.1.5) becomes

$$(M_B\delta_{ij} + M_{ij})\frac{dU_j}{dt} = f_{Bi}. \tag{5}$$

Hence no force is required to maintain steady motion. This result is referred to as D'Alembert's paradox, since it contradicts experience. The fallacy lies not in the direct neglect of viscous forces, but rather in the assumption that there is no vorticity in the fluid outside the body. A body moving through a real fluid has behind it a wake containing vorticity.[1]

A torque must be applied in general to maintain the motion. Put

$$\mathbf{x} = \mathbf{U}t + \mathbf{x}' \tag{6}$$

into (4.3.1), where $\mathbf{x}'$ is measured relative to the centre of volume. Then

$$\mathbf{A}_B = t\mathbf{U} \times \mathbf{I}_B + \int \phi\,\mathbf{x}' \times \mathbf{n}\,dS_B. \tag{7}$$

The integral in (7) is constant when the body is in steady motion without rotation. Likewise, the angular momentum of the body is constant. Hence the torque to be applied by external forces is given by (4.3.3) and is[2]

$$\mathbf{G} = \mathbf{U} \times \mathbf{I}_B. \tag{8}$$

This torque is about the origin, but since the resultant force is zero, it is also the torque about any point.

The torque vanishes when $\mathbf{I}_B$ is parallel to $\mathbf{U}$. This occurs when the velocity is parallel to the principal axes of the virtual mass tensor. If any two of the principal axes are equal, any direction in the plane formed by these two is a principal axis such that motion parallel to it is free of torque. If all three principal axes are equal, then there is no torque for motion in any direction. Thus, for example, there is no torque on a cube in steady motion.

The considerations of this section apply to both two- and three-dimensional motion, with the understanding that, in the former case, forces are per unit length, and there is no circulation about the body. The results

---

[1] The present considerations are likely to be most applicable when either the body is streamlined, so that there is only a thin wake containing vorticity, or the body is in rapid oscillatory motion so that the boundary layers produce vorticity of opposite sign, which cancels.

[2] The torque exerted by the fluid on the body is $-\mathbf{G}$.

are significantly different when there is circulation in two dimensions, as will be shown below.

For a sphere and cylinder of radius a, the virtual mass tensors are $M'\delta_{ij}$, where M', called the virtual mass, is

$$M' = \tfrac{2}{3}\pi a^3, \quad M' = \pi a^2, \tag{9}$$

respectively. For a circular disc of radius a, the principal values are

$$\left(\tfrac{8}{3}a^3, \ 0, \ 0\right), \tag{10}$$

the non-zero term being obviously for motion normal to the plane of the disc. Lamb [1932 §114] gives the formulae for an ellipsoid of general shape.

The body can be replaced by its equivalent system of bound vorticity and the present argument applies equally well to a vortex, showing that no net force is required to maintain a steady vortex, such as the Hill spherical vortex, although in general a net couple must be applied. This also applies to a combination of vortex and body, such as the Föppl flow past a cylinder with two stationary free vortices, or a stationary vortex behind a sphere.

Suppose the body moves at a small angle to one of the principal axes, which is supposed to be the x-axis, the other co-ordinate axes being aligned along the other principal axes. If the velocity is (u, v, w), where $v, w \ll u$, the virtual momentum is $\mathbf{I}_B = (Au, Bv, Cw)$ and the torque exerted by the fluid on the body is $(0, (C - A)uw, (A - B)uv)$ to first order, where (A, B, C) are the principal coefficients of the virtual mass tensor. If $A > C$ and $A > B$, this torque would tend to rotate the body so that the velocity is aligned along the x-axis. That is, the free motion would in the first instance tend to decrease v and w from positive values. We infer that the motion should be stable if it is in the direction of greatest coefficient of virtual mass and is otherwise unstable. This is in accord with the observed tendency of discs to set themselves normal to the flow and for prolate ellipsoids to be unstable when moving without spin parallel to the axis of symmetry. For a prolate ellipsoid, $A/B = A/C \to 0$ as the eccentricity tends to 1, with A taken along the axis of symmetry (see Lamb [1932 chap. V]).[3]

A simple application of the virtual mass concept is the initial acceleration of a buoyant or heavy rigid body of mass M released from rest in a fluid of arbitrary viscosity. At $t = 0+$, there is no vorticity in the fluid, and

---

[3] Greenhill's demonstration of the stabilizing effect of spin is discussed by Lamb [1932 §128]. It must of course be remembered that a discussion of stability assuming potential flow past the body is only physically realistic while the body is streamlined; i.e., there is negligible vorticity in the flow.

hence at the initial instant

$$\frac{d}{dt}(M\mathbf{U} + \mathbf{M} : \mathbf{U}) = \mathbf{B}, \tag{11}$$

where $\mathbf{B}$ is the buoyancy force, $\mathbf{M}$ is the inertial mass tensor, and $\mathbf{U}$ is the velocity of the centre of mass, assumed to coincide with the centre of hydrostatic pressure, about which there is no torque. The initial acceleration then follows from the inversion of

$$M\frac{d\mathbf{U}}{dt} + \mathbf{M} : \frac{d\mathbf{U}}{dt} = \mathbf{B}. \tag{12}$$

5.2 Attraction of vortices or bodies to walls

If a body or vortex has a trajectory that brings it close to a fixed surface, on which $\mathbf{u} \cdot \mathbf{n} = 0$, it follows from (4.1.8) that the impulse of the vortex or total momentum (real plus apparent) of the body will acquire a component in the direction towards the surface. If $\mathbf{I}_v$ is generally in the same sense as $\mathbf{U}$ (the velocity of the vortex), then the vortex or body will appear to be attracted to the body. If $\mathbf{I}_v$ and $\mathbf{U}$ are in opposite senses, then there will be an apparent repulsion. Both cases are possible, although the former is usual.[4] According to Kelvin [1868], the attraction may be easily observed by projecting a smoke ring past an ordinary 12-inch globe hung by a thin cord.

Note that the virtual mass tensor still exists and is given by (1.4), but the potentials Φ^j and χ_α are affected by the presence of the solid wall, and hence the coefficients of the virtual mass tensor will be functions of the position of the body.[5] Consider, for example, a solid sphere moving towards a plane wall at right angles. By expanding the velocity potential in powers of a/X (a = sphere radius, X = distance from the wall) it can be shown the the virtual momentum taken positive when directed towards the wall is (see Lamb [1932 §98])

$$I_B = \tfrac{2}{3}\pi a^3 \left(1 + \frac{a^3}{4X^3}\right) U, \tag{1}$$

where $U = -\dot{X}$.

---

[4] The second case could be produced by projecting a solid torus with sufficiently large cyclic motion in the direction opposite to that of the flow through the torus or a vortex ring with strong axial flow.

[5] In addition to the time dependence of the M_{ij}, the equation of motion of the body is changed by the necessity of including the entire right-hand side of (4.1.8).

On the other hand, it can be shown by calculation of the pressure, or by energy methods (Lamb [1932 §137]), that to the same order of approximation

$$\left(m + \tfrac{2}{3}\pi a^3\right)\dot{U} = \frac{-3\pi a^6 U^2}{8X^4} + F, \tag{2}$$

where m is the mass of the sphere and F is the external force directed towards the wall. Thus, when $F = 0$, the sphere appears to be repelled from the wall. However, there is no contradiction with (4.1.6), because the decrease in I_B due to a decrease in U is more than compensated for by the increase due to the decrease in X (when $U > 0$). Indeed, to this order of approximation

$$\frac{d}{dt}(mU + I_B) = \frac{\pi a^6 U^2}{8X^4}. \tag{3}$$

The right-hand side is the contribution to (4.1.8) from the integral over the wall.

The fact that increase of speed may mean decrease of impulse or virtual momentum is even more striking for a vortex ring. From (3.8.1), the impulse of a thin cored vortex ring is approximately $\pi \Gamma R^2$, where Γ is its circulation and R is the radius of the ring. The speed is given by (10.2.1). As the ring approaches a wall, its radius changes, owing to convection in the velocity field of its image, which clearly causes the ring to expand, with Γ kept constant, leading to a decrease in velocity but an increase in the impulse, in accordance with (3.7.14). It is interesting to note, as stated by Kelvin [1881], that the integral of the pressure over the plane surface is zero, so that the total force on the plane is zero. This result follows immediately from considerations of the real momentum in a hemisphere of radius R_∞. The motion is unsteady, and at large distances the velocity is that of a pair of equal and opposite vortex rings (a 'ring dipole') and is $O(r^{-4})$. The velocity potential and hence also the pressure are $O(r^{-3})$. Therefore there is no momentum flux across the hemisphere at infinity.[6] The momentum in the hemisphere is constant, because the centre of mass of the fluid is fixed. No external forces are applied. Hence the integral of the pressure variations over the plane wall is zero. Note that the argument depends on the wall being plane and infinite to ensure the velocity is $O(r^{-4})$ at infinity. If the wall were replaced by a circular cylinder, the result would not be true. In two dimensions, the calculation can be carried out explicitly

---

[6] Alternatively note that the impulse of the ring plus image is zero from symmetry and use the asymptotic expressions of §3.2.

for a pair of equal and opposite rectilinear line vortices approaching a wall, and the result can be demonstrated directly.

The result naturally had serious implications for Kelvin's [1867] theory of vortex atoms, as the vortex atoms cannot now communicate momentum to a wall by collisions like the atoms and molecules of kinetic theory. Kelvin [1881] claimed that the result failed if the wall were at the end of a finite tube so that the vortex could return along the tube wall, but this statement seems to be in error. Suppose the wall is at the end of a semi-infinite circular tube of radius b. At infinity, it is a consequence of potential theory that $\phi = O(e^{-kx})$, where k is the smallest root of $J_0(kb) = 0$. Hence there is no momentum flux to infinity and the pressure on the end wall is zero at all times.

It can be shown (Saffman [1979]) that a symmetrical pair of two-dimensional vortices of finite size incident normally on a rigid wall approach monotonically without rebound. Suppose the wall is $y = 0$ and the vortices are symmetrical about $x = 0$ and approaching from above. The total vorticity Γ in the first quadrant is

$$\Gamma = \int_0^\infty \int_0^\infty \omega \, dx \, dy. \tag{4}$$

This is conserved in the absence of viscosity. The co-ordinates $(\bar{x}, \bar{y})$ of the centroid of vorticity in the quadrant are related to the components (I_x, I_y) of hydrodynamic impulse

$$I_x = \int_0^\infty \int_0^\infty y\omega \, dx \, dy, \quad I_y = -\int_0^\infty \int_0^\infty x\omega \, dx \, dy, \tag{5}$$

by the relations

$$\bar{x} = -I_y/\Gamma, \quad \bar{y} = I_x/\Gamma. \tag{6}$$

It follows from the Euler equations that

$$\frac{dI_x}{dt} = -\int_0^\infty \int_0^\infty \left\{ \frac{\partial}{\partial x} (u\omega y) + \frac{\partial}{\partial y} (v\omega y) - v\omega \right\} dx \, dy, \tag{7}$$

where (u, v) are the x and y components of velocity. Integrating by parts and using the boundary conditions

$$u = 0 \quad \text{on } x = 0, \qquad v = 0 \quad \text{on } y = 0, \tag{8}$$

it follows that

$$\frac{d\bar{y}}{dt} = -\frac{1}{2\Gamma} \int_0^\infty v_0^2 \, dy. \tag{9}$$

Similarly,

$$\frac{d\bar{x}}{dt} = \frac{1}{2\Gamma} \int_0^\infty u_0^2 \, dx. \tag{10}$$

Here,

$$v_0 = v(0, y), \quad u_0 = u(x, 0). \tag{11}$$

Suppose that the vortex pair is initially at a large distance from the wall $x = 0$. Then (9) predicts that $\bar{y}$ decreases at a rate depending upon the velocity field on the axis of symmetry. As the pair approaches the wall, the vortices begin to separate under the action of the image vorticity in the wall. This is shown by (10), according to which $\bar{x}$ increases when u_0 is not negligibly small on the wall. Eventually, the vortices will be so far apart that v_0 is negligible. The centroids of the vortices will then have reached an asymptotic distance from the wall and continue parallel to it. Since the right-hand side of (9) is always negative, inviscid dynamics predicts that the vorticity centroids approach the wall monotonically, however large the vortices or their deformation may be, and asymptote to a line parallel to the wall. It cannot, however, be asserted that the final motion of each vortex is steady. It is possible that the final state is a periodic or aperiodic oscillation for vortices of finite size. (See §9.3 for further comment.)

5.3 Force on a body in a non-uniform stream

The flow in the centre of a parallel-sided wind tunnel is not exactly uniform because of the growth of boundary layers on the tunnel walls, which produce an acceleration along the centreline and a negative longitudinal pressure gradient. There is no force on a body at rest in a uniform stream, but we can expect that when the stream is not uniform a force will act on the body. This problem was studied by Kelvin for spherical particles. An analysis for bodies of general shape was given by Taylor [1928]; see also Lamb [1932 §143]. For general results, it is necessary to assume that the body is small compared with the scale of inhomogeneity of the stream. Then the free stream in the vicinity of the body (assumed to be at rest) can be approximated by the form

$$u_i = u_i^{(0)} + \alpha_{ij} x_j = u_i^{(0)} + u_i^{(1)}, \quad \text{say}, \tag{1}$$

where the origin is taken in the body, and $\mathbf{u}^{(0)}$ and the α_{ij} are constants with $\alpha_{ij} = \alpha_{ji}$ since the free stream is assumed to be irrotational.

The body can be replaced by a corresponding distribution of bound vorticity

$$\omega_i = \omega_i^{(0)} + \omega_i^{(1)}. \tag{2}$$

The drag force on the body is (see (3.7.14) or (4.1.8)),

$$\mathbf{D} = \int_B (\mathbf{u}^{(0)} + \mathbf{u}^{(1)}) \times \boldsymbol{\omega} \, d\mathbf{x} = \int_B \mathbf{u}^{(1)} \times \boldsymbol{\omega}^{(0)} \, d\mathbf{x} + O(\alpha_{ij}^2), \tag{3}$$

since $\mathbf{u}^{(0)}$ is constant and the volume integrals of $\boldsymbol{\omega}^{(0)}$ and $\boldsymbol{\omega}^{(1)}$ separately vanish. The integral in (3) will be related to the apparent momentum of the body when it is moving steadily through unbounded fluid and hence to the apparent mass. First we note that for a closed distribution of vorticity inside a volume V,

$$\int_V (x_i\omega_j + x_j\omega_i) \, d\mathbf{x} = - \int_S x_ix_j \, \boldsymbol{\omega} \cdot \mathbf{n} \, dS = 0, \tag{4}$$

and hence

$$\int_B x_i\omega_j^{(0)} \, d\mathbf{x} = \epsilon_{ijk}I_{Vk} \tag{5}$$

relates the impulse of the image vorticity for the body in a uniform stream to the integral on the right-hand side of (3). The relation between $\mathbf{I}_V$ and the apparent mass follows from (4.2.3)

$$I_{Vi} = -M_{ij}u_j^{(0)} - Vu_i^{(0)}, \tag{6}$$

where V is the volume of the body and M_{ij} is its virtual mass tensor. The negative sign is present because the speed of the body relative to the fluid is $-\mathbf{u}^{(0)}$. Substituting for $\mathbf{u}^{(1)}$ in (3), we obtain

$$D_i = \alpha_{ij}(M_{jk} + V\delta_{jk})u_k^{(0)}, \tag{7}$$

which is the result obtained by Taylor and Lamb in rather different ways. It is to be noted that the pressure gradient due to the acceleration of the free stream is near the body

$$\partial p / \partial x_i = -\alpha_{ij}u_j^{(0)}, \tag{8}$$

and if the force on the body were calculated as if it were a statical problem, the force would be simply $V\alpha_{ij}u_j^{(0)}$ which is incorrect.

5.4 Force on a body in an accelerated irrotational stream

Suppose that relative to the body, the speed of the fluid is $\mathbf{u}_E = \mathbf{u}^{(0)}(t) + \mathbf{u}^{(1)}(\mathbf{x}, t) \approx \mathbf{u}^{(0)}(t) + \alpha(t) : \mathbf{x}$. The bound distribution of vorticity is $\omega = \omega^{(0)} + \omega^{(1)}$. The impulse $\mathbf{I}_v$ of the bound vorticity is given by (3.6). We have

$$\frac{d\mathbf{I}_v}{dt} = \int \mathbf{F} \, dV_B + \int \mathbf{u}_E \times \omega \, dV_B, \tag{1}$$

where $\mathbf{F}$ is the force that binds the vorticity. Then

$$\int \mathbf{F} \, dV_B = -\mathbf{M} : \dot{\mathbf{u}}^{(0)} - V\dot{\mathbf{u}}^{(0)} + \alpha : \mathbf{I}_v + O(\alpha^2). \tag{2}$$

If $\mathbf{D}$ is the force exerted on the body by the fluid,

$$\int \mathbf{F} \, dV_B = -\mathbf{D} + V\dot{\mathbf{W}}, \tag{3}$$

where $\mathbf{W}$ is the speed of the body relative to an inertial frame. Relative to instantaneous inertial co-ordinates $\mathbf{x}' = \mathbf{x} + \mathbf{W}t$, the speed of the fluid is

$$\mathbf{u}^{(0)} + \alpha : (\mathbf{x}' - \mathbf{W}t) + \mathbf{W}. \tag{4}$$

It follows from (3) and (2) that

$$\mathbf{D} = \mathbf{M} : \dot{\mathbf{u}}^{(0)} + V(\dot{\mathbf{W}} + \dot{\mathbf{u}}^{(0)}) + \alpha : (\mathbf{M} : \mathbf{u}^{(0)} + V\mathbf{u}^{(0)}). \tag{5}$$

If the body accelerates through fluid otherwise at rest, $\alpha = 0$ and $\mathbf{u} = -\mathbf{W}$, so we recover the previous result.

Further, it follows from (4) that the pressure gradient in the vicinity of the body is

$$\begin{aligned}
-\nabla p &= \dot{\mathbf{u}}^{(0)} + \dot{\mathbf{W}} - \alpha : \mathbf{W} + (\mathbf{u}^{(0)} + \mathbf{W}) : \alpha \\
&= \dot{\mathbf{u}}^{(0)} + \dot{\mathbf{W}} + \alpha : \mathbf{u}^{(0)}.
\end{aligned} \tag{6}$$

Thus the terms in the expression (5) for the drag $\mathbf{D}$ which involve the volume V of the body can be interpreted in terms of the hydrostatic pressure force.

5.5 Rotating bodies

We now consider the motion of rigid bodies in incompressible, inviscid fluid allowing for the possibility that the orientation of the body is not fixed and the body may rotate with angular velocity Ω in addition to its translation with velocity **U**. The velocity potential now takes the form (summation over repeated symbols is implied)

$$\phi = \Phi^j U_j + \Gamma_\alpha \chi_\alpha + \Psi^j \Omega_j, \tag{1}$$

where Φ^j and χ_α are defined as in §1 and the Ψ^j are harmonic functions satisfying

$$\partial \Psi^j / \partial n = \epsilon_{ijk} x'_k n_i \quad \text{on } S_B. \tag{2}$$

The co-ordinate $\mathbf{x}'$ is position relative to the centre of volume of the body. The virtual momentum of the body is

$$\mathbf{I}_B = \mathbf{I}_R^j U_j + \Gamma_\alpha \mathbf{I}^\alpha + \mathbf{I}_R^j \Omega_j \tag{3}$$

where

$$\mathbf{I}_R^j = \int \Psi^j \mathbf{n} \, dS_B. \tag{4}$$

The components of the vectors $\mathbf{I}_R^j$ form a tensor which is not necessarily symmetrical.

The vectors $\mathbf{I}_B^j$, $\mathbf{I}^\alpha$, $\mathbf{I}_R^j$ are constant relative to axes fixed in the body but not relative to axes fixed in space. Thus to obtain equations of motion, it is best to refer the velocity and angular velocity to axes rotating with angular velocity $\mathbf{\Omega}$. The equation of motion in unbounded fluid with external applied force **f** is then

$$\mathbf{I}_B^j \dot{U}_j + \mathbf{I}_R^j \dot{\Omega}_j + \mathbf{\Omega} \times \mathbf{I}_B + M d\mathbf{U}/dt = \mathbf{f}. \tag{5}$$

The symbol $^\cdot$ denotes a time derivative relative to the rotating axes.

A similar but more complicated equation can be obtained for the rate of change of angular momentum, but it is simpler to proceed otherwise for the general case via the Lagrangian equation of dynamical motion as discussed, for example, by Lamb [1932 chap. 6]. We note, however, that if the body is not translating and the cyclic constants Γ_α are zero, the apparent angular momentum is

$$\mathbf{A}_B = \Omega_j \int \Psi^j \mathbf{x} \times \mathbf{n} \, dS_B. \tag{6}$$

If the body has cubic symmetry, the tensor whose coefficients arise from the integral in (6) will be proportional to the unit tensor. The apparent angular momentum will then be proportional to the angular velocity,[7] and no torque will be required to maintain steady rotation.

[7] The constant of proportionality is not necessarily zero; the sphere is a special case for which the Ψ^j vanish.

5.6 Torque on a body in a non-uniform stream

From (3.7.19), a body held at rest in a non-uniform stream will be subject to a torque

$$\mathbf{T} = \int \mathbf{x} \times (\mathbf{u}_E \times \boldsymbol{\omega}) \, dV_B$$

$$= \int \mathbf{x} \times (\mathbf{u}^{(0)} \times \boldsymbol{\omega}^{(0)}) \, dV_B + \int \mathbf{x} \times (\mathbf{u}^{(1)} \times \boldsymbol{\omega}^{(0)}) \, dV_B \qquad (1)$$

$$+ \int \mathbf{x} \times (\mathbf{u}^{(0)} \times \boldsymbol{\omega}^{(1)}) \, dV_B + O(\alpha_{ij}^2),$$

on substituting (3.1) and (3.2) for the external velocity and bound vorticity. The first expression on the right-hand side of (1) can be evaluated in terms of the impulse of the body moving through fluid at rest.

First it is easily verified with the aid of (3.4) that

$$\int \mathbf{x} \times (\mathbf{u}^{(0)} \times \boldsymbol{\omega}^{(0)}) \, dV_B = -\mathbf{I}_v \times \mathbf{u}^{(0)}. \qquad (2)$$

Since $\mathbf{I}_v - \mathbf{I}_B$ is parallel to $\mathbf{u}^{(0)}$ and a body at rest in a uniform stream of velocity $\mathbf{u}^{(0)}$ is equivalent to the same body moving with speed $-\mathbf{u}^{(0)}$ through fluid at rest, this provides an alternative derivation of (1.8).

For the second term in (1), substitution of (3.1) gives the expression

$$\alpha_{ij} \int_B x_j x_k \omega_k^{(0)} \, d\mathbf{x} - \alpha_{jk} \int_B \omega_i^{(0)} x_j x_k \, d\mathbf{x} \qquad (3)$$

in terms of the distribution $\boldsymbol{\omega}^{(0)}$ of bound vorticity for the body at rest in a uniform stream.

The last term in (1) can be written, with the aid of (3.4), (4.2.2) and (3.2.15), and noting that the velocity produced by $\boldsymbol{\omega}^{(1)}$ must have equal and opposite normal component to $\mathbf{u}^{(1)}$ on S_B,

$$-\mathbf{u}^{(0)} \times \int \mathbf{u}^{(1)} \, dV_B - \mathbf{u}^{(0)} \times \int \phi^{(1)} \mathbf{n} \, dS_B, \qquad (4)$$

where $\phi^{(1)}$ is the solution of Laplace's equation satisfying on S_B,

$$\partial \phi^{(1)} / \partial n + \mathbf{u}^{(1)} \cdot \mathbf{n} = 0, \qquad (5)$$

and the first term in (4) is zero if the origin is taken at the centre of volume.

Taylor [1928] considered the example of a sphere held at a point off centre in a curved stream. Let us take $\mathbf{u}^{(0)} = (U, \, 0, \, 0)$, $\alpha_{12} = \alpha_{21} = \alpha$ and all other $\alpha_{ij} = 0$. The centre of curvature of the streamlines then lies in the positive y-direction. Suppose the centre of the sphere is at the point $(0, Y, Z)$. The torque about the x-axis is given by (3) and is

$$T_1 = \alpha \int (y^2 \omega_y^{(0)} + yz\,\omega_z^{(0)} - xy\,\omega_x^{(0)})\,dV_B.$$ (6)

It is easily shown from the known solution for potential flow past a sphere (for instance, we take for $\omega^{(0)}$ the equivalent vortex sheet on the surface of the sphere) that

$$T_1 = \alpha Z \int (y - Y)\omega_z^{(0)}\,dV_B = -2U\pi a^4 \alpha Z.$$ (7)

Thus there are two equilibrium positions if the sphere is free to rotate about the x-axis, both with the centre lying in the plane of the streamlines, the one with $Y > 0$ being stable, and the other with $Y < 0$ being unstable. As noted by Taylor, this answer can be verified directly because the force on the sphere is necessarily through its centre and is $2U\pi a^3 \alpha$. Taylor argued that an elongated slender body of revolution with centre line bent into the arc of a circle, looking like a small banana, would have the same type of symmetry as the sphere held off centre,[8] the centre of curvature of the centre line being in the opposite direction of that towards the sphere's point of support. The stable equilibrium position, when supported and free to rotate about the axis, is that with the centres of curvature of streamlines and centre line of body on opposite sides. This was verified experimentally. Bodies of revolution with fins were also considered by Taylor.

5.7 Self-propulsion of a deformable body

Consider the example of a deformable body of mass M and constant volume, whose centroid moves with velocity U in the x-direction. The shape of the body is assumed to be symmetrical about the x-axis. In the absence of external forces,

$$MU + I_B(t) = 0,$$ (1)

where $I_B(t)$ is the (in general) time-dependent x-component of the virtual momentum. From the linearity of Laplace's equation and the symmetry, it follows that

$$I_B(t) = \hat{M}(t)\,U + S(t),$$ (2)

[8] The more streamlined the body, the more is potential theory likely to be applicable.

where $\hat{M}(t)$ is a function of the instantaneous shape and $S(t)$ is a function of the rate of deformation. Then

$$U = -\frac{S(t)}{M + \hat{M}(t)}. \tag{3}$$

Note that even if the shape is a periodic function of time, so that $\overline{\hat{M}} = \overline{S} = 0$, where $\overline{}$ denotes an average over the period, it does not follow that $\overline{U} = 0$, and hence a deformable body can move through a perfect fluid, although not unidirectionally (Saffman [1967]).[9]

5.8 Buoyant vortex rings

It is essential for most of the applications involving the concepts of bound vorticity and vortex force that the density be constant. Although the Helmholtz and Kelvin theorems remain valid for barotropic fluid – that is, $\rho = \rho(p)$, which includes compressible flow provided no shocks are present – the analysis and mathematical transformations rely heavily on $\nabla\rho = 0$ and div $\mathbf{u} = 0$.

The theory can, however, be applied to stratified flow in a gravitational field if the Boussinesq approximation is allowed. We consider the case of constant gravity. The inviscid equations of motion are

$$\rho D\mathbf{u}/Dt = -\nabla p - \rho g\mathbf{k}, \tag{1}$$

$$D\rho/Dt = -\rho\,\mathrm{div}\,\mathbf{u}. \tag{2}$$

An equation of state for the density is required. Let us suppose that

$$\rho = \rho_0\,(1 + c), \tag{3}$$

where ρ_0 is a constant density and c is some scalar concentration obeying a diffusion equation

$$\partial c/\partial t + \mathbf{u}\cdot\nabla c = 0. \tag{4}$$

The Boussinesq approximation is the limit $c \to 0$, $g \to \infty$, $cg \to \theta$, say. The equations become

$$D\mathbf{u}/Dt = -\nabla p/\rho_0 - \theta\mathbf{k}, \tag{5}$$

$$\mathrm{div}\,\mathbf{u} = 0, \tag{6}$$

---

[9] Benjamin and Ellis [1990] have used self-propulsion to explain the erratic motions of small bubbles in liquids irradiated by sound and have corrected algebraic errors in Saffman's work.

and

$$\partial\theta/\partial t + \mathbf{u} \cdot \nabla\theta = 0, \tag{7}$$

where θ can be identified with $(\rho - \rho_0)g$ for applications.

A relatively simple example is afforded by the buoyant vortex ring of radius R propagating vertically. Suppose the fluid inside the cores has density ρ_1, and the ambient density is ρ_0. If the circulation around the ring is κ, Kelvin's circulation theorem implies that κ is constant. The hydrodynamic impulse I is given by

$$I \approx \rho_0 \kappa R^2. \tag{8}$$

The buoyancy force on the ring is $(\rho_0 - \rho_1)2\pi gRa^2$, where a ($\ll R$) is the core radius. Entrainment or mixing between the core and the ambient fluid is neglected, so $\rho_1 = $ const. and $Ra^2 = $ const. Momentum conservation gives

$$dI/dt = 2\rho_0\kappa RdR/dt = (\rho_0 - \rho_1)2\pi gRa^2. \tag{9}$$

Write $V = 2\pi^2 Ra^2$ to denote the constant core volume. Then

$$R^2 = R_0^2 + \frac{2Vt}{\kappa\pi\rho_0}(\rho_0 - \rho_1)g \tag{10}$$

gives the evolution with time of the ring radius.

The behaviour depends on the signs of κ and $\rho_0 - \rho_1$. For a light (heavy) ring propagating upwards (downwards) – that is, $\kappa(\rho_0 - \rho_1) > 0$ – the ring radius increases with time and the ring slows down, since the speed $\propto 1/R$. But the ring does go to ∞ since $\int dt/\sqrt{t}$ diverges. For the opposite case, R approaches zero in a finite time as the ring accelerates, but the approximations become inconsistent when $R \not\gg a$. Experimentally, a cascading instability is seen in this case.

The motion of two-dimensional vortex pairs in a stratified atmosphere is of considerable interest in connection with the descent of aircraft trailing vortices. Hill [1975a] has carried out a numerical study of the problem.

6

CREATION OF VORTICITY

6.1 Formation of a vortex sheet

In perfect barotropic fluid, acted upon by conservative forces with a single-valued potential, the first Helmholtz law (§1.5) says that it is not possible to endow a fluid particle with vorticity, and Kelvin's circulation theorem (§1.6) shows that the circulation around a material circuit is zero if initially zero. The question arises whether vorticity can be created without violating these theorems, and without invoking viscosity, non-conservative forces or baroclinic effects. There is no a priori reason why they are not important in subsequent motion if present initially, and so one wishes to know if vorticity can be created without appeal to these effects.

Klein [1910] addressed this question with his Kaffeelöffel experiment. (See also Betz [1950].) The conclusion is that the Helmholtz and Kelvin theorems preclude the generation of piece-wise continuous vorticity, but do not prevent the formation of vortex sheets or the generation of circulation. Consider Klein's experiment. A two-dimensional plate of width $2a$ is set in motion through a perfect incompressible fluid with velocity U normal to the plate. We introduce the complex potential $w(z) = \phi + i\psi$, $z = x + iy$. The boundary conditions are $\psi = Uy$ on $x = 0$, $|y| < a$ (the axes are taken to coincide instantaneously with the plate with the y-axis along the plate and the x-axis in the direction of motion), and $w \sim 0$ as $z \to \infty$ (circulation at infinity is not allowed). By standard methods (see, e.g., Batchelor [1967 §6.5]) or by inspection, one obtains

$$w = U(z - \sqrt{a^2 + z^2}).$$ (1)

On the plate,

$$\phi = \mp U\sqrt{a^2 - y^2}, \qquad x = \pm 0,$$ (2)

$$v = \frac{\partial \phi}{\partial y} = \pm \frac{Uy}{\sqrt{a^2 - y^2}}, \quad x = \pm 0. \tag{3}$$

The virtual momentum is in the x-direction and its magnitude is

$$I_B = \oint \phi \mathbf{n} \cdot \mathbf{i} \, ds = 4U \int_0^a \sqrt{a^2 - y^2} \, dy = \pi U a^2. \tag{4}$$

The kinetic energy is

$$\iint \tfrac{1}{2} (\nabla \phi)^2 \, dx \, dy = \tfrac{1}{2} \oint \phi \, \partial\phi/\partial n \, ds = -2 \int_0^a \phi \, \partial\phi/\partial x \, dy = \tfrac{1}{2} \pi U^2 a^2. \tag{5}$$

The circulation around an instantaneous streamline from A to A', say, is

$$\int_A^{A'} \mathbf{u} \cdot d\mathbf{s} = \phi(A') - \phi(A) = 2U \sqrt{a^2 - y^2}. \tag{6}$$

The plate is therefore equivalent to a vortex sheet of strength

$$\kappa = 2Uy/\sqrt{a^2 - y^2}. \tag{7}$$

The bound vorticity $\omega(x, y)$ is therefore

$$\omega = 2Uy \, \delta(x)/\sqrt{a^2 - y^2}, \quad \text{for } y < a. \tag{8}$$

The hydrodynamic impulse of the bound vorticity is

$$\mathbf{I_V} = \int \mathbf{r} \times \mathbf{k} \omega \, dx \, dy = \mathbf{i} \oint \kappa y \, dy = \mathbf{i} \int_0^a \frac{4Uy^2}{\sqrt{a^2 - y^2}} \, dy = \pi U a^2 \, \mathbf{i}. \tag{9}$$

Note that $I_B = I_V$, because the volume of the body is zero.

The kinetic energy of the bound vorticity is

$$\tfrac{1}{2} \int \omega \psi \, dx \, dy = \tfrac{1}{2} \int_{-a}^a Uy\kappa \, dy = \tfrac{1}{2} \pi U^2 a^2. \tag{10}$$

The difference in the kinetic energy of the bound vorticity and the flow about the body is the kinetic energy of the fluid flow inside the body, which in this case is zero as the velocity is bounded and the volume is zero. Thus the values given by (5) and (10) are the same. (A general expression for the kinetic energy of the irrotational flow caused by a body moving with velocity $\mathbf{U}$ through fluid at rest is $\tfrac{1}{2} \mathbf{I}_B \cdot \mathbf{U}$.)

Klein argued that if the plate is removed (by pulling out the coffee spoon) or dissolved (certainly possible in a 'gedanken' experiment), then a vortex sheet is left in the fluid. Non-zero circulation has been created around circuits drawn as shown in Figure 6.1-1. There is no contradiction with the Helmholtz–Kelvin laws, because no fluid particle has acquired

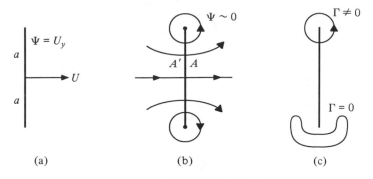

6.1-1 Klein's Kaffeelöffel experiment. (a) Motion of plate. (b) Streamlines.
(c) Contours of zero and non-zero circulation.

vorticity, and no circulation has been generated about *closed contours* lying entirely within the fluid. The new contours with circulation were not closed beforehand; they intersected the body. This is an example in which vorticity has been created as a sheet (i.e., a singular distribution) by changing the topological properties of the flow.

6.2 Leading edge suction on the sheet

The further dynamics of the sheet is of considerable interest. The velocity of the sheet is found by the procedures described in §2.2. The velocity of the sheet along its normal at station y is

$$\int_{-a}^{a} \frac{2uy'}{\sqrt{a^2 - y'^2}} \frac{dy'}{y' - y} = U, \qquad |y| < a. \tag{1}$$

Thus the sheet appears to continue forward unchanged, except that the sheet strength and induced velocity are singular at the ends $y = \pm a$, where a principal value cannot be taken. Dynamically, the singularity requires an external force T, which can be called a leading edge suction, to be applied at the tips of the plate. To see this, consider the momentum balance for the flow in an infinitesimal circle of radius ϵ centred at $y = a$, see Figure 6.2-1.

We make the flow steady by superposing a velocity $-U$ in the x-direction. In the neighbourhood of the tip

$$w = -U\sqrt{a^2 + z^2} \sim -\sqrt{2a}\, U\, (i\zeta)^{1/2}, \tag{2}$$

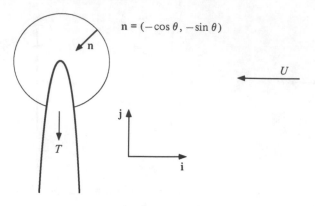

6.2-1 Flow near the tip of the plate.

where $\zeta = z - ia$. If $T\mathbf{j}$ is the external force applied to the fluid, corresponding to a tension T in the plate near the tip,

$$T\mathbf{j} + \oint_{|\zeta|=\epsilon} p\mathbf{n}\,ds + \oint_{|\zeta|=\epsilon} \mathbf{u}\,(\mathbf{u}\cdot\mathbf{n})\,ds = 0. \tag{3}$$

The velocity components (u, v) follow from

$$u - iv = dw/dz \sim \sqrt{a/2}\,U\,(i/\zeta)^{1/2}. \tag{4}$$

Thus

$$u = -U\sqrt{a/2\epsilon}\cos(\pi/4 - \theta/2), \qquad v = -U\sqrt{a/2\epsilon}\sin(\pi/4 - \theta/2). \tag{5}$$

and

$$\mathbf{u}^2 = aU^2/2\epsilon, \qquad \mathbf{u}\cdot\mathbf{n} = -U\sqrt{a/2\epsilon}\cos(\pi/4 + \theta/2). \tag{6}$$

Using Bernoulli's equation to give the pressure, substituting into (3) and evaluating the integrals, we find

$$T = \tfrac{1}{2}\pi U^2 a. \tag{7}$$

The plate can move through the fluid if it can withstand the tension described by (7). If the plate were not sufficiently strong, it would be torn apart. The vortex sheet has, however, no means of withstanding the tension, and the uniform translation described by (1) is dynamically inconsistent.

This result can also be found from vortex force considerations. We work in the frame of reference in which the flow is at rest at infinity. Consider that part of the sheet between y_1 and y_2. It has hydrodynamic impulse $\mathbf{I}_v = \int_{y_1}^{y_2} \mathbf{r} \times \mathbf{k}\,\kappa\,dy$. The equation of motion (3.7.14) for this region of vorticity is ($d\mathbf{r}/dt = U\mathbf{i}$)

$$d\mathbf{I}_v/dt = -U\mathbf{j} \int_{y_1}^{y_2} \kappa \, dy = \mathbf{f} + \int_{y_1}^{y_2} \int \mathbf{u}_e \times \boldsymbol{\omega} \, dx \, dy, \tag{8}$$

where $\mathbf{f} = (T_2 - T_1)\mathbf{j}$ is the external force. In the last term in (8), we can replace $\mathbf{u}_e$ by $\mathbf{u}$ since the self-induced velocity makes no contribution to the vortex force. Now for $y_1 > -a$ and $y_2 < a$, $\mathbf{u} = U\mathbf{i}$ on the sheet. Hence, taking the y-component of (8) and remembering that $\omega = \kappa(y)\,\delta(x)$, we have $\mathbf{f} = 0$; that is, $T_1 = T_2 = T$, and the tension is constant along the sheet. To find T, we apply (8) to the large semicircle $y \geq 0$, $x^2 + y^2 < R^2$ and let $R \to \infty$. Then

$$-U \int_0^a \kappa \, dy = -T + \iint_{\substack{y \geq 0 \\ x^2+y^2 \leq R^2}} (\mathbf{u} \times \boldsymbol{\omega})_y \, dx \, dy. \tag{9}$$

Now

$$\iint (\mathbf{u} \times \boldsymbol{\omega})_y \, dx \, dy = \oint (\tfrac{1}{2}\mathbf{u}^2 \mathbf{n} - \mathbf{u}(\mathbf{u} \cdot \mathbf{n}))_y \, ds = -\int_{-\infty}^{\infty} \tfrac{1}{2} u^2 \, dx, \tag{10}$$

since the contribution from the semicircle at infinity vanishes. Substituting into (10),

$$u(x, 0) = \partial\phi/\partial x = U(1 - |x|/\sqrt{a^2 + x^2}), \tag{11}$$

we find

$$\int_{-\infty}^{\infty} \tfrac{1}{2} u^2 \, dx = (2 - \tfrac{1}{2}\pi)U^2 a. \tag{12}$$

Since $\int_0^a \kappa \, dy = 2Ua$, we recover (7) from (9).

6.3 Approximate development of the sheet into a vortex pair

The vortex sheet created by the removal of the plate is not an equilibrium solution. The longitudinal velocity profile at $x = 0$ is shown in Figure 6.3-1. The singularity in the force on the bound vorticity is manifested by the singularity in the induced velocity field. The subsequent evolution of the sheet is a problem of great interest. The appropriate equations are obtained in §8.1 and the structure of the solution described in §8.4. Here we shall confine ourselves to a qualitative discussion.

The problem is important in connection with the wake of an aircraft. To a first approximation, the flow at a station Z downstream of a rectangular,

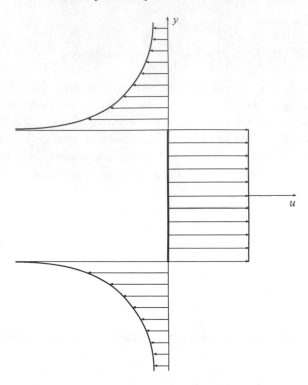

6.3-1 Longitudinal velocity profile created by motion of the plate. Vortex sheet on the plate rolls up into two counter-rotating cores.

elliptically loaded wing of span $2a$ in a stream of velocity W at small angle of attack α is that of the developing vortex sheet at time $t = Z/W$ after the plate has been withdrawn, with the identification $U = W\alpha$ (see Figure 6.3-2).

The vortex sheet is pulled out by the infinite velocity at the tips, where the vorticity is largest, into a spiral which keeps on growing as it rolls up by entraining more and more of the sheet. In the absence of viscosity, the sheet remains infinitely thin, but the roll-up continues to give a spiral in which the distance between consecutive turns goes to zero and the sheet is continuously stretched so that its strength (i.e., the jump in velocity) decreases to zero. The spiral is approximately circularly symmetric. The asymptotic nature of the velocity field is not definite but probably approaches a continuous, nowhere differentiable distribution.

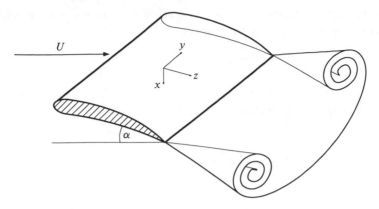

6.3-2 Vortex field behind a wing.

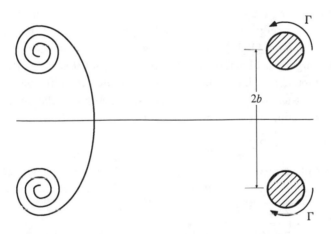

6.3-3 Intermediate and asymptotic stages of vortex sheet roll-up.

In this case, we can estimate properties of a coarse-grained approximation.[1] We assume, on the basis of the calculations of vortex sheet roll-up, that the coarse-grained limit of the sheet roll-up is two, approximately circular, counter-rotating vortices of strengths $\pm\Gamma$, whose centroids are distance b apart (see Figure 6.3-3). The roll-up conserves hydrodynamic impulse (this is true even in the presence of viscosity (§3.6)). It is assumed

[1] In a real physical situtation, coarse-graining, to give a smooth vorticity field and smooth out the velocity jumps, will be accomplished by viscosity (see Moore and Saffman [1973]). Theoretically, we can maintain the fiction of an inviscid fluid, provided the artifice of a mathematical smoothing is introduced.

further that circulation is conserved (i.e., there is no vorticity diffusion across the centreline $y = 0$). Then

$$\Gamma = \int_0^a \kappa \, dy = 2Ua \tag{1}$$

gives the vortex strengths. The separation b is found from equating the x-components of hydrodynamic impulse of the original sheet and the vortex pair;

$$\int_{-a}^a y\kappa(y) \, dy = \pi U a^2 = \Gamma b. \tag{2}$$

Then

$$b = \tfrac{1}{2}\pi a. \tag{3}$$

It remains to estimate the size of the cores, and the distribution of vorticity inside them. We give here (for historical reasons) an argument of Prandtl [1919] based on energy conservation, although it has been superseded by an argument of Betz [1932] based on angular impulse conservation (see §8.4). These arguments contradict the erroneous belief that the vortex sheet rolls up asymptotically into a pair of line vortices. Prandtl assumed that the vorticity ω was constant inside the cores, which had radius δ, and that the core deformation could be neglected. Then

$$\pi\delta^2\omega = \Gamma = 2Ua. \tag{4}$$

To this order of approximation, the velocity inside the cores is tangential of magnitude $\Gamma r/2\pi\delta^2$. Then the kinetic energy of the fluid inside the cores is

$$\int_0^\delta 2\pi r \left(\frac{\Gamma r}{2\pi\delta^2} \right)^2 dr = \frac{\Gamma^2}{8\pi}. \tag{5}$$

For the motion outside the cores, we have the stream function ψ given by

$$\psi = -\frac{\Gamma}{2\pi} \log\left(\frac{r}{r'}\right), \tag{6}$$

where r and r' are the distances to the centres of the cores. The kinetic energy of the fluid external to the cores is

$$-\tfrac{1}{2} \oint_{\text{cores}} \psi \, \partial\psi/\partial n \, ds. \tag{7}$$

On the first core,

$$\psi \approx (\Gamma/2\pi) \log(b/\delta) + O(\delta/b), \qquad \oint -\partial\psi/\partial n \, ds = \Gamma. \tag{8}$$

Hence the kinetic energy of the fluid external to the cores is, approximately, $(\Gamma^2/2\pi) \log(b/\delta)$. The initial kinetic energy of the sheet is given by (1.5). Following Prandtl in assuming that the kinetic energy is conserved during the roll-up, we have (see also Durand [1934 vol. II p. 328])

$$\frac{\Gamma^2}{8\pi} + \frac{\Gamma^2}{2\pi} \log \frac{b}{\delta} = \tfrac{1}{2}\pi U^2 a^2 = \frac{\pi\Gamma^2}{8}, \tag{9}$$

from which follows

$$\delta \approx 0.11b. \tag{10}$$

If the core radius is identified as the distance at which the tangential velocity is a maximum (see §2.1), the estimate (10) is somewhat larger than typically measured values, and Prandtl's model seems inadequate. (It is, of course, possible to calculate the kinetic energy of the assumed vorticity distribution more accurately. Spreiter and Sachs [1951] calculated exactly the kinetic energy of uniform vortices bounded by rigid cylinders. But the surfaces of the vortices are then vortex sheets, and the relevance of the calculation is not clear.) The assumption of uniform vorticity in the cores appears to be the cause of the discrepancy (see §8.4).

6.4 Formation of a vortex ring

The process described above for two-dimensional flow can be applied in axisymmetric three-dimensional flow. Suppose a circular disc of radius a is moving normal to its plane with velocity U. The velocity potential can be found by standard methods of analysis (Lamb [1932 §108], Batchelor [1967 §6.8]). The streamline pattern in the meridional plane is similar to that for the two-dimensional case. On the disc, the velocity potential is (employing cylindrical polar co-ordinates x, r, θ)

$$\phi = \mp \frac{2U}{\pi} \sqrt{a^2 - r^2} \quad \text{on } x = \pm 0, \ y^2 + z^2 - r^2 < a^2. \tag{1}$$

If the disc is dissolved, there remains a circular vortex sheet, with vortex lines that are circles about the axis. The strength and vorticity of the sheet are, respectively,

$$\kappa(r) = \frac{4U}{\pi} \frac{r}{\sqrt{a^2 - r^2}}, \qquad \boldsymbol{\omega} = \kappa\,\boldsymbol{\theta}\,\delta(x). \tag{2}$$

By symmetry, the hydrodynamic impulse is in the x-direction, of magnitude

$$I = \tfrac{1}{2}\int (\mathbf{x} \times \boldsymbol{\omega})_x \, dV = \tfrac{1}{2}\int_0^a 2\pi r^2 \kappa \, dr = 8Ua^3/3. \tag{3}$$

The kinetic energy is

$$E = \tfrac{1}{2} \int \phi \, \partial\phi/\partial n \, dS = -\tfrac{1}{2} U \int_0^a 2\pi r[\phi] \, dr$$
$$= 4U^2 \int_0^a r\sqrt{a^2 - r^2} \, dr = 4U^2 a^3/3, \tag{4}$$

since $\partial\phi/\partial n = -U$ on the disc. The circulation around a circuit beginning and ending at the centre of the disc and enclosing the disc is

$$\Gamma = \int_0^a \kappa \, dr = [\phi]_{r=0} = 4Ua/\pi. \tag{5}$$

The circular vortex sheet is again not a steady solution because of the unbalanced leading edge suction at the edge, and it rolls up into a vortex ring. Taylor [1953] examined the properties of the uniform vortex ring formed by coarse-graining the rolled up sheet into a circular vortex ring of radius R, core radius c, and circulation Γ. The hydrodynamic impulse of the ring is $\Gamma\pi R^2$. Assuming that the impulse (as well as the circulation) are conserved, we have

$$R = \sqrt{2/3} \, a. \tag{6}$$

Taylor (following Prandtl's argument) determined the core radius and the speed of the ring by assuming that energy was conserved and that the vorticity in the core was uniform. We can proceed more generally using formulae for the properties of rings with non-uniform vorticity. We need the following results for the speed V and kinetic energy E of a ring (to be deduced in §10.3) (see also Fraenkel [1970, 1972], Saffman [1970])

$$V = \frac{\Gamma}{4\pi R} \left[\log \frac{8R}{c} - \frac{1}{2} + \int_0^c \left(\frac{\Gamma(s)}{\Gamma} \right)^2 \frac{ds}{s} + o\left(\frac{c}{R}\right) \right], \tag{7}$$

$$E = \frac{1}{2}\Gamma^2 R \left[\log \frac{8R}{c} - 2 + \int_0^c \left(\frac{\Gamma(s)}{\Gamma} \right)^2 \frac{ds}{s} + o\left(\frac{c}{R}\right) \right]. \tag{8}$$

Here, $\Gamma(s)$ is the circulation about a circle of radius s lying in the meridional plane centred on the axis. For a uniform core (Kelvin), $\Gamma(s)/\Gamma = s^2/c^2$, $\int (\Gamma(s)/\Gamma)^2 \, ds/s = 1/4$, and we obtain Kelvin's formula (10.2.1) for the speed of a vortex ring. For a hollow core, $\Gamma(s) = \Gamma H(c-s)$, and the integral is zero, giving Hicks's [1885] result (10.2.2) for the speed of a hollow ring.

Assuming that energy is conserved, we obtain from (4) and the elimination of the integral between (7) and (8),

$$\frac{V}{U} = \frac{1}{4} + \frac{1}{\pi^2} \left(\frac{3}{2}\right)^{3/2} = 0.44, \tag{9}$$

independent of the distribution of $\Gamma(s)$.

The equation for the core radius c is

$$\log \frac{8R}{c} + \int_0^c \left(\frac{\Gamma(s)}{\Gamma}\right)^2 \frac{ds}{s} = 2 + \frac{\pi^2}{6} \left(\frac{3}{2}\right)^{1/2} = 4.01. \tag{10}$$

For a uniform core, $c/R = 0.19$; for a hollow core $c/R = 0.14$. $\Gamma(s)$ for vortex rings produced in nature is difficult to determine.

6.5 Creation of circulation about a body

The Klein Kaffeelöffel experiment and its extensions show how vorticity is created in a fluid by a fundamental topological change. A solid surface was dissolved, thereby putting the bound vorticity into the fluid. But if just the creation of circulation is desired, it suffices to change the topology of the body. For instance, the coffee spoon could be shrunk into two cylinders of radius c distance $2b$ apart by creating an infinitesimal slit, which then grows. If no external forces are applied, the impulse would remain constant and this would determine the speed of the body. In general, work would be done in the process, and the kinetic energy of the new flow would be different from that of the initial configuration. Similarly, a disc could be deformed into a ring.

Consider the two-dimensional case and suppose $c \ll b$, so that the cylinders are infinitesimal. The upper cylinder has a virtual momentum

$$\mathbf{I}_B = -\Gamma X \mathbf{j} + \Gamma b \mathbf{i}, \tag{1}$$

where $X(t)$ is the x-co-ordinate of the mid-point. Then the drag force on the upper cylinder is (from §4.4)

$$\mathbf{D} = \int \omega \, \mathbf{u}_E \times \mathbf{k} \, dS - d\mathbf{I}_B/dt. \tag{2}$$

In (2) we can put $\mathbf{u}_E = (\Gamma/4\pi b)\mathbf{i}$. Then

$$\mathbf{D} = -\Gamma b \mathbf{i} + \Gamma \mathbf{j} (\dot{X} - \Gamma/4\pi b). \tag{3}$$

If by some mechanical means, the distance b is changed (and no transverse force is required when $\dot{X} = \Gamma/4\pi b$), a drag force $\Gamma \dot{b}$ is then generated on

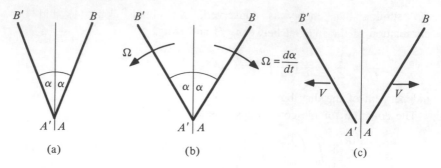

6.5-1 Clap and fling. (a) Start of beat. (b) Rotation of wing. (c) Fling stage.

each part. This means of communicating a force to a body in a perfect fluid has been proposed as the basis of an insect lift mechanism by Weis-Fogh [1973] and Lighthill [1973].

The process is called 'clap and fling' as a hovering insect moves its wings in a horizontal plane. Consider Figure 6.5-1. Stage (*a*) is the start of a beat. The wings are touching at a vertex, and the angle between them starts to increase. This generates a circulation $\pm\Gamma = \Omega a^2 g(\alpha)$ about the *open* curves ABA and $A'B'A'$. The function $g(\alpha)$ can be calculated by standard potential theory (see Lighthill). The fling then starts with the wings separating at the vertex AA' and moving sideways with velocities $\pm V$. This produces a lift $V\Gamma h(X, \alpha)$ where $h(X) \to 1$ as $X \to \infty$ to give the Kutta lift result for a body in unbounded flow. For $X < \infty$, $h(X, \alpha) < 1$, because each wing is in the downwash of the other.

Lift is produced while the angle between the wings opens prior to the fling, because

$$\mathbf{I}_B = -\mathbf{j}a^3 h(\alpha)\,\Omega,\tag{4}$$

and since $\mathbf{D} = -d\mathbf{I}_B/dt$ in unbounded flow, there is a lift

$$L = a^3\,\frac{d}{dt}\,(\Omega\,h(\alpha))\tag{5}$$

and the total contribution $\int L\,dt \neq 0$ if the fling starts (i.e., the wings separate) while $\Omega > 0$.

The clap then reverses the fling, bringing the wing back to stage (*a*) and completing the cycle. However, the clap requires the existence of a different physical mechanism. If it were the exact reverse, negative lift would occur, which would exactly cancel the lift generated in the fling (see §5.7). Within the context of potential theory (i.e., neglecting the possibility of circulation generation by separation), the clap process is made possible by three-dimensional effects. The wings are not moving in a plane, but around a cylinder with the insect's body as the vertical axis. The continuation of the fling through $180°$ brings the wings together at the end of half a beat. The circulation is then annulled by stopping the rotation and the fling repeats with the wing returning with opposite-sign circulation, giving lift.[2] In three dimensions, the bound vortex lines in the wing will have to close in the fluid by creation of vortex sheets. The energy of the vortex sheets manifests itself as an induced drag, so that work must be done to produce the lift. Lift requires a downward flux of momentum to balance the weight, and the induced drag can also be interpreted as the kinetic energy of the downwash produced by the vorticity in the fluid (see §14.1).

6.6 Generation of vorticity by flow separation

Consider the flow of an inviscid incompressible fluid past a body started from rest by conservative forces or motion of boundaries. The motion is not unique, unless the flow is continuous. In this case, the Helmholtz–Kelvin theorems show that the vorticity in the flow is zero and the velocity is the gradient of a single-valued potential. If discontinuous flow is allowed, for example, by allowing vortex sheets to enter the fluid from the body, the unsteady flow of a perfect fluid does not have a unique solution.

Consider the flow produced by a flat plate moving normal to itself. The unique potential flow solution is that discussed in §1 in connection with the Kaffeelöffel experiment. This flow is continuous at all points *inside* the fluid. As Figure 6.6-1(a) demonstrates an alternative is a separated flow with vortex sheets springing from the sharp edges (Anton [1939]). The velocity field is now not continuous.

A second example is the flow out of a tube. Either potential flow can exist, or the flow separates at the edge of the tube and vortex sheets appear which roll-up into vortex rings (see Figure 6.6-1(b)).

A third example is the starting flow past an airfoil. The continuous potential flow appears below in Figure 6.7-1(a). Figure 6.6-1(c) shows the separated flow with a vortex sheet leaving the trailing edge and a separation bubble at the nose.

---

[2] This could be analysed in two dimensions by considering a periodic infinite array of wings.

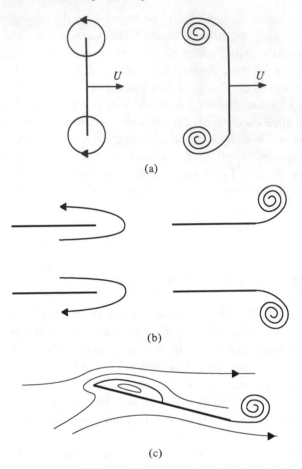

6.6-1 Generation of vorticity by flow separation. (a) Motion of a plate. (b) Outflow
from an orifice. (c) Starting vortex caused by acceleration of an airfoil.

The potential flows are continuous but not bounded, the velocity being
infinite at the sharp edges. The separated flows are discontinuous, but the
velocities are bounded. It is speculated that they are therefore more likely
to be limits of solutions of the Navier–Stokes equations as $\nu \to 0$, with t
fixed. However, within the framework of a purely inviscid theory there is
no way to choose between the flows.

Moreover, there is no known reason why separation should not occur for
flow past a smooth body, as sketched in Figure 6.6-2, in an infinite number
of ways depending upon the separation point, in the sense that each flow

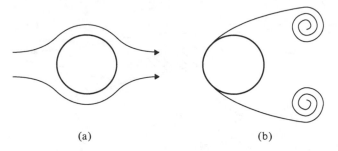

(a)   (b)

6.6-2 Speculative generation of circulation by separation of flow past a smooth body.
(a) Irrotational flow, no separation. (b) Flow with shed vortex sheets.

is a solution of the Euler equations.[3] The creation of vortex sheets by flow
separation does not violate the Helmholtz–Kelvin theorems. The sheets
are infinitely thin and do not contain any fluid particles that were in the
fluid at $t = 0$; the vorticity comes entirely from the boundaries. Also,
the circulation around any material circuit lying entirely within the fluid at
$t = 0$ is zero for all $t > 0$. Circulation can appear around circuits lying
entirely inside the fluid, but these circuits are not the evolution of any
initial material circuit.

For steady flow past a body, there are obviously infinitely many possible
solutions of the Euler equations, depending upon the distribution of vorticity
(e.g., Föppl-type flows with vortices of finite size).

6.7   Accelerated flow past a wing

For the case of bodies with sharp edges, the lack of uniqueness of
Euler flow can be removed by imposition of the so-called Kutta condition
that requires the velocity to be bounded, but not necessarily continuous.
We consider the problem of a two-dimensional airfoil set into motion
(Wagner [1925], Sedov [1965], Karman and Burgers [1934]), and calculate
the growth of circulation Γ around the wing. The problem can be analysed
with the approximation that the roll-up of the vortex sheet shed from the
trailing edge is negligible and that the vortex sheet is convected with the
free stream.

---

[3] We emphasise that the issue of which flow is the $\nu \to 0$ limit of a Navier–Stokes
solution is a separate question. It should also be remembered that the limit $\nu \to 0$ may
also include viscosity-dependent boundary conditions, so a unique limit may not exist
(Saffman [1981]).

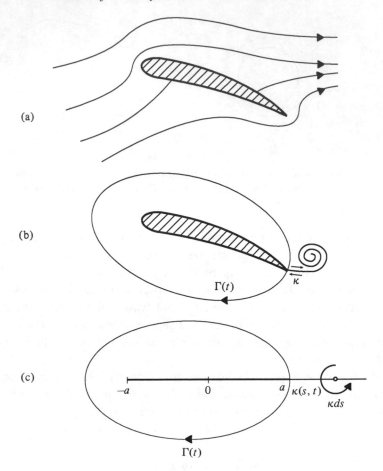

6.7-1 Accelerated flow past a wing. (a) Initial potential flow. (b) Circulation and vortex wake for $t > 0$. (c) Mathematical model.

For simplicity, consider a flat plate of length $c = 2a$ at small angle of attack α, set instantaneously into motion with constant velocity V (see Figure 6.7-1). At time t, the circulation around the wing is $\Gamma(t)$ and the vortex sheet has strength $\kappa(s, t)$, where s is the distance from the trailing edge. With the assumption that the self-induced motion of the sheet is negligible, the sheet strength satisfies the equation

$$\frac{\partial \kappa}{\partial t} + V \frac{\partial \kappa}{\partial s} = 0. \tag{1}$$

From conservation of total circulation around the wing and sheet,

$$\frac{d\Gamma}{dt} - V\kappa(0,t) = 0, \quad \text{or} \quad \Gamma = \int_0^{Vt} \kappa(s,t)\, ds. \tag{2}$$

To find the flow, we need to solve two potential flow problems. The first is the flow past a flat plate in a uniform stream V at angle of attack α (see Batchelor [1967 §6.7]). The complex potential is (see §7.2)

$$w = \tfrac{1}{2} V(\zeta\, e^{-i\alpha} + (a^2/\zeta)\, e^{i\alpha}) + i\gamma_0/2\pi \,\log \zeta, \tag{3}$$

where $\zeta = z + \sqrt{z^2 - a^2}$ and $-\gamma_0$ (real) is the circulation around the plate. The flat plate $-a < \Re z < a$ has been mapped into the circle $|\zeta| = a$.

The second potential problem is the flow induced by a line vortex of strength $\kappa\, ds$ at the point $(s + a,\ 0)$.[4] This is

$$w = \frac{i\, ds}{2\pi} \left[-\kappa \log\{\zeta - (s + a + \sqrt{s^2 + 2as})\} + \gamma_1 \log \zeta \right.$$
$$\left. + \kappa \log\left\{\zeta - \frac{a^2}{s + a + \sqrt{s^2 + 2as}}\right\} \right], \tag{4}$$

where γ_1 (real) is a function of time. The induced circulation about the wing due to the element ds of the sheet is $-(\kappa + \gamma_1)\, ds$.

The bound circulations around the wing are determined from imposition of the Kutta condition. Since $u - iv = dw/dz = (dw/d\zeta)(d\zeta/dz)$, and $d\zeta/dz = \infty$ at $z = a$, the condition of finite velocity at the trailing edge requires $dw/d\zeta = 0$ at $\zeta = a$. From (3) we obtain

$$\gamma_0 = 2\pi a V \sin \alpha \sim \pi V c \alpha. \tag{5}$$

From (4),

$$\gamma_1 = -\kappa\sqrt{\frac{s + c}{s}}. \tag{6}$$

The total bound vorticity inside the wing at time t is $-\Gamma(t)$, where

$$\Gamma(t) = \gamma_0 + \int_0^{Vt} (\kappa + \gamma_1)\, ds. \tag{7}$$

Using (2), we have

$$\gamma_0 = V\pi c \alpha = -\int_0^{Vt} \gamma_1\, ds = \int_0^{Vt} \kappa\sqrt{\frac{s + c}{s}}\, ds. \tag{8}$$

---

[4] The y-component is actually $s\alpha$, but this is neglected.

Now from (1), $\kappa = K(Vt - s)$. Putting $\sigma = Vt - s$, we have on changing variables in (8),

$$V\pi c\alpha = \int_0^{Vt} K(\sigma)\sqrt{\frac{Vt + c - \sigma}{Vt - \sigma}}\, d\sigma, \tag{9}$$

which is a Volterra integral equation for the sheet strength $K(\sigma)$. Note that $K(0)$ is the value at the end of the sheet, and $K(Vt)$ is the value at the trailing edge.

The formal solution of (9) is readily obtained using the Laplace transform.

$$\frac{V\pi c\alpha}{p} = \tilde{K}(p)\overline{\left(\frac{\sigma + c}{\sigma}\right)^{1/2}}, \tag{10}$$

where $\overline{}$ denotes the Laplace transform in the variable p with respect to σ. We have written $\tilde{K}(p) = \overline{K(\sigma)}$.

Exact inversion is not possible, but we can use asymptotic techniques for small and large times. For small times, σ is small and we are interested in the form of the transform for large p. Then

$$\overline{\left(\frac{\sigma + c}{\sigma}\right)^{1/2}} \sim c^{1/2} \int_0^{\infty} \frac{e^{-p\sigma}}{\sigma^{1/2}}\, d\sigma = \sqrt{\frac{\pi c}{p}}. \tag{11}$$

It follows that

$$\tilde{K}(p) \sim V\alpha\sqrt{\pi c/p}, \tag{12}$$

which in turn implies

$$K(\sigma) \sim V\alpha\sqrt{c/\sigma}, \quad \text{i.e.,} \quad \kappa(s, t) \sim V\alpha\sqrt{c/(Vt - s)}. \tag{13}$$

Note the square root singularity at the end of the sheet. The circulation $\Gamma(t)$ is given from (2) by

$$\Gamma \sim 2V\alpha\sqrt{cVt} \tag{14}$$

and thus increases initially like the square root of the time.

For large times, we are interested in small p. Then

$$\int_0^{\infty} \sqrt{1 + \frac{c}{\sigma}}\, e^{-p\sigma}\, d\sigma \sim \frac{1}{p}[1 + p\log pc], \tag{15}$$

and

$$\tilde{K}(p) \sim V\pi c\alpha(1 - p\log pc). \tag{16}$$

The limiting circulation Γ_{∞} as $t \to \infty$ is

$$\Gamma_{\infty} = \int_0^{\infty} K(\sigma)\, d\sigma = \tilde{K}(0) = V\pi c\alpha. \tag{17}$$

Wagner's approximate inversion of the Laplace transform shows that $\Gamma = 0.9\Gamma_\infty$, when $t = 7c/V$.

The lift and drag on the wing can be found by using the unsteady form of Bernoulli's equation for potential flow to evaluate the pressure (taking account of the suction at the leading edge), but we shall use this problem as an example of vortex force and bound vorticity. The impulse of the bound vorticity in the wing ($\int_W$ denotes integration inside the wing and $\oint_W$ is integration around the wing) is

$$\mathbf{I} = \int_W \omega\, \mathbf{r} \times \mathbf{k}\, dS. \tag{18}$$

Then,

$$\frac{d\mathbf{I}}{dt} = \int_W \frac{\partial\omega}{\partial t}\, \mathbf{r} \times \mathbf{k}\, dS$$

$$= \int_W \mathbf{F}\, dS - \int_W (\mathbf{r} \times \mathbf{k})(\mathbf{u} \cdot \nabla\omega)\, dS + \oint_W (\mathbf{r} \times \mathbf{k})(\mathbf{n} \times \mathbf{F} \cdot \mathbf{k})\, ds \tag{19}$$

$$= -\mathbf{D} + \int_W \mathbf{u} \times \mathbf{k}\,\omega\, dS + \oint_W (\mathbf{u} \cdot \mathbf{n})(\mathbf{k} \times \mathbf{r})\,\omega\, ds$$

after a further integration by parts and requiring that $\mathbf{F}$ satisfies (3.7.4) on the wing surface, where $-\mathbf{D}$ is the external force required to keep the wing stationary so that $\mathbf{D}$ is the force exerted on the body by the fluid, and $\mathbf{F} \times \mathbf{n}$ is taken to be zero on the boundary of the wing. The last term in (19) does not in general vanish, because of the flux of vorticity out of the wing into the vortex sheet, but we can make this term vanish by choosing the origin at the trailing edge of the wing.[5] Then

$$\mathbf{D} = \int_W \mathbf{u} \times \mathbf{k}\,\omega\, dS - \frac{d\mathbf{I}}{dt}. \tag{20}$$

The first term on the right-hand side of (20) is the vortex force. The contribution from the free stream gives a lift $V\Gamma(t)$ in the direction perpendicular to the free stream. The contribution to the vortex force from the velocity induced by the vorticity in the trailing sheet gives a component $O(\alpha^2)$ parallel to the free stream. This is the 'induced drag' associated with the generation of the kinetic energy in the trailing vortex system.

It remains to calculate the last term in (20). There are contributions from the bound vorticity which is the image of the trailing vorticity and the bound vorticity induced by the Kutta condition. (The bound vorticity induced by the free stream is constant.) Consider first the image of the

---

[5] The apparent lack of Galilean invariance is due to the circulation around the wing not being zero.

trailing vorticity. From (18), the y-component which contributes to the lift is

$$I_y^T = - \int_{-a}^{a} x \left[\int \omega \, dy \right] dx = \int_{-a}^{a} x \left[u(x, 0+) - u(x, 0-) \right] dx = - \oint_W xu \, dx,$$

$$(21)$$

where u is the velocity induced by the trailing vorticity and can be taken as given by (4) with $\gamma_1 = -\kappa$, so that the total vorticity of the image system of the trailing vorticity is taken to be zero. The position of the origin is then unimportant. On the wing, $u = dw/dz$, $x = z$, $dx = dz$ and hence the contribution δI_y^T from the trailing element of vorticity $\kappa \, ds$, is given by

$$\frac{\delta I_y^T}{\kappa \, ds} = - \oint z \frac{dw}{dz} \, dz = \oint \frac{1}{2} \left(\zeta + \frac{a^2}{\zeta} \right) \frac{dw}{d\zeta} \, d\zeta$$

$$= \oint \zeta \frac{dw}{d\zeta} \, d\zeta, \quad \text{since } dw \text{ is real and } \Re\zeta = \Re(a^2/\zeta)$$

$$= \frac{i}{2\pi} \oint \left(\frac{\zeta}{\zeta - r_1} - \frac{\zeta}{\zeta - r_2} + 1 \right) d\zeta, \quad r_{1,2} = s + a \pm \sqrt{s^2 + 2as},$$

$$= s + a - \sqrt{s^2 + 2as}. \tag{22}$$

Hence,

$$I_y^T = \int_0^{Vt} \kappa(s, t) \left(s + a - \sqrt{s^2 + 2as} \right) ds. \tag{23}$$

Finally, we have the contribution from the bound vorticity with circulation $\Gamma(t)$ induced by the Kutta condition. Remembering that the origin is at the trailing edge, we have

$$I_y^K = \int_{-a}^{a} \left[\int \omega \, dy \right] (a - x) \, dx = \oint u(a - x) \, dx = -\Gamma a, \tag{24}$$

since $u = \Gamma/(2\pi a \sin \theta)$, when $x = a \cos \theta$. Using (2), we obtain from (23) and (24),

$$I_y = \int_0^{Vt} \kappa(s, t) \left(s - \sqrt{s^2 + sc} \right) ds + \text{const.} \tag{25}$$

Now, $\kappa(s, t) = K(Vt - s)$. Differentiating with respect to time, and integrating by parts, we obtain

$$\frac{dI_y}{dt} = \int_0^{Vt} K(\sigma) \left[V - \frac{V^2 t - \sigma V + \frac{1}{2} Vc}{\{V^2 t^2 - 2\sigma Vt + \sigma^2 + Vtc - \sigma c\}^{1/2}} \right] d\sigma$$

$$\to 0 \quad \text{as } t \to \infty. \tag{26}$$

So $D \to \pi V^2 \alpha c$.

For small t, we have from (25), using (13),

$$I_y \sim - \int_0^{Vt} \kappa \sqrt{sc} \, ds = - \int_0^{Vt} V\alpha c \sqrt{\frac{s}{Vt - s}} \, ds = -\tfrac{1}{2} \pi V^2 \alpha c t. \tag{27}$$

Then from (20), since $\Gamma(0) = 0$, the lift L is given by

$$L = \tfrac{1}{2} \pi V^2 \alpha c \quad \text{as } t \to 0. \tag{28}$$

Thus for the impulsively generated motion, the initial lift on the wing is one-half the final lift. This behaviour (called the Wagner effect) has been verified experimentally by Walker [1931]. The final lift is attained after a time $O(c/V)$. The shed vorticity then rolls up into a starting vortex. If the wing suddenly stops moving, the Kutta condition is violated, and vorticity will then be shed until the circulation around the wing is reduced to zero. This trailing vorticity rolls up into an ending vortex. As $t \to \infty$, the starting and ending vortices form a pair of equal and opposite vortices with a hydrodynamic impulse equal to the lift integrated over the time of flight.

7

DYNAMICS OF LINE VORTICES IN TWO-DIMENSIONAL FLOW

7.1 Stability of a polygonal array

For two-dimensional motion of an incompressible fluid, the vorticity $\omega(x, y, t)$ and stream function $\psi(x, y, t)$, where $u = \partial\psi/\partial y$, $v = -\partial\psi/\partial x$, are related by

$$\psi(x, y, t) = \frac{-1}{2\pi} \int \omega(x', y', t) \log r \, dx' \, dy' + \Psi(x, y, t), \tag{1}$$

where $r = \sqrt{(x - x')^2 + (y - y')^2}$, and $\nabla^2 \Psi = 0$. If the fluid is unbounded and at rest at infinity, and the vorticity distribution is compact, then $\Psi = 0$. In the absence of viscosity, the vorticity satisfies (3.10.7).

As discussed in §2.3, the point or line vortex is a singular distribution $\omega = \kappa\delta(\mathbf{r} - \mathbf{R})$ which moves with the fluid in the absence of external forces. The velocity field is

$$\mathbf{u} = \frac{\kappa}{2\pi} \frac{\mathbf{k} \times (\mathbf{r} - \mathbf{R})}{|\mathbf{r} - \mathbf{R}|^2} + \mathbf{u}_F(\mathbf{r}, t), \tag{2}$$

where $\mathbf{u}_F$ is non-singular in the vicinity of $\mathbf{R}$. Note that $\mathbf{u}_F$ need not be irrotational. The singular vortex could be immersed in a field of continuous vorticity, although this is usually supposed not to be the case. The motion of the vortex is given by

$$\frac{d\mathbf{R}}{dt} = \mathbf{u}_F(\mathbf{R}, t), \qquad \kappa = \text{const.} \tag{3}$$

For an assembly of N vortices, strengths κ_j, located at $\mathbf{R}_j(t)$, the stream function is

$$\psi = -\sum_{j=1}^{N} \frac{\kappa_j}{2\pi} \log |\mathbf{r} - \mathbf{R}_j| + \Psi, \tag{4}$$

where Ψ is non-singular in the vicinity of the vortices and vanishes if the fluid is unbounded and at rest at infinity. The velocity field is

$$\mathbf{u} = -\mathbf{k} \times \nabla \psi, \tag{5}$$

and the evolution is described by $2N$ non-linear ordinary differential equations which result from

$$\frac{d\mathbf{R}_j}{dt} = \sum_k' \frac{\kappa_k}{2\pi} \frac{\mathbf{k} \times (\mathbf{R}_j - \mathbf{R}_k)}{|\mathbf{R}_j - \mathbf{R}_k|^2} - \mathbf{k} \times \nabla \Psi(\mathbf{R}_j, t), \tag{6}$$

where the $'$ denotes that infinite terms are excluded from the sum.

Given that Ψ can be determined (usually it is due to the presence of bodies and requires for its calculation the Green's function of Laplace's equation with Dirichlet boundary conditions), the evolution of the point vortices reduces to the solution of a set of ordinary differential equations and is an exercise in dynamical system theory.

The simplest example is the vortex pair. The counter-rotating vortex pair consists of two vortices strengths $\pm\kappa$ separated by a distance b. The pair moves with velocity $\kappa/2\pi b$ in the direction normal to the line of centres.

The co-rotating pair consists of equal vortices of strength κ separated by distance b rotating about their midpoint with angular velocity $\kappa/\pi b^2$.

The equations for three vortices are integrable, but the motion of more than three vortices is, in general, unsteady and irregular (see Aref [1983] for a review of this topic).

Some special equilibrium conditions are of interest. Consider the case of N vortices at the vertices of a regular polygon (Kelvin [1878], Thomson [1883]). We follow here the treatment of Havelock [1931]. Clearly there is an equilibrium in which the polygon rotates without change of shape, with angular velocity $\Omega = \kappa(N-1)/4\pi a^2$, where a is the radius of the circle on which the vortices lie. We consider the linear stability of the equilibrium configuration to infinitesimal disturbances. We suppose the nth vortex is displaced radially to $a + r_n$ and the angular co-ordinate is $2n\pi/N + \Omega t + \theta_n$. The velocities are given by (6). Substituting and linearising in r_n and θ_n, one obtains after some algebra for the $n = 1$ vortex,

$$\dot{r}_1 = \frac{-\kappa}{4\pi a} \sum_1^{N-1} \frac{\theta_{n+1} - \theta_1}{1 - C_n}, \tag{7}$$

$$a\dot{\theta}_1 = -\Omega r_1 + \frac{\kappa}{4\pi a} \sum_1^{N-1} \left\{ \frac{C_n}{1 - C_n} \frac{r_1}{a} - \frac{1}{1 - C_n} \frac{r_{n+1}}{a} \right\}, \tag{8}$$

where $C_n = \cos(2n\pi/N)$. Using $\sum_1^{N-1}(1 - C_n)^{-1} = \frac{1}{6}(n^2 - 1)$, we can rewrite these equations as

$$(4\pi a/\kappa)\,\dot{r}_1 = A\theta_1 - \sum_1^{N-1} c_n\theta_{n+1}, \tag{9}$$

$$(4\pi a^3/\kappa)\,\dot{\theta}_1 = Br_1 - \sum_1^{N-1} c_n r_{n+1}, \tag{10}$$

where $c_n = 1/(1 - C_n)$, $A = \frac{1}{6}(N^2 - 1)$, $B = \frac{1}{6}(N - 1)(N - 11)$. There are similar equations for the other vortices. For the mth vortex,

$$(4\pi a/\kappa)\dot{r}_m = A\theta_m - \sum_1^{N-1} c_n\theta_{n+m}, \tag{11}$$

$$(4\pi a^3/\kappa)\dot{\theta}_m = Br_m - \sum_{n=1}^{N-1} c_n r_{n+m}, \tag{12}$$

with the convention that $n + m = (n + m) \bmod N$ in the suffix.

From the invariance under rotation, we expect eigenfunctions of the form

$$r_m = \alpha(t)e^{2km\pi i/N}, \quad \theta_m = \beta(t)e^{2km\pi i/N} \tag{13}$$

for $0 \le k \le N - 1$, corresponding to N independent eigenfunctions. We can regard k as a subharmonic wave number. When $k = 0$, all vortices behave in the same way. There is a wavelength N/k when this ratio is an integer. Substitution of (13) into (11) and (12) shows that solutions of the form (13) exist if

$$(4\pi a/\kappa)\dot{\alpha}(t) = A\beta - S\beta, \quad (4\pi a^3/\kappa)\dot{\beta} = B\alpha - S\alpha, \tag{14}$$

where

$$S = \sum_1^{N-1} \frac{e^{2kn\pi i/N}}{1 - \cos 2n\pi/N} = \frac{1}{6}(N^2 - 1) - k(N - k).$$

It follows now that α and β are proportional to $e^{\sigma t}$ where

$$\sigma^2 = \left(\frac{\kappa}{4\pi a^2}\right) k(N - k)\{k(N - k) - 2(N - 1)\}. \tag{15}$$

When $k = 0$, it follows that $\sigma = 0$. In this case, the displacement consists of a small change in the ring radius and a small change in the angular velocity, corresponding to a transition to a steady state near the original one. For the other eigenfunctions, stability depends on the sign of

$$D = k(N - k) - 2(N - 1). \tag{16}$$

If $D > 0$ for any k $(0 < k \leq N-1)$, the system is unstable with disturbances growing exponentially in t. If $D < 0$, for all k, the system is linearly stable. (Stability in this case to finite amplitude disturbances is another question.) If $D = 0$, there are unstable disturbances which grow linearly in time. If $N > 7$, it is easy to show that there exists values of k for which $D > 0$. If $N < 7$, $D < 0$ for all k. If $N = 7$, $D < 0$ except for $k = 3$ or 4, for which $D = 0$. Havelock considers further cases, including the addition of an arbitrary tangential velocity $V(r)$ (which need not be irrotational) and the presence of outer or inner boundaries and a second ring of vortices. Further studies are reviewed by Aref et al. [1988].

7.2 Vortices near walls

The case of a single line vortex near a plane wall is easily solved by the method of images. The addition of an image vortex of opposite strength satisfies the kinematic boundary condition of zero normal velocity at the wall, and the pair moves with velocity $\kappa/4\pi h$ parallel to the wall, where κ is the vortex strength and h is the distance from the wall. Note that the vortex can be reduced to rest by superposition of a uniform stream of magnitude $\kappa/4\pi h$ parallel to the wall.

A more complicated problem is the counter-rotating vortex pair in a parallel sided channel.[1] The image system is a doubly infinite row, and it is preferable to employ the techniques of complex variables (Batchelor [1967 §6.5]). We introduce the complex potential

$$w(z) = \phi + i\psi, \quad z = x + iy, \tag{1}$$

which is an analytic function of z and such that the velocity components (u, v) are given by its derivative

$$dw/dz = u - iv. \tag{2}$$

For a single line vortex of strength κ at z_0, the complex potential is

$$w = -\frac{i\kappa}{2\pi} \log(z - z_0). \tag{3}$$

Now consider a counter-rotating pair, symmetrically placed in a parallel sided channel of width b, distance $b - 2h$ apart. Take the origin on the bottom wall and the x-axis parallel to the walls. Then

$$w = \frac{i\kappa}{2\pi} \log \frac{z - i(b - h)}{z - ih} + W(z), \tag{4}$$

[1] This configuration has applications to the dynamics of liquid helium II.

where W is analytic in the channel, vanishes at $x \to \pm\infty$, and is such that $\Im(w) = 0$ on the walls.[2] To find W, we map the channel conformally into the upper half plane by the map

$$\zeta = e^{\pi z/b}, \quad z = b/\pi \, \log \zeta. \tag{5}$$

The vortices map into the points $e^{i\pi h/b}$ and $-e^{-i\pi h/b}$. In the ζ-plane, images at the complex conjugate points can be used, to give

$$w = -\frac{i\kappa}{2\pi} \log \frac{\zeta - e^{i\pi h/b}}{\zeta - e^{-i\pi h/b}} + \frac{i\kappa}{2\pi} \log \frac{\zeta + e^{-i\pi h/b}}{\zeta + e^{i\pi h/b}}. \tag{6}$$

The velocity (U, V) of the vortex at $z = ih$ is given by the other vortex plus the contributions from the images, and is

$$U - iV = \frac{d}{dz} \left[\frac{i\kappa}{2\pi} \log(z - i(b - h)) + W(z) \right]_{z=ih}$$
$$= \lim_{z \to ih} \frac{d}{dz} \left(w + \frac{i\kappa}{2\pi} \log(z - ih) \right) \tag{7}$$

where w is given by (6) as a function of ζ. Noting that $w + i\kappa/2\pi \log(\zeta - e^{i\pi h/b})$ is analytic, we have

$$U - iV = \frac{d}{d\zeta} \left(w + \frac{i\kappa}{2\pi} \log(\zeta - e^{i\pi h/b}) \right) \bigg/ \frac{dz}{d\zeta} + \frac{i\kappa}{2\pi} \frac{d}{dz} \log \left(\frac{z - ih}{\zeta - e^{i\pi h/b}} \right), \tag{8}$$

with the right-hand side evaluated at $\zeta = e^{i\pi h/b}$. Since

$$\frac{z - ih}{\zeta - e^{i\pi h/b}} = \frac{dz}{d\zeta} + \frac{1}{2} \frac{d^2 z}{d\zeta^2} (\zeta - e^{i\pi h/b}),$$

where the derivatives are evaluated at $\zeta = e^{i\pi/b}$, the last term in (8) is

$$\frac{i\kappa}{4\pi} \frac{d^2 z/d\zeta^2}{(dz/d\zeta)^2} \bigg|_{\zeta=e^{i\pi h/b}}. \tag{9}$$

Substitution and evaluation gives

$$U = \frac{\kappa}{2b \tan(2\pi h/b)}, \quad V = 0. \tag{10}$$

It is easily verified that the correct limits are obtained for $h \to 0$ and $h \to b/2$. The vortices are stationary when $h = b/4$. The complex potential (6) can be expressed as

$$w = \frac{i\kappa}{2\pi} \log \frac{\sinh \frac{\pi}{b} (z - ih)}{\sinh \frac{\pi}{b} (z - i(b - h))}. \tag{11}$$

---

[2] An arbitrary constant or function of time can be added.

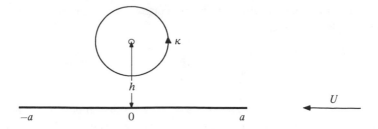

7.2-1 Vortex standing over a flat plate parallel to a uniform stream.

Another example of interest is the Föppl vortex pair behind a circular cylinder of radius a in a uniform stream. If the vortices are at z_0 and $\bar{z}_0$, with strengths $-\kappa$ and κ respectively, the image vortices are at $a^2/\bar{z}_0$ and a^2/z_0 and of strengths κ and $-\kappa$. The complex potential is

$$w = U\left(z + \frac{a^2}{z}\right) + \frac{i\kappa}{2\pi} \log\left(\frac{z - z_0}{z - \bar{z}_0} \frac{z - a^2/z_0}{z - a^2/\bar{z}_0}\right). \tag{12}$$

This has the appropriate singularities and behaviour at infinity, and it is readily verified that $\Im w = 0$ on $r = a$. The vortices are at rest if

$$\frac{d}{dz}\left[w - i\kappa \log(z - z_0)\right]_{z=z_0} = 0. \tag{13}$$

There are two equations for three real unknowns, κ, x_0, y_0. After reduction, it is found that (13) is satisfied if the equations (2.4.7) and (2.4.8) are satisfied.

Föppl also discussed the corresponding case of flow past a flat plate normal to the flow. (see also Villat [1930 pp. 162–171, 182–14]; Smith and Clark [1975]). It was shown that it is impossible to find a position and strength for the vortices which satisfies the Kutta condition at the edge of the plate; that is, there is no flow for which the velocity on the plate is bounded. Roy [1972] has shown that the Kutta condition can be satisfied by superimposing a sink on the vortex.[3] The claim is made that this flow is stable.

Another class of interesting flows is the asymmetric configurations of bodies and vortices. Consider the case of a vortex of strength κ at a distance h above the midpoint of a flat plate of length $2a$ (see Figure 7.2-1) parallel to a stream of velocity U in the negative x-direction. The conformal map

$$\zeta = z + (z^2 - a^2)^{1/2}, \quad z = \tfrac{1}{2}(\zeta + 1/\zeta), \tag{14}$$

---

[3] This theory is a model of flow past a slender delta wing at small angle of attack.

maps the plate into a circle with the vortex at $\zeta = ih + i\sqrt{a^2 + h^2}$. Then the complex potential is[4]

$$w = -\frac{U}{2}\left(\zeta + \frac{a^2}{\zeta}\right) - \frac{i\kappa}{2\pi}\log(\zeta - ih') + \frac{i\kappa}{2\pi}\log(\zeta - ia^2/h') - \frac{i\gamma}{2\pi}\log\zeta,$$

(15)

where $h' = h + \sqrt{a^2 + h^2}$ and γ (real) is arbitrary and is the strength of a bound line vortex at the centre of the circle. The circulation about the plate is

$$\Gamma = \gamma - \kappa. \tag{16}$$

It is possible in this flow to choose h, κ and γ so that the vortex is stationary and the Kutta condition is satisfied at the ends of the plate. Since the mapping is singular at $z = \pm a$, the Kutta condition is

$$dw/d\zeta = 0, \quad \text{on } \zeta = \pm a. \tag{17}$$

This gives on evaluation

$$\frac{\gamma}{\kappa} = \frac{h'^2 - a^2}{h'^2 + a^2}. \tag{18}$$

The vortex in the free stream is stationary if

$$\frac{d}{d\zeta}\left(w + \frac{i\kappa}{2\pi}\log(\zeta - ih')\right) + \frac{i\kappa}{4\pi}\frac{d^2z/d\zeta^2}{(dz/d\zeta)^2} = 0 \tag{19}$$

when $\zeta = ih'$. This gives on evaluation ($a \le h' < \infty$)

$$\gamma = \pi U(h'^2 - a^2)^2(h'^2 + a^2)/2a^2h'^3. \tag{20}$$

The lift on the plate is the vortex force on the vortex in the free stream and the bound vorticity and is[5]

$$L = U(\kappa + \gamma - \kappa) = U\gamma. \tag{21}$$

The stability of the flow is a difficult problem, as it involves the unsteady Kutta condition.

---

[4] If w(z) is analytic for $|z| \le R$, $(R > a)$, then $w(z) + \overline{w}(a^2/z) - i\kappa/2\pi \log z$ is analytic for $a \le z \le R$, and has constant imaginary part on $|z| = a$, where $a^2/\overline{z} = z$.

[5] Saffman and Sheffield [1977] consider the case of a plate at incidence. Sheffield [1978] discusses the case of a Joukowski airfoil. Saffman and Tanveer [1984a] and Tanveer [1983, 1985] examine the case of a point vortex over a wing with a forward-facing flap. All configurations show considerable lift enhancement. Note, however, that geometrical constraints must sometimes be satisfied in order for a solution to exist. These calculations are relevant to the study of the Kasper wing, a high-lift wing with a trapped vortex (Kasper [1979], Sunderland [1976].)

7.3 Kirchhoff–Routh path function

Suppose we have N line vortices of strengths $\kappa_1, \ldots, \kappa_n$ at points $z_1, \ldots, z_n$ in a region bounded by surfaces in prescribed motion, that is, surfaces at rest or in given motion. (Free surfaces on which the pressure is given or obeys some relation are excluded.) Then the complex potential has the form

$$w = f(z) + \sum_{j=1}^{N} \kappa_j G(z; z_j) = \phi + i\psi, \tag{1}$$

where $f(z)$ is the irrotational flow generalized by motion of the boundaries and cyclic conditions. The function G is the unique complex Green's function such that $\Im(G) = 0$ on all bounding surfaces, and

$$g(z; z_j) = G(z; z_j) + \frac{i}{2\pi} \log(z - z_j) \tag{2}$$

is an analytic and single-valued function of z in the region. Note that $g(z; z_j)$ is not in general analytic in the second variable z_j. However, a reciprocity property holds,

$$\Im g(z; z_j) = \Im g(z_j; z). \tag{3}$$

To see this, let $\Im G(z; z_1) = \psi_1(z)$, $\Im G(z; z_2) = \psi_2(z)$. Both ψ_1 and ψ_2 are harmonic and single valued; hence

$$\oint \left(\frac{\partial \psi_1}{\partial n} \psi_2 - \frac{\partial \psi_2}{\partial n} \psi_1 \right) ds = 0.$$

Choose a circuit consisting of the boundaries and infinitesimal circles around z_1 and z_2. It follows that $\psi_1(z_2) = \psi_2(z_1)$. The function g can be thought of as the complex potential induced by the images, in the boundaries, of the vortex at z_j. If unbounded, $g = 0$ and

$$G(z : z_j) = \frac{-i}{2\pi} \log(z - z_j). \tag{4}$$

Consider now the velocity (u_j, v_j) of the jth vortex. Differentiating (1) with respect to z at $z = z_j$, we have

$$\kappa_j(u_j - iv_j) = \kappa_j f'(z_j) + \sum_{k=1}^{N}{}' \kappa_j \kappa_k \frac{\partial G}{\partial z_j}(z_j; z_k) + \kappa_j^2 \frac{\partial}{\partial z} g(z; z_j) \bigg|_{z=z_j}, \tag{5}$$

where $\sum'$ means that the term with $k = j$ is omitted. Define $\Psi(x_1, y_1, \ldots, x_N, y_N)$ by

$$\Psi = \Im\left[\sum_j \kappa_j f(z_j) + \frac{1}{2}\sum_{j \neq k}\sum \kappa_j \kappa_k \, G(z_j; z_k) + \sum_j \frac{\kappa_j^2}{2} g(z_j; z_j)\right] \tag{6}$$

It follows from the symmetry of $\Im G$ that

$$\kappa_j(u_j - iv_j) = \frac{\partial \Psi}{\partial y_j} + i\frac{\partial \Psi}{\partial x_j}. \tag{7}$$

The function Ψ, which acts like a multidimensional stream function, will be called the Kirchhoff–Routh function. If the domain is unbounded,

$$\Psi = -\frac{1}{4\pi}\sum_{j \neq k}\sum \kappa_j \kappa_k \, \log \, |z_j - z_k|. \tag{8}$$

In general, if the boundaries are moving, f, G and g will be functions of time.

If the boundaries are fixed,

$$\frac{d}{dt}\Psi = \sum_j \left(\frac{\partial \Psi}{\partial x_j}\frac{dx_j}{dt} + \frac{\partial \Psi}{\partial y_j}\frac{dy_j}{dt}\right) = 0, \tag{9}$$

by (7). Thus Ψ is an invariant of the motion for fixed boundaries. In this case, when $N = 1$ the function $\Psi(x, y)$ is called the Kirchhoff–Routh path function, and the trajectory of the vortex is given by $\Psi = \text{constant}$.

Equations (7) are of Hamiltonian form. Define (no summation)

$$q_j = \sqrt{\kappa_j}\, x_j, \qquad p_j = \sqrt{\kappa_j}\, y_j, \qquad\qquad \kappa_j > 0 \tag{10}$$

$$q_j = \sqrt{-\kappa_j}\, y_j, \qquad p_j = \sqrt{-\kappa_j}\, x_j, \qquad\qquad \kappa_j < 0, \tag{11}$$

then (7) can be written

$$\frac{d\mathbf{q}}{dt} = \frac{\partial H}{\partial \mathbf{p}}, \qquad \frac{\partial \mathbf{p}}{\partial t} = -\frac{\partial H}{\partial \mathbf{q}}, \qquad H = \Psi(\mathbf{p}, \mathbf{q}). \tag{12}$$

The co-ordinates of the vortices are the phase space configuration of a dynamical system with Hamiltonian H.

If the boundaries are at rest, $\partial H/\partial t = 0$, and H can be identified with the finite part of the energy. To see this, note that for fixed boundaries $f(z) = 0$ and the stream function $\psi(x, y)$ is given by

$$\psi(x, y) = \sum_1^N \kappa_j \psi_j, \tag{13}$$

where

$$\psi_j = -\frac{1}{2\pi} \, \log \, r_j + \Im \, g(z; z_j) = \Im G(z; z_j), \tag{14}$$

and r_j denotes the distance to the jth vortex. The kinetic energy is

$$K.E. = \lim_{\epsilon \to 0} \frac{1}{2} \int_{|z-z_j|>\epsilon} (\nabla\psi)^2 dS = \lim_{\epsilon \to 0} -\frac{1}{2} \sum_j \oint_{|z-z_j|=\epsilon} \psi \frac{\partial\psi}{\partial n} \, ds. \tag{15}$$

Substituting (14), and remembering that the g are finite so that $\partial\psi_j/\partial n = -1/2\pi\epsilon + O(1)$ and $\oint ds = 2\pi\epsilon$, we obtain

$$K.E. \sim \frac{1}{2} \sum\sum_{j\neq k} \kappa_j\kappa_k \, \Im G(z_j; z_k) - \frac{1}{2} \sum_j \kappa_j^2 \lim_{\epsilon \to 0} \oint \psi_j \frac{\partial\psi}{\partial n} \, ds$$

$$\sim -\sum \frac{\kappa_j^2}{4\pi} \, \log \epsilon + \frac{1}{2} \sum\sum_{j\neq k} \kappa_j\kappa_k \, \Im G(z_j; z_k)$$

$$+ \frac{1}{2} \sum \kappa_j^2 \, \Im g(z_j; z_j)$$

$$= \Psi - \log \epsilon \sum \frac{\kappa_j^2}{4\pi}. \tag{16}$$

If unbounded, $g = 0$ and there is an extra divergent contribution $\left(\sum \kappa_j\right)^2 \log R/4\pi$ from the circle of radius R at infinity.

Other invariants are the linear impulse $\mathbf{I}$ and angular impulse A.

$$\mathbf{I} = \int \omega \, \mathbf{r} \times \mathbf{k} \, dS = \int (\omega y, -\omega x) \, dS$$

$$= \left(\sum \kappa_j y_j, -\sum \kappa_j x_j\right), \tag{17}$$

$$A = -\frac{1}{2} \int \omega r^2 \, dS = -\frac{1}{2} \sum \kappa_j(x_j^2 + y_j^2). \tag{18}$$

These invariants behave in the usual way. For a finite group of N vortices,

$$\frac{d\mathbf{I}_N}{dt} = \int \omega \, \mathbf{u}_{\rm E} \times \mathbf{k} \, dS = \left(\sum_1^N \kappa_j v_{\rm E}(\mathbf{r}_j), -\sum_1^N \kappa_j u_{\rm E}(\mathbf{r}_j)\right), \tag{19}$$

$$\frac{dA_N}{dt} = \mathbf{k} \cdot \int \omega \mathbf{r} \times (\mathbf{u}_{\rm E} \times \mathbf{k}) \, dS = -\sum_1^N \kappa_j(\mathbf{r}_j \cdot \mathbf{u}_{\rm E}(\mathbf{r}_j)). \tag{20}$$

Here $\mathbf{u}_{\rm E}(\mathbf{r})$ is the external velocity, that is, the actual velocity field with the Biot–Savart contribution of the N vortices subtracted. (The images of the vortices contribute to $\mathbf{u}_{\rm E}$.) In unbounded fluid, $\mathbf{u}_{\rm E} = 0$ and $\mathbf{I}$ and A are constant.

For the case when the total vorticity is non-zero, that is, $\sum_j \kappa_j \neq 0$, we can define a two-dimensional centroid by (see §3.10)

$$\bar{x} = \frac{\sum \kappa_j x_j}{\sum \kappa_j}, \quad \bar{y} = \frac{\sum \kappa_j y_j}{\sum \kappa_j}. \tag{21}$$

We define the angular impulse A_G about the centroid as

$$A_G = -\frac{1}{2} \sum_j \kappa_j [(x_j - \bar{x})^2 + (y_j - \bar{y})^2]. \tag{22}$$

Then

$$\frac{dA_G}{dt} = -\sum_j \kappa_j (\mathbf{r}_j - \bar{\mathbf{r}}) \cdot (u_E(\mathbf{r}_j) - \bar{\mathbf{u}}) \tag{23}$$

when $\bar{\mathbf{u}} = (d\bar{x}/dt, d\bar{y}/dt)$ is the velocity of the centroid. The expression (23) is the moment of the vortex force about the centroid. If this is zero, then $A_G = \text{constant}$.

7.4 Conformal mapping and the Kirchhoff–Routh path function

The behaviour of the Kirchhoff–Routh function under conformal mapping is of interest. Consider an assembly of N vortices moving inside a fixed domain D in the z-plane. The motion is described by a complex potential $w(z)$. Let $z = Z(z')$ be the conformal map of D into a region D' in the z'-plane. Then $w(z) \to w'(z')$, where w' is analytic in z', $\Im w' = 0$ on the boundaries, and $w' \sim -\kappa_j/2\pi \log r'_j$ near the image of each vortex.

For the domain D there is a Kirchhoff–Routh function $\Psi(\mathbf{z})$, where $\mathbf{z} = (z_1, z_2, \ldots, z_N)$, and for D' there is a Kirchhoff–Routh function $\Psi'(\mathbf{z}')$ which describes the motion of actual vortices in D'. In general

$$\Psi(\mathbf{z}) \neq \Psi'(\mathbf{z}'), \tag{1}$$

and the trajectories in the two domains are not the conformal maps of each other under the mapping Z. But there is a relation between them. We have

$$u_j - iv_j = \frac{d\bar{z}_j}{dt} = \frac{d}{dz} \left[w + \frac{i\kappa_j}{2\pi} \log(z - z_j) \right]_{z=z_j} = \frac{1}{\kappa_j} \left[\frac{\partial \Psi}{\partial y_j} - i \frac{\partial \Psi}{\partial x_j} \right] \tag{2}$$

and

$$u'_j - iv'_j = \frac{d\bar{z}'_j}{dt} = \frac{d}{dz'}\left[w' + \frac{i\kappa_j}{2\pi}\log(z'-z'_j)\right]_{z'=z'_j} = \frac{1}{\kappa_j}\left[\frac{\partial\Psi'}{\partial y'_j} - i\frac{\partial\Psi'}{\partial x'_j}\right].$$

(3)

It follows from the chain rule (see §2), that

$$\frac{d\bar{z}_j}{dt} = \frac{d\bar{z}'_j}{dt} \bigg/ \frac{dZ}{dz'} + \frac{i\kappa_j}{4\pi}\frac{d^2Z/dz'^2}{(dZ/dz')^2},$$

(4)

where the derivatives of Z are evaluated at z'_j. Now

$$d\Psi = \sum_j\left(\frac{\partial\Psi}{\partial x_j}dx_j + \frac{\partial\Psi}{\partial y_j}dy_j\right) = \Im\sum_j\kappa_j(u_j - iv_j)(dx_j + idy_j)$$

$$= \Im\sum_j\kappa_j\frac{d\bar{z}_j}{dt}dz_j = \Im\sum_j\kappa_j\frac{d\bar{z}'_j}{dt}dz'_j + \Im\sum_j\frac{i\kappa_j^2}{4\pi}\frac{d^2Z/dz'^2}{dZ/dz'}dz'_j$$

$$= d\Psi' + \frac{1}{4\pi}\Im\sum_j i\kappa_j^2\, d\left(\log\frac{dZ}{dz'}\right)_{z'_j}$$

$$= d\Psi' + \frac{1}{4\pi}\sum_j\kappa_j^2 d\left(\log\left|\frac{dZ}{dz'}\right|\right)_{z'_j}.$$

(5)

Hence,[6]

$$\Psi = \Psi' + \sum_{j=1}^{N}\frac{\kappa_j^2}{4\pi}\log\left|\frac{dZ}{dz'}\right|_{z'_j}.$$

(6)

As an example of the application of the path function, consider the motion of a symmetrically placed counter-rotating vortex pair of strength κ approaching perpendicularly a slit in a wall (Karweit [1975]), see Figure 7.4-1. The conformal map $z = \sinh\zeta$, $z = x + iy$, $\zeta = \xi + i\eta$, maps the upper half-plane into a strip of width $\pi/2$ in the ζ-plane.[7] The complex potential induced by a vortex in a channel is given in §2. Thus

$$w = -\frac{i\kappa}{2\pi}\log\left(\sinh\left(\frac{\zeta - \zeta_0}{\zeta - \bar{\zeta}_o}\right)\right)$$

---

[6] The result was given for $N = 1$ by Routh [1881]. The generalization to N vortices is due to Lin [1941, 1943].

[7] The Schwarz–Cristoffel theorem ensures that there exists a transformation reducible to quadrature which maps the interior of polygons (which may have a vertex at infinity) into each other (see Batchelor [1967 §6.5])

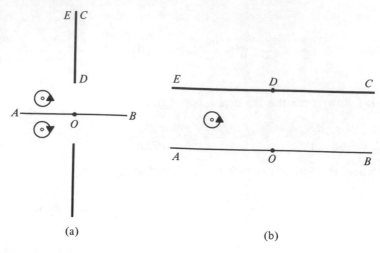

(a)   (b)

7.4-1 Vortex pair approaching a slit in a wall. (a) Physical plane. (b) Mapped plane.

$$= -\frac{i\kappa}{2\pi} \log \left(\frac{z\sqrt{1+z_0^2} - z_0\sqrt{1+z^2}}{z\sqrt{1+\bar{z}_0^2} - \bar{z}_0\sqrt{1-z^2}} \right), \tag{7}$$

where z_0 is the position of the vortex in the upper half-plane. To find the trajectories, we first find the path function in the ζ-plane. From §2, we have for the velocity of the vortex

$$u_0' = -\frac{\kappa}{2\pi \tan 2\eta_0} = \frac{1}{\kappa} \frac{\partial \Psi'}{\partial \eta_0}. \tag{8}$$

Hence

$$\Psi' = \frac{\kappa^2}{4\pi} \log(\sin 2\eta). \tag{9}$$

Now $dz/d\zeta = \cosh\zeta$, and it follows that the path function in the original plane is

$$\Psi = \frac{\kappa^2}{4\pi} \log \sin 2\eta + \frac{\kappa^2}{4\pi} \log \left(\cosh^2\xi - \sin^2\eta \right)^{1/2}. \tag{10}$$

The trajectories $\Psi = $ constant are sketched in Figure 7.4-2. On the dividing trajectory, $\partial\Psi/\partial\eta = 0$, which happens when $\tan\eta = 1/\sqrt{3}$. The equation of the dividing streamline is therefore

$$\sin 2\eta \left(\cosh^2\xi - \sin^2\eta \right)^{1/2} = 4/3\sqrt{3}. \tag{11}$$

It is easily seen from (11) that a pair coming from infinity will pass through

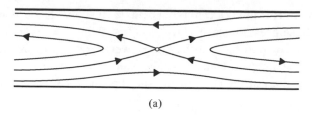

(a)

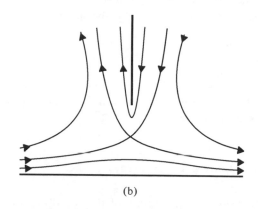

(b)

7.4-2 Trajectories. (a) Mapped plane. (b) Physical plane.

the slit of width 2 if the separation is less than $4/3\sqrt{3}$. Otherwise they will be reflected.[8]

There are difficulties with the solution just obtained. The Kutta condition is violated, which suggests it would be appropriate to study a flow in which vortex sheets are shed from the edges of the slit. Also, there is a pressure jump across the slit. From (7), the velocity potential at infinity asymptotes

$$\phi(+\infty) = -\frac{\kappa}{2\pi} i(\bar{\zeta}_0 - \zeta_0) = -\frac{\eta_0}{\pi}, \tag{12}$$

$$\phi(-\infty) = -\frac{\kappa}{2\pi} i(\zeta_0 - \bar{\zeta}_0) = \frac{\eta_0}{\pi}. \tag{13}$$

It follows from the unsteady Bernoulli equation, since η_0 depends upon t, that

$$p(+\infty) \neq p(-\infty). \tag{14}$$

Moreover, the solution is not unique, as we can add a potential $w_Q = Q\zeta = Q\sinh^{-1} z$, ($Q$ real) to the potential (7), corresponding to a flux $Q\pi$

---

[8] Other cases of vortex pairs moving near orifices have been studied by Sheffield [1977].

through the slit. The flux Q can be chosen so that the Kutta condition is satisfied. It suffices to make $dw/d\zeta - Q = 0$ at $\zeta = i\pi/2$, that is, $Q = -(\kappa/2\pi)\sin 2\eta_0/(\cos 2\eta_0 - \cosh 2\xi_0)$. However, $Q = Q(t)$ produces an infinite pressure at infinity. Other solutions can be constructed by adding multipole singularities at the edge of the slit. For instance, we can add

$$w_m = m/(1 + e^{2\zeta})^n, \quad n = 1, 2, \ldots. \tag{15}$$

The corresponding stream function ψ_m vanishes on $\eta = 0$ and $\eta = \pi/2$. The Kutta condition cannot be satisfied by the addition of (15), but the velocity potential ϕ_m satisfies

$$\phi_m(\infty) = 0, \quad \phi_m(-\infty) = m, \tag{16}$$

so choosing $m = -2\eta_0/\pi$ ensures that $p(+\infty) = p(-\infty)$. The calculation of the particle trajectories is now, however, a more difficult problem as the path function cannot be employed.[9]

7.5   Stability of infinite periodic arrays

In §1, we considered the stability of the equilibrium configuration of vortices at the vertices of a regular polygon. We now consider the stability of an infinite row of vortices of equal strength (Lamb [1932 §156]). The equilibrium configuration consists of vortices of strength κ at the points $x = ma$, $y = 0$ $(m = 1, 2 \ldots)$. The velocity at any point induced by the vortices is a conditionally convergent sum. To obtain definite results, we follow Lamb and adopt the convention of using a 'principal value' by considering the limit as $N \to \infty$ for $-N \le m \le N$.[10]

The complex potential of the undisturbed row is ($\sum'$ means $m = 0$ is excluded)

$$w = -\frac{i\kappa}{2\pi} \lim_{N\to\infty} {\sum_{-N}^{N}}' \log\left(\frac{z - ma}{-ma}\right)$$

$$= -\frac{i\kappa}{2\pi} \sum_1^\infty \left(1 - \frac{z^2}{m^2 a^2}\right) - \frac{i\kappa}{2\pi} \log z$$

---

[9]   A similar lack of uniqueness holds in three dimensions for the problem of a vortex ring approaching a wall with a hole (Miloh and Shlien [1977]). The solution is again arbitrary to the addition of a flux Q through the hole, but in this case the flux can be chosed to ensure $p(\infty) = p(-\infty)$. This solution still violates the Kutta condition, and flows with higher-order singularities at the edge of the hole can be added.

[10]   The velocity as $y \to \pm\infty$ is $\pm \frac{1}{2}\kappa/a$. The row is a discrete model of an infinite straight vortex sheet of strength κ/a.

$$= -\frac{i\kappa}{2\pi} \log\left(\sin\frac{\pi z}{a}\right), \tag{1}$$

since $\sin z / z = \prod_1^\infty (1 - z^2/m^2\pi^2)$. (The infinite constant $\sum' \log(1/-ma)$ was added to produce a finite result.) Now $d/dz \log(\sin(\pi z/a)/(z - ma)) = 0$, when $z \to ma$. Thus all the vortices are at rest. (This is obvious from the symmetry.)

We now study the stability to infinitesimal displacements of the vortices. The mth vortex is assumed to be displaced from $z_m = ma$ to $z_m + z'_m$. The new complex potential is (neglecting squares of primed quantities)

$$
\begin{aligned}
w &= -\frac{i\kappa}{2\pi} \sum_{-\infty}^{\infty}{}' \log\left(1 - \frac{z}{ma} + \frac{z'_m}{ma}\right) - \frac{i\kappa}{2\pi} \log(z - z'_0) \\
&= -\frac{i\kappa}{2\pi} \sum_{1}^{\infty} \log\left(1 - \frac{z^2}{m^2 a^2} + \frac{z(z'_m + z'_{-m})}{m^2 a^2} + \frac{z'_m + z'_{-m}}{ma}\right) \\
&\quad - \frac{i\kappa}{2\pi} \log(z - z'_0). \tag{2}
\end{aligned}
$$

The velocity of the $m = 0$ vortex is obtained by differentiating the sum with respect to z and evaluating at $z = z'_0$. To first order,

$$w = -\frac{i\kappa}{2\pi} \log\left(\frac{a}{\pi z} \sin\frac{\pi z}{a}\right) - \frac{i\kappa z}{2\pi a^2} \sum_{-\infty}^{\infty}{}' \frac{z'_m}{m^2} - \frac{i\kappa}{2\pi} \log(z - z'_0), \tag{3}$$

and hence

$$\frac{dz'_0}{dt} = -\frac{i\kappa\pi}{6a^2} z'_0 - \frac{i\kappa}{2\pi a^2} \sum_{-\infty}^{\infty}{}' \frac{z'_m}{m^2}. \tag{4}$$

The first term on the right-hand side of (4) is the velocity field of the other vortices in the undisturbed row near the equilibrium position of the $m = 0$ vortex. The second term is the contribution to the velocity of the $m = 0$ vortex from the displacement of the other vortices.

Since $\sum_1^\infty 1/m^2 = \frac{1}{6}\pi^2$, we can write (4) as

$$\frac{dz'_0}{dt} = \frac{i\kappa}{2\pi a^2} \sum_{-\infty}^{\infty}{}' \frac{z'_0 - z'_m}{m^2}. \tag{5}$$

From the symmetry, it is clear that this equation holds for the jth vortex, with the $m = j$ term then omitted from the sum.

We have an infinite system of equations. We examine disturbances of the form

$$z'_m = \zeta(t) e^{2\pi i m p}. \tag{6}$$

Note that without loss of generality, we can suppose $0 \le p \le 1$. The number p can be regarded as a dimensionless subharmonic wave number. If $p = \frac{1}{2}$, the disturbance has wavelength $2a$, in the sense that every other vortex moves in the same way, so the basic unit is the motion of vortex pairs. If p is very small, the disturbance is a long-wave modulation. If $p = 0$, we can talk of a superharmonic disturbance in which every vortex moves in the same way. In the present case, it is obvious that the superharmonic displacement is neutrally stable, being the trivial displacement of the row as a whole.

Substituting (6) into (5), we see that we have a solution if

$$\frac{d\bar{\zeta}}{dt} = \frac{i\kappa\zeta}{2\pi a^2} \sum_{-\infty}^{\infty}{}' \frac{1 - e^{2\pi i m p}}{m^2}. \tag{7}$$

It can be shown that, for $0 \le \alpha \le 2\pi$,

$$\sum_{-\infty}^{\infty}{}' \frac{e^{im\alpha}}{m^2} = \frac{\pi^2}{3} - \frac{1}{2}\alpha(2\pi - \alpha).$$

Taking the complex conjugate of (7) and differentiating with respect to time, we obtain $(0 \le p \le 1)$

$$\frac{d^2\zeta}{dt^2} = \sigma^2\zeta, \quad \text{where } \sigma = \frac{\kappa\pi}{a^2} p(1 - p). \tag{8}$$

Thus the disturbance grows exponentially in time, the growth rate being a maximum when $p = \frac{1}{2}$; that is, the pairing instability is most unstable. For p very small, the wavelength λ of the disturbance is approximately a/p and

$$\sigma \sim \pi\kappa \, a\lambda. \tag{9}$$

(Compare the result in §8.2 for the stability of a vortex sheet.)

When $\zeta \propto e^{\sigma t}$, it follows from (7) that $\zeta - i\bar{\zeta} = 0$, that is, $x_0' = y_0'$. From (4), the local velocity field of the undisturbed row near the $m = 0$ vortex is $\kappa\pi/6a^2 \, (y, x)$. Thus the vortices move in the direction of the positive principal rate of strain when the disturbance corresponds to the growing eigenvalue.

There is an exact solution for the finite amplitude evolution of the array for $p = \frac{1}{2}$. Let neighbouring vortices have co-ordinates $z_0(t)$, $z_1(t)$. We have two rows, with vortices at $z = z_0 + 2ma$, $z = z_1 + 2ma$, $-\infty < m < \infty$. From expression (1) for the complex potential of a single row, we have for the complex potential

$$w = -\frac{i\kappa}{2\pi} \log \sin\left(\frac{\pi(z-z_0)}{2a}\right) - \frac{i\kappa}{2\pi} \log \sin\left(\frac{\pi(z-z_1)}{2a}\right). \tag{10}$$

From the symmetry, $z_0 + z_1 = a$. Hence,

$$\frac{d\bar{z}_0}{dt} = -\frac{i\kappa}{4a} \cot \frac{\pi}{2a}(z_0 - z_1). \tag{11}$$

Now $z_1 - z_0 = a - 2z_0$, and writing $\zeta = z_1 - z_0$ for the relative displacement of the two vortices, we have from (11)

$$\frac{d\bar{\zeta}}{dt} = -\frac{i\kappa}{2a} \cot\left(\frac{\pi\zeta}{2a}\right). \tag{12}$$

This equation can be integrated in closed form, but an easier way to determine the relative orbit is to note that (12) is the equation for the path of a fluid particle in the velocity field of a row of equal vortices distance $2a$ apart (C. Williamson [1989 private communication]). Hence the relative positions lie on the streamlines of the velocity with complex potential $-i(\kappa/\pi) \log \sin(\pi\zeta/2a)$, that is,

$$\left|\sin \frac{\pi\zeta}{2a}\right| = \text{constant}. \tag{13}$$

For the limiting (heteroclinic) orbit, the constant is 1, and the locus is

$$\sinh \frac{\pi\eta}{2a} = \cos \frac{\pi\xi}{2a}. \tag{14}$$

The distance of closest approach occurs when $\xi = 0$ and is $2a/\pi \sinh^{-1} 1 = 0.56a$.

The stability of doubly infinite arrays was investigated by Tkachenko [1966]. It was shown that the arrangement in which the vortices are at the vertices of an infinite triangular lattice is stable to infinitesimal disturbances, but the square and hexagonal (honeycomb) lattices are unstable (see Aref et al [1988]).

7.6 The Karman vortex street

The double row of staggered vortices behind a bluff body that exists in a range of Reynolds number around 100 is a well-known observation. Lamb [1932 §370a] refers to descriptions by Ahlborn [1902], Bénard [1908], Karman and Rubach [1913]. Lugt [1979 p.7] shows a ninth-century

B.C. Assyrian relief with a staggered double row, and it is clearly depicted in some of Leonardo da Vinci's sketches.

The idealisation is an infinite double row of counter-rotating point vortices of strength $-\gamma$ at the points $z_m = ma$, and vortices of strength γ at $z = (m + \frac{1}{2})a + ih$, $-\infty < m < \infty$. The complex potential is

$$w = \frac{i\gamma}{2\pi} \log \left(\sin \frac{\pi z}{a} \right) - \frac{i\gamma}{2\pi} \log \left(\sin \frac{\pi}{a} (z - \tfrac{1}{2}a - ih) \right). \tag{1}$$

Note that $w \to 0$ as $y \to \pm\infty$. The velocity of the vortex at $z = 0$ is

$$W = -\frac{i\gamma}{2a} \cot \left(-\frac{\pi}{2} - \frac{i\pi h}{a} \right) = \frac{\gamma}{2a} \tanh \frac{\pi h}{a}. \tag{2}$$

From symmetry, each vortex moves with the same velocity W relative to the flow at infinity.

The stability to infinitesimal disturbances was discussed by Karman. We follow here the treatment by Lamb [1932 §156]. The vortices in the bottom row are displaced in the x and y directions by amounts $(\alpha(t), \beta(t)) e^{2\pi imp}$, respectively, and those in the top row to $(\alpha'(t), \beta'(t)) e^{2\pi imp}$. Without loss of generality, we can take $0 \le p \le 1$, and again regard p as a dimensionless subharmonic wavenumber.

By an analysis similar to that in §5, we obtain the equations, defining $k = h/a$ as the aspect ratio of the street,

$$\frac{2\pi a^2}{\gamma} \frac{d\alpha}{dt} = -A\beta - B\alpha' - C\beta', \tag{3}$$

$$\frac{2\pi a^2}{\gamma} \frac{d\beta}{dt} = -A\alpha - C\alpha' + B\beta', \tag{4}$$

$$\frac{2\pi a^2}{\gamma} \frac{d\alpha'}{dt} = A\beta' - B\alpha + C\beta, \tag{5}$$

$$\frac{2\pi a^2}{\gamma} \frac{d\beta'}{dt} = A\alpha' + C\alpha + B\beta, \tag{6}$$

where

$$A = \pi^2 \{ 2p(1 - p) - \operatorname{sech}^2 k\pi \}, \tag{7}$$

$$B = i\pi^2 \left\{ 2p \frac{\cosh k\pi(1 - 2p)}{\cosh k\pi} + \frac{\sinh 2k\pi p}{\cosh^2 k\pi} \right\}, \tag{8}$$

$$C = \pi^2 \left\{ \frac{\cosh 2\pi kp}{\cosh^2 k\pi} - 2p \frac{\cosh k\pi(1 - 2p)}{\cosh k\pi} \right\}. \tag{9}$$

There are for each p two possible modes of disturbance; symmetric modes with $\alpha = \alpha'$, $\beta = -\beta'$, and antisymmetric modes with $\alpha = -\alpha'$,

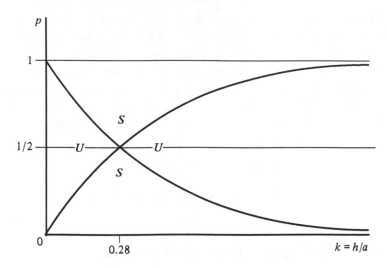

7.6-1 Stability boundaries in the k-p plane.

$\beta = \beta'$. Both modes have the same stability properties. The solutions are proportional to $e^{\sigma t}$, where

$$\sigma = \mp B + (A^2 - C^2)^{1/2}. \tag{10}$$

Take $-B$ for the symmetrical mode, and $+B$ for the antisymmetrical mode. Since A and C are real, stability requires $A^2 \le C^2$.

In Figure 7.6-1, we plot in the k-p plane the curves $A^2 = C^2$. The symmetry about $p = \frac{1}{2}$ follows from the invariance under the translation $x \to x + \frac{1}{2} a$. It is clear from the figure that there is, for each value of k, a range of wavenumbers for which the disturbance is unstable except for the special value k_c where the curves cross. For $k \ne k_c$, not all wavenumbers are unstable, but only one is required for instability. The critical value of k_c is found on putting $p = \frac{1}{2}$ in (7) and (9), and is

$$k_c = \frac{1}{\pi} \sinh^{-1} 1 = 0.2801 \ldots . \tag{11}$$

For this particular value of k_c, the street is stable to all infinitesimal disturbances, but it does not follow that it is stable to finite amplitude disturbances. Kochin, Kibel and Roze [1964] demonstrate by application of the Lyapunov method that there exist finite-amplitude unstable disturbances.

An elegant demonstration was given by Domm [1956]. Consider the $p = \frac{1}{2}$ pairing instability mode. The basic cell of the disturbance consists of four vortices, two in the top row and two in the bottom row, and the array breaks up into four rows, each consisting of vortices separated by $2a$ and giving rise to a complex potential like (1). Thus the eight co-ordinates satisfy a system of eight first-order non-linear, ordinary differential equations. Let the complex displacements of the four vortices relative to their equilibrium positions in a frame of reference moving with the array, that is, with speed W relative to the flow at infinity, be $\zeta_1, \zeta_2, \zeta_3, \zeta_4$, where ζ_1 and ζ_2 are in the top row and ζ_3 and ζ_4 are in the bottom row. The exact equations possess an integral

$$\zeta_1 + \zeta_2 - \zeta_3 - \zeta_4 = K. \tag{12}$$

This is a consequence of the invariance of impulse. Write

$$\Phi = \zeta_1 - \zeta_2 + \zeta_3 - \zeta_4, \qquad \Psi = \zeta_1 - \zeta_2 - \zeta_3 + \zeta_4. \tag{13}$$

Domm shows, retaining terms of the second order, that

$$\frac{d\Phi}{dt} = \frac{\gamma\pi}{4a^2}\left(1 + \frac{\pi K}{2a}\right)\Psi, \qquad \frac{d\Psi}{dt} = -\frac{\gamma\pi}{4a^2}\left(1 - \frac{\pi K}{2a}\right)\Phi \tag{14}$$

when $\sinh k\pi = 1$. Then Φ and Ψ are proportional to $e^{\sigma t}$, where

$$\sigma^2 = -\frac{\gamma^2\pi^2}{64a^4}\left(1 - \frac{\pi^2}{4a^2}K\overline{K} + \frac{\pi}{2a}(K - \overline{K})\right). \tag{15}$$

Then if the initial displacements are such that $\Im K \neq 0$ (i.e., the rows are not displaced sideways by equal amounts), then the disturbance grows exponentially, but at a rate proportional to the amplitude of the disturbance. The Karman vortex street is therefore unstable for all aspect rations k, but the growth rate depends on the amplitude of the disturbance when $k = k_c$.

Karman [1911, 1912] and Karman and Rubach [1913] modeled the unsteady wake behind a bluff body and obtained from momentum considerations a relation between the drag on the body and the properties of the street. Assume that it is possible in a body-fixed frame of reference to separate the wake at some station downstream of the body into regions of vorticity of strength $\pm\Gamma$, separated by irrotational fluid, whose centroids are at a transverse distance h apart and move downstream with speed $U_0 - U_s$ relative to the body, where U_0 is the speed of the free stream. The flow is then periodic with period τ, where

$$\tau = \frac{a}{U_0 - U_s}, \tag{16}$$

and a is the downstream separation of the vortices. Consider now the momentum balance in a region $\mathcal{A}$ fixed with respect to the body, see

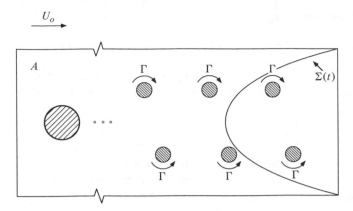

7.6-2 Momentum balance for the Karman vortex street.

Figure 7.6-2, whose downstream boundary $\Sigma(t)$ moves periodically in time. If $\mathbf{I}$ is the hydrodynamic impulse in the region (including the body which is replaced by a distribution of bound vorticity and an external force density $-\mathbf{F}$), so that the drag on the body $\mathbf{D} = (D, L)$ is

$$\mathbf{D} = \int_A \mathbf{F} \, dA, \tag{17}$$

then (see §3.10),

$$\frac{d\mathbf{I}}{dt} = -\mathbf{D} + \int_A \mathbf{u} \times \omega \, dA + \int_\Sigma \mathbf{u} \cdot \mathbf{n} \, \omega \times \mathbf{r} \, ds - \int_\Sigma \mathbf{q} \cdot \mathbf{n} \, \omega \times \mathbf{r} \, ds, \tag{18}$$

where $\mathbf{q}$ is the velocity of $\Sigma(t)$. The first integral is equal to

$$\oint_{\partial A} \left\{ \tfrac{1}{2} \mathbf{u}^2 \mathbf{n} - \mathbf{u}(\mathbf{u} \cdot \mathbf{n}) \right\} \, ds. \tag{19}$$

The last two are zero if Σ lies entirely in irrotational fluid. We now take the time average of (18) over a period τ, and write $\mathbf{u}' = \mathbf{u} - \mathbf{U}_0$ for the velocity relative to the free stream. We can replace $\mathbf{u}$ in (19) by $\mathbf{u}'$, since $\int_A \omega \, dA = 0$. We can thus neglect the contributions to (19) from the front and sides. Further, (19) is independent of the contour while the contour is in irrotational fluid. We now suppose that $\mathbf{q}$ is such that $\Sigma(t)$ is in irrotational fluid, except when it jumps instantaneously (with infinite velocity $\mathbf{q}$) to allow the passage of two vortices out of A to compensate for the creation of new vortices at the body. Then (the contribution from the second integral in (18) being zero as the integrand is non-zero only for an infinitesimal time)

$$\tau\overline{\mathbf{D}} = \tau \int_{\Sigma} \left\{ \tfrac{1}{2}\mathbf{u}'^2\,\mathbf{n} - \mathbf{u}'\,(\mathbf{u}' \cdot \mathbf{n}) \right\} ds + \int dt \int_{\Sigma} \mathbf{q} \cdot \mathbf{n}\,\mathbf{r} \times \omega\, ds. \tag{20}$$

The last term is $\gamma h\mathbf{i}$, where h is the vertical displacement of the vortex centroids. The first term can be written $-\tfrac{1}{2}\int\{(u^2-v^2)\,dy-2uv\,dx\}$. Hence

$$\overline{D} = \frac{\gamma h}{a}(U_0 - U_s) - \tfrac{1}{2}\Im \int_{-i\infty}^{i\infty} \overline{\left(\frac{dw}{dz}\right)^2}\, dz. \tag{21}$$

Here, w is the complex potential in the irrotational flow.

For the special case of $w(z)$ given by the Karman vortex street of point vortices, the last term is found on evaluation (Goldstein [1965 §243]) to be

$$\frac{\gamma^2}{2a^2}\left(\frac{a}{\pi} - h\tanh\frac{\pi h}{a}\right),$$

and then

$$\overline{D} = \frac{\gamma^2}{2a\pi} + \frac{\gamma h}{a}(U_0 - 2W) \tag{22}$$

on putting $U_s = W$.

Saffman and Schatzman [1982b] have studied an inviscid model of a vortex street wake of finite core cortices to investigate effects of core size on the relation between drag, aspect ratio and vortex strength.

7.7   Statistical mechanics of assemblies of line vortices

Consider N line vortices in a simply connected region of area V bounded by walls at rest. It was shown in §3 that the motion of the vortices is described by a Hamiltonian $H(\mathbf{p},\mathbf{q})$, which is the Kirchhoff–Routh path function. When $N \gg 1$, and $V < \infty$, Onsager [1949] suggested that the ideas of statistical mechanics be employed. The motion of the vortices can be described by a trajectory in phase space with co-ordinates $\mathbf{p}$, $\mathbf{q}$, and since these are proportional to the physical co-ordinates, the total volume of phase space is bounded. The element of phase space $d\Omega$, say, is

$$d\Omega = dp_1\,dq_1\,\ldots\,dp_N\,dq_N = \left(\prod_{1}^{N}|\kappa_j|\right) dx_1\,dy_1\,\ldots\,dx_N\,dy_N \tag{1}$$

and

$$\int d\Omega = \prod_{1}^{N}|\kappa_j|\,V^N. \tag{2}$$

The trajectory lies on the surface $H = $ constant. In general, there are no other invariants. The linear impulse $\mathbf{I}$ is not conserved because pressure is exerted on the walls. The angular impulse A is conserved if the region is bounded by a circular wall.

An ensemble of solutions is described by motion of a cloud of points in phase space. Liouville's theorem that phase space is conserved is satisfied because of the Hamiltonian structure; that is,

$$\frac{D}{Dt} d\Omega = \operatorname{div} \mathbf{Q} \; d\Omega = 0, \tag{3}$$

since

$$\operatorname{div} \mathbf{Q} = \sum_j \left(\frac{\partial}{\partial p_j} \left(\frac{dp_j}{dt} \right) + \frac{\partial}{\partial q_j} \left(\frac{dq_j}{dt} \right) \right) = 0. \tag{4}$$

Consider

$$W(E) = \int_{H<E} d\Omega, \qquad W(-\infty) = 0, \qquad W(\infty) = \prod_1^N |\kappa_j| \, V^N. \tag{5}$$

By construction, W is a monotonically increasing function of E, so that $W' = dW/dE > 0$. Then W' is the 'density' of configurations in phase or configuration space, and $W' \, dE$ is the volume of phase space in which H lies between E and $E+dE$. Define entropy (see Landau and Lifshitz [1958]) S by

$$S = \log W'. \tag{6}$$

For given volume, the state is characterized by its 'energy' E. The temperature Θ is defined by

$$\frac{1}{\Theta} = \frac{dS}{dE} = \frac{d^2W/dE^2}{dW/dE}. \tag{7}$$

Since the volume of phase space is finite,[11] $W'(-\infty) = W'(\infty) = 0$, and W' has a maximum (possibly more than one) at a point of inflexion E_m where $d^2W(E_m)/dE^2 = 0$. Suppose for simplicity, there is only one maximum. For $E < E_m$, $\Theta > 0$; for $E > E_m$, $\Theta < 0$ and we have a state of negative temperature.[12]

---

[11] For a physically sensible system, we require that the circulation around the boundary be finite. Therefore in the limit $N \to \infty$ of statistical mechanics, we require that the strength κ_j of the vortices $\to 0$ like $1/N$. It is unlikely that equilibrium states exist in the limit $V \to \infty$.

[12] These are known to occur for various physical systems, e.g., paramagnetic dielectrics, Landau and Lifshitz [1958 §70].

Onsager states that in a system with positive Θ, the most probable state will be that in which the vortices are well mixed and the state is disorganized. But when the temperature is negative, clustering is favoured and like-signed vortices will clump into 'organized structures'.[13] Onsager's ideas apply to an assembly of point vortices. Suppose, however, that we have a smooth distribution of vorticity in a region of area V bounded by walls at rest. We can identify this vorticity distribution with an assembly of N line vortices which has the same invariants, that is, $H = T$ (and equal angular impulse if in a circular region), where T is the kinetic energy of the smooth distribution and is a discretisation in the sense that as $N \to \infty$, the circulation about any circuit in the region is the same. Let us assume that the thermodynamic equation of state $S = S(E)$ is independent of the particular form of the discretisation as $N \to \infty$. Then we can identify a temperature with a smooth distribution. Unfortunately, we have no idea at present if this assumption is correct. Let us continue, however, and suppose the continuous distribution evolves with time under the Euler equations. At some later time t, we can again identify discretisations and an equation of state. The question arises whether the equation of state is invariant. They will be the same if phase space is ergodic and only the energy of the system matters. The work of Goodman, Hou and Lowengrub [1990] suggests that the discretisations are on the same trajectory in phase space. They prove that in the limit $N \to \infty$, the trajectories of the line vortices coincide with the paths of the particles at the initial positions of the vortices, as determined by the smooth distribution of vorticity. If these ideas are relevant, then it becomes important to estimate the temperature of a smooth distribution of vorticity and the most likely state to which it will evolve.

---

[13] Joyce and Montgomery [1972, 1973] provide some numerical evidence for the existence of negative temperature states. There exist theoretical studies of the problem of calculating the entropy and determining the vortex configuration in the most probable state; e.g., Kida [1975], Novikov [1976], Pointin and Lundgren [1976].

8

VORTEX SHEETS IN TWO DIMENSIONS

8.1 The Birkhoff–Rott equation

Consider a vortex sheet in uniform fluid of constant density[1] and consider how it moves. Suppose at some instant of time the locus of the sheet can be described parametrically by the equations $x = X(s,t)$, $y = Y(s,t)$, $z = x + iy = Z(s,t)$, where s is arc length. Let $\kappa(s,t)$ denote the strength of the sheet, that is, the jump in tangential velocity. Then (see §2.2), the velocity field induced by the sheet is

$$u - iv = \frac{-i}{2\pi} \int \frac{\kappa(s',t)\,ds'}{z - Z(s',t)}. \tag{1}$$

When the field point z is on the sheet, we define the averaged induced velocity on the sheet at station s as the principal value

$$\tilde{U} - i\tilde{V} = \frac{-i}{2\pi} \fint \frac{\kappa(s',t)\,ds'}{z - Z(s',t)}. \tag{2}$$

where $\fint$ denotes principal value. This can be shown to be the arithmetic mean of the two limits of (1) as $z \to Z(s,t)$ from the two sides of the sheet.

Suppose that in addition to the velocity induced by the sheet, there is an external velocity field $U_E - iV_E$ produced by bodies, continuous distributions of vorticity and other vortex sheets and singularities, which is continuous on the sheet.[2] We define the velocity of the sheet to be

$$\frac{d\bar{Z}}{dt} = U - iV = \tilde{U} - i\tilde{V} + U_E - iV_E. \tag{3}$$

[1] As mentioned in §2.2, vortex sheets which are also surfaces of density discontinuity are of great importance, but these flows are outside the scope of the present work.

[2] It should be remembered that smooth vorticity distributions can exist in the fluid adjacent to the sheet; i.e., the sheet can separate vortical regions.

We now define $\Gamma(P_1, P_2)$ as the integrated sheet strength or circulation between points P_1 and P_2 in the sheet. It was shown in §2.2 that, in the absence of external forces on the sheet, Γ is conserved; that is,

$$\frac{d\Gamma}{dt} = 0, \tag{4}$$

when the points P_1 and P_2 move with the velocity given by (3).

Now change variables, expressing both the equation of the sheet and its strength by giving the parametric equation in terms of Γ and t; that is,

$$z = Z(\Gamma, t) \tag{5}$$

is the locus of the sheet and $\kappa = |\partial Z/\partial\Gamma|^{-1}$. Also $\kappa\, ds = d\Gamma$, and (2) and (4) can be expressed as

$$\frac{\partial \overline{Z}}{\partial t}(\Gamma, t) = \frac{-i}{2\pi} \int \frac{d\Gamma'}{Z(\Gamma, t) - Z(\Gamma', t)} + U_E - iV_E. \tag{6}$$

This non-linear, singular integro-differential equation is called the Birkhoff–Rott equation. It is given explicitly by Birkhoff [1962] and implied in the work of Rott [1956]. It describes the evolution of the sheet, given its initial shape and strength. Note that $U_E - iV_E$ may depend implicitly on the sheet shape and strength through image vorticity. The equation contains the full description, as both the kinematic condition of continuous normal velocity and continuity of pressure are satisfied. It is an elegant formulation, but is not necessarily the most suitable for numerical computation of sheet evolution.[3]

8.2 Kelvin–Helmholtz instability

One of the important features of vortex sheet dynamics is the instability to infinitesimal two-dimensional disturbances of the plane infinite uniform[4] vortex sheet. In terms of the formulation[5] of §1, the undisturbed sheet is

$$Z = \frac{\Gamma}{U}, \quad \kappa = U. \tag{1}$$

---

[3] In §3, we present evidence by Moore [1979, 1984] that the evolution of an initially smooth vortex sheet may be ill posed, and that a singularity in the shape of the sheet develops in a finite time.

[4] See Hocking [1964, 1965] for a discussion of the stability of non-uniform vortex sheets.

[5] See Batchelor [1967 §7.1] for a simpler formulation in terms of velocity potentials and a physical explanation of the instability.

We consider an infinitesimal periodic disturbance which perturbs the sheet into the curve

$$Z = \frac{\Gamma}{U} + \sum_{-\infty}^{\infty} a_n(t) e^{in\Gamma\alpha}, \tag{2}$$

where the coefficients $a_n(t)$ are assumed to be infinitesimal. Note that

$$Z\left(\Gamma + \frac{2\pi}{\alpha}\right) = Z(\Gamma) + \frac{2\pi}{\alpha U}, \tag{3}$$

so the physical wavelength is $\lambda = 2\pi/\alpha U$.

Substituting into the Birkhoff–Rott equation (1.6), we have

$$\sum_{-\infty}^{\infty} \frac{d\bar{a}_n}{dt} e^{-in\Gamma\alpha} = \frac{-i}{2\pi} \int_{-\infty}^{\infty} \frac{d\Gamma'}{\dfrac{\Gamma - \Gamma'}{U}\left[1 + \sum_{-\infty}^{\infty} a_n \dfrac{e^{in\Gamma\alpha} - e^{in\Gamma'\alpha}}{(\Gamma - \Gamma')/U}\right]},$$

$$\text{put} \quad \theta = \Gamma' - \Gamma,$$

$$= \frac{iU}{2\pi} \int_{-\infty}^{\infty} \frac{d\theta/\theta}{1 - \sum_{-\infty}^{\infty} U a_n e^{in\Gamma\alpha}(1 - e^{in\theta\alpha})/\theta} \tag{4}$$

$$= \frac{iU}{2\pi} \int_{-\infty}^{\infty} \frac{d\theta}{\theta} \left(1 + \sum_{-\infty}^{\infty} U a_n e^{in\Gamma\alpha} \left(\frac{1 - e^{in\theta\alpha}}{\theta}\right) + O(a_n^2)\right).$$

The principal value integral refers to both $\theta = 0$ and $\theta = \infty$, so $\int_{-\infty}^{\infty} d\theta/\theta = 0$. Further, $\int(1 - e^{in\theta\alpha}) d\theta/\theta^2 = \pi n\alpha$. Hence to first order in the amplitude of the disturbance,

$$\sum_{-\infty}^{\infty} \frac{d\bar{a}_n}{dt} e^{-in\Gamma\alpha} = \frac{iU^2\alpha}{2} \sum_{-\infty}^{\infty} n a_n e^{in\Gamma\alpha}. \tag{5}$$

Equating coefficients of $e^{-in\Gamma\alpha}$, we have

$$\frac{d\bar{a}_n}{dt} = \frac{-i\pi nU}{\lambda} a_{-n}, \quad -\infty < n < \infty. \tag{6}$$

The nth Fourier coefficients, a_n and a_{-n}, thus grow like $e^{\sigma t}$, where

$$\sigma = \pm \pi nU/\lambda. \tag{7}$$

The sheet is therefore unstable. Moreover, the smaller the wavelength λ or the higher the mode n, the faster the growth rate. This suggests that the motion of a vortex sheet may be an ill-posed problem. We discuss this in §3 below.

The Kelvin–Helmholtz instability may, however, be suppressed by stretching the sheet in unsteady flow. Consider for instance, a heuristic argument (Saffman [1974]) and suppose that the vortex sheet lies along the x-axis in an externally imposed irrotational field $U_E = \beta(t)x$, $V_E = -\beta(t)y$. The wavelength of the disturbance is now not constant in time and the undisturbed state is not an equilibrium solution because of the pressure jump, but provided $U >> \beta$, the amplitude A of the disturbance grows like

$$\frac{1}{A}\frac{dA}{dt} = \frac{\pi U}{\lambda}, \quad \text{where} \quad \frac{d\lambda}{dt} = \beta\lambda. \tag{8}$$

If β is an increasing function of time, the amplitude A saturates. The wave slope $\propto A/\lambda$ decays as $t \to \infty$.

A rigorous analysis has been carried out by Moore and Griffith-Jones [1974] for the stability of an expanding circular vortex sheet. The undisturbed unsteady flow is (using polar co-ordinates r, θ)

$$u_r = \frac{\dot{R}R}{r}; \quad u_\theta = 0, \ r < R(t); \quad u_\theta = \frac{\Gamma}{2\pi r}, \ r > R(t); \tag{9}$$

where $R(t)$ is to be given, and $\Gamma = \text{constant}$. This field describes a circular vortex sheet of radius R and strength $\Gamma/2\pi R$ expanding due to a source at the origin.

A disturbance is imposed that deforms the circle into the curve

$$r = R + \epsilon(t)e^{is\theta}, \tag{10}$$

where $|\epsilon| << R$ and s is an integer. The flow is irrotational both inside and outside the vortex, with velocity potentials

$$\phi = R\dot{R}\log r + A(t)r^s \, e^{is\theta}, \quad r < R + \epsilon \, e^{ist}, \tag{11}$$

$$\phi = R\dot{R}\log r + \frac{\Gamma\theta}{2\pi} + B(t)r^{-s} \, e^{is\theta}, \quad r > R + \epsilon \, e^{ist}. \tag{12}$$

The boundary conditions that the sheet is a material surface and the pressure is continuous give

$$\frac{D}{Dt}(R + \epsilon e^{is\theta} - r) = 0, \quad \frac{\partial}{\partial\theta}\left[\frac{\partial\phi}{\partial t} + \frac{1}{2}(\nabla\phi)^2\right] = 0, \quad \text{on } r = R + \epsilon e^{ist},$$

$$\tag{13}$$

where [] refers to the jump across the vortex sheet. Substituting and linearising leads to the equation for the amplitude $\epsilon(t)$

$$\ddot{\epsilon} + \dot{\epsilon}\left(\frac{2\dot{R}}{R} + \frac{is\Gamma}{2\pi R^2}\right) + \epsilon\left(\frac{\ddot{R}}{R} - \frac{s(s-1)\Gamma^2}{8\pi^2 R^4}\right) = 0. \tag{14}$$

A common problem in unsteady flow is the precise definition of stability. Moore and Griffith-Jones discuss this question and adopt the criterion that

ϵ be bounded. They consider first the case $R(t) = R_0(1 + at)^n$, $a > 0$, and examine the non-trivial cases $s > 2$. Defining

$$\beta = \Gamma(s^2 - 2s)^{1/2}/2\pi a R_0^2, \quad \tau = 1 + at, \quad p = \frac{1}{2 - 4n}, \tag{15}$$

they show that as $\tau \to \infty$,

$$|\epsilon(t)| \sim \exp(\beta|p|\tau^{1-2n}), \quad n < \frac{1}{2} \tag{16}$$

$$|\epsilon(t)| \sim \tau^{\frac{1}{2}[1+\sqrt{1+\beta^2}]}, \quad n = \frac{1}{2} \tag{17}$$

$$|\epsilon(t)| \sim \tau^{1-n}, \quad n > \frac{1}{2}. \tag{18}$$

Then, according to their criterion, the expanding sheet is stable if $n \geq 1$, and otherwise unstable, but it is pointed out that the real change occurs at $n = \frac{1}{2}$. For $n > \frac{1}{2}$, the disturbances grow weakly and in a way independent of the wavelength $2\pi R/s$, whereas for $n < \frac{1}{2}$ the disturbances grow exponentially and the short waves with large s grow fastest, as in the Kelvin–Helmholtz instability which is recovered for $n = 0$. For general $R(t)$, the WKB method is applied for large s, and gives results consistent with those obtained from the special case.[6]

8.3 The ill-posedness of vortex sheets

By linear superposition, it follows from the analysis of §2 that

$$y = \epsilon \sum_{1}^{\infty} A_n \sin \frac{n\pi x}{\lambda} \exp \frac{n\pi Ut}{\lambda} \tag{1}$$

describes the evolution of a periodic infinitesimal disturbance for arbitrary values of the coefficients A_n. As pointed out by Birkhoff and Fisher [1959] (see also Birkhoff [1962], Saffman and Baker [1979]), the choice

$$A_n = \exp(-|n|^{1/2} - n\pi Ut_0/\lambda) \tag{2}$$

gives a smooth (i.e., infinitely differentiable) wave for $t \leq t_0$, but (1) converges nowhere for $t > t_0$. This result is suggestive, but since the equations are non-linear, the possibility exists that non-linear interactions may produce exponential decay like e^{-n^2}, say, and the sheet would remain smooth for all time.

---

[6] Moore [1976] analyses the behaviour of short wavelength disturbances on an arbitrary vortex sheet.

Moore [1979, 1984] has examined this problem for solutions of the Birkhoff–Rott equation with initial condition

$$Z(\Gamma,0) = \Gamma + i\epsilon \sin \Gamma, \quad -\infty < \Gamma < \infty, \tag{3}$$

for $\epsilon \ll 1$. His method of solution is to write

$$Z(\Gamma,t) = \Gamma + 2i \sum_{1}^{\infty} A_n(t) \sin n\Gamma, \quad A_n(0) = \tfrac{1}{2}\epsilon\,\delta_{n1}. \tag{4}$$

Substitution into (2.4) and expansion in powers of the A_n, as for the examination of the Kelvin–Helmholtz instability, leads to an infinite system of infinitely many non-linear equations in infinitely many unknowns. Moore notes that the expansion is of row-echelon type, that is, $A_n = \epsilon^n A_{n0} + \epsilon^{n+2} A_{n2} + \cdots$, and the A_{n0} can be determined recursively. That is, the equations are of the form

$$\frac{d}{dt}\,\overline{A}_{n0} = f_n(A_{10}, \ldots A_{n0}) \tag{5}$$

where the f_n are polynomials, which may depend logarithmically upon ϵ. By a series of ingenious steps, Moore obtains the asymptotic expression for large n,[7]

$$\epsilon^n A_{n0} \sim t^{-1}(2\pi)^{-1/2}(1+i)n^{-5/2}\,\exp\{n(1 + \tfrac{1}{2}t + \log \tfrac{1}{4}\epsilon t)\}. \tag{6}$$

Thus the exponential decay of the coefficients is lost when $t = t_c$, where

$$1 + \tfrac{1}{2}t_c + \log t_c = \log(4/\epsilon). \tag{7}$$

The shape of the sheet develops a singularity at $\Gamma = 2n\pi$,

$$Z(\Gamma,t) = \Gamma + \frac{2\sqrt{3}}{3t}(1+i)\left\{(1 - e^{i\Gamma}\epsilon\,\Theta)^{2/3} - (1 - e^{-i\Gamma}\epsilon\,\Theta)^{3/2}\right\}, \tag{8}$$

plus less singular terms, where $\Theta = \tfrac{1}{4}t\exp(\tfrac{1}{2}t+1)$. As $t \to t_c-$, $\epsilon\,\Theta \to 1$, the slope stays finite but an infinite curvature develops.[8] The sheet is like the curve $y' = |x'|^{3/2}$ at the singularity, where $y' = 0$ is the local tangent. The strength of the sheet,

$$\kappa(\Gamma) = 1 - \frac{\sqrt{3}}{t_c}(t_c - t + ((t_c - t)^2 + 4\Gamma^2)^{1/2})^{1/2} + O(1/t_c^2), \tag{9}$$

---

[7] Meiron, Baker and Orszag [1982] have carried out further studies of the system of o.d.e.'s for the coefficients A_n and confirm Moore's results with minor quantitative changes.

[8] Caflisch and Orellena [1986] prove existence using an alternative formulation by Moore in terms of the solution of a pair of non-linear hyperbolic equations.

remains bounded as $t \to t_c$, but the distribution has a cuspidal form[9] with $\kappa - \kappa_c \propto x^{\frac{1}{2}}$. Note that the singularity appears first at the points where the sheet is compressed most as described by the linear evolution. The sheet is stretched most at the points $\Gamma = (2n + 1)\pi$.

The representation $Z = Z(\Gamma, t_c)$ is not analytic in Γ at $t = t_c$. That is, the function Z cannot be continued analytically for complex Γ into a strip $|\Im\Gamma| < \delta$. It is therefore implied that no solution exists of the Birkhoff–Rott equation for $t > t_c$, which can be regarded as a bounded continuation of the solution for $t < t_c$. However, this question is still open and the existence of a solution with singularities (e.g., a spiral) for $t > t_c$ cannot be ruled out (see §5).

8.4 Roll-up of a semi-infinite vortex sheet: The Kaden spiral

In §6.1, we discussed the creation of a vortex sheet in the Klein Kaffeelöffel experiment and its roll-up to form two coarse-grained vortices. The initial strength of the sheet is

$$\kappa = 2Ux/\sqrt{a^2 - x^2}. \tag{1}$$

The strength has an infinite square root at the ends. This distribution can also be interpreted as the vortex sheet strength behind an elliptically loaded wing which produces an infinite discontinuity in the velocity at the tip.

Kaden [1931] considered the case of a semi-infinite vortex sheet, initially extending along the positive x-axis. The initial strength and circulation are

$$\kappa = \gamma x^{-1/2}, \quad \Gamma = 2\gamma x^{1/2}, \qquad 0 < x < \infty, \tag{2}$$

and the initial shape for the Birkhoff–Rott equation is

$$Z = \frac{\Gamma^2}{4\gamma^2}, \quad 0 < \Gamma < \infty. \tag{3}$$

This problem enables us to determine the nature of the singularity at the end of the sheet.

Since there is no externally imposed length scale, $Z = Z(\Gamma, t, \gamma)$ for $t > 0$, and we infer from dimensional analysis that

$$Z(\Gamma, t, \gamma) = (\gamma t)^{2/3} \zeta(\tau), \quad \text{where } \tau = \Gamma/\gamma^{4/3} t^{1/3}. \tag{4}$$

[9] Schwartz [1981] examines the initial problem for the roll-up of a finite vortex sheet with a strength that vanishes at the end so that the velocity is well defined everywhere at the initial time, and shows that a singularity forms at the tip after a finite time.

Substitution into the Birkhoff–Rott equation gives the singular ordinary integro-differential equation for $\zeta(\tau)$,

$$2\bar{\zeta} - \tau\frac{d\bar{\zeta}}{d\tau} = \frac{-3i}{2\pi}\int_0^\infty \frac{d\tau'}{\zeta(\tau) - \zeta(\tau')}, \quad 0 \le \tau < \infty. \tag{5}$$

The initial instant $t = 0$ corresponds to $\tau = \infty$, and we have a boundary condition

$$\zeta(\tau) \sim \tfrac{1}{4}\tau^2 \quad \text{as } \tau \to \infty. \tag{6}$$

Note that (6) is actually an exact solution of (4) for $0 < \tau < \infty$. The problem is that it does not provide a solution for $\tau = 0$, since $\int_0^\infty d\tau/\tau^2 = \infty$. The failure is due to leading edge suction, as described in §6.2. We conclude that $\zeta_0 \equiv \zeta(0) \neq 0$, and the solution should satisfy the criteria

$$|\zeta_0| < \infty, \quad \left|\int_0^\infty \frac{d\tau'}{\zeta_0 - \zeta(\tau')}\right| < \infty. \tag{7}$$

It is not known if (6) and (7) are sufficient to ensure existence of a unique solution of (5), but proceeding on the assumption that a solution exists, simple geometrical arguments can be given to determine the shape of the vortex sheet for $t > 0$.

Following Kaden, it is assumed that the sheet rolls up into a spiral which ends at $Z_0 = \zeta_0(\gamma t)^{2/3}$ and that in the vicinity of the tip the spiral is tightly wound and the turns are approximately circular and have negligible radial motion.[10] By dimensional considerations, the circulation around a circle of radius r is

$$\Gamma(r) = 2\gamma(\lambda r)^{1/2}, \tag{8}$$

where λ is some unknown coefficient. The tangential velocity v_θ at radius r is given by

$$v_\theta = \Gamma/2\pi r = \frac{\gamma\lambda^{1/2}}{\pi r^{1/2}}. \tag{9}$$

Consider now a fluid particle on the sheet with circulation co-ordinate Γ_P. For this particle, its radial and tangential co-ordinates r_P, θ_P follow from (8) and (9), using $v_\theta = r_P\dot{\theta}_P$,

$$r_P = \frac{\Gamma_P^2}{4\gamma^2\lambda}, \quad \theta_P = \frac{\gamma\lambda^{1/2}}{\pi r_P^{3/2}} t + \text{const.} \tag{10}$$

---

[10] This assumption is consistent with numerical calculations of the roll-up (Moore [1974]) and numerical analysis of the Birkhoff–Rott equation (Moore [1981]), and asymptotic analysis of (4) by Moore [1975].

For small r, or large t, the second of these relations gives the asymptotic equation of the spiral, dropping the suffix P,

$$r \sim \left(\frac{\gamma^2 \lambda}{\pi^2} \right)^{1/3} \left(\frac{t}{\theta} \right)^{2/3}. \tag{11}$$

It is clear from (11) that as $\theta \to \infty$, the spiral consists of tightly wound circular turns verifying the provisional assumptions on which the argument was based. The spiral is infinitely long, since the element of length ds is given by

$$ds = r d\theta \sqrt{1 + \frac{1}{r^2} \left(\frac{dr}{d\theta} \right)^2} \sim \frac{d\theta}{\theta^{2/3}}, \quad \text{and} \quad \int ds = \int^\infty \frac{d\theta}{\theta^{2/3}} = \infty. \tag{12}$$

Returning to the dimensionless formulation for the shape expressed in the form $\zeta = \zeta_0 + f(\tau) e^{i\phi(\tau)}$, we have asymptotically as $\tau \to 0$,

$$f(\tau) = r(\gamma t)^{-\frac{2}{3}} \sim \frac{\tau^2}{4\lambda}, \tag{13}$$

$$\phi(\tau) = \theta \sim \frac{8\lambda^2}{\pi \tau^3} + \epsilon, \tag{14}$$

where ϵ is some unknown constant. γ_0, λ and ϵ cannot be determined locally by consideration of the vortex sheet dynamics near the tip of the spiral, but depend globally upon the entire solution of (5). The tangential velocity $v_\theta \propto r^{-1/2} \to \infty$ as $r \to 0$, so the velocity field is singular at the centre of the spiral. However, the tip velocity is finite, since

$$\int_0 \frac{d\tau}{\zeta - \zeta_0} \sim \int_0 \frac{d\tau}{\tau^2 e^{8i\lambda^3/\pi\tau^3}} < \infty. \tag{15}$$

The strength of the sheet is given by

$$\frac{1}{\kappa} = \left| \frac{dZ}{d\Gamma} \right| = \frac{t^{1/3}}{\gamma^{2/3}} \left| \frac{d\zeta}{d\tau} \right| \sim \frac{6\lambda^3}{\pi\tau^2} \frac{t^{1/3}}{\gamma^{2/3}}, \tag{16}$$

i.e.,

$$\kappa \propto \frac{\tau^2}{t^{1/3}} \propto \frac{\Gamma^2}{t} \propto \frac{r}{t} \propto \frac{t^{1/3}}{\theta^{2/3}}. \tag{17}$$

Thus the sheet strength decreases as the centre is approached, or at a fixed r as $t \to \infty$, representing the fact that the sheet is being continuously stretched as the spiral tightens. The sheet strength vanishes as $r \to \infty$, owing to the initial condition. Thus there will be a value of τ at which the sheet strength is a maximum at a given time. The stability of the Kaden spiral is an open question. It is possible that the stretching stabilizes the sheet against the Kelvin–Helmholtz instability (see §2). If unstable, the instability may first appear where the sheet strength is a maximum (Moore [1974]).

Moore [1975] has investigated higher-order corrections to (13) and (14). Writing

$$\zeta = \zeta_0 + F(\tau)e^{i8\lambda^2/\pi\tau^3 + i\epsilon}, \tag{18}$$

he shows from an asymptotic analysis of (5) that[11]

$$F(\tau) \sim \frac{\tau^2}{4\lambda} + \left(Ae^{\frac{16i\lambda^2}{\pi\tau^3}} + Be^{\frac{-16i\lambda^2}{\pi\tau^3}} \right) \tau^{3n+2}, \tag{19}$$

where $n = \frac{1}{3}\left(\sqrt{13} - 1\right) = 0.87$. The turns of the spiral are therefore elliptical, and the excess axis ratio at distance r from the centre of the spiral is $O(\tau^{3n}) = O(r^{3n+2}/(\gamma t)^n)$. The constants A and B are not determined by the local analysis but depend upon the global structure.

As pointed out by Moore, the elliptical correction to the inner turns of the Kaden spiral can be anticipated by a simple argument which also leads to the value of n. Neglecting the detailed spiral structure (i.e., coarse graining) we replace the tightly wound turns by a circular vortex core of radius R with a tangential velocity field $k/r^{1/2}$. The outer part of the spiral will effectively impose an irrotational straining field, that is, will perturb the stream function by a term proportional to $e^{2i\theta}$. Thus the stream function inside the core will be perturbed into the form $\psi = 2kr^{1/2} + \delta g(r)\cos 2\theta$, where $\delta \ll 1$. For this to be a quasi-steady solution, the Jacobian of ψ and $\nabla^2\psi$ must vanish. Substituting $g(r) \propto r^N$ and linearising in δ, gives $N = \frac{1}{2}\sqrt{13}$. Thus the core is deformed into an ellipse with excess axis ratio $R^{N-\frac{1}{2}} = R^{\frac{3}{2}n}$.

Moore's analysis of the higher-order corrections suggests that the equations are sufficient to determine an isolated (i.e., locally unique) solution and that the three unknown parameters $(\lambda, \zeta_0, \epsilon)$ describing the core structure will be determined by the equations, assuming that a solution exists. There is, however, a method of estimating λ proposed by Betz [1932]. It is based on the hypothesis that the vortex force exerted on the vorticity that forms the core inside a circle of radius R has no torque about the centroid of this vorticity. That is, if

$$\omega(r) = \frac{1}{r}\frac{\partial}{\partial r}(rv_\theta) = \frac{1}{2}\frac{\gamma\lambda^{1/2}}{\pi r^{3/2}} \tag{20}$$

---

[11] Guiraud and Zeytounian [1977] have rederived Moore's results using a multiscale expansion of the velocity potential for the flow.

is, using (9), the coarse-grained vorticity in the core, which comes from the length X of the initial sheet whose strength is $\gamma x^{-1/2}$,

$$2\pi \int_0^R r\omega dr = \int_0^X \frac{\gamma}{x^{1/2}} dx. \tag{21}$$

From the hypothesis of conservation of angular momentum

$$-\pi \int_0^R r^3 \omega dr = \int_0^X \frac{\gamma}{x^{1/2}} (x - \bar{x})^2 dx, \tag{22}$$

where

$$\bar{x} = \int_0^X \gamma x^{1/2} dx \bigg/ \int_0^X \frac{\gamma}{x^{1/2}} dx. \tag{23}$$

Substituting (20), we find from (21) and (22),

$$X = \lambda R, \quad \lambda = \frac{3}{2}. \tag{24}$$

The value of λ can be regarded as the degree of roll-up or tightening of the sheet. The vorticity in a length X of the sheet moves into a circle of radius $X/\lambda < X$ if $\lambda > 1$. So tightening does occur, but it is not large.[12] The idea that sheets roll-up into very thin filaments is fallacious (unless there is significant three-dimensional stretching).

The location of the centre of the spiral can be estimated from the conservation of impulse. At time t, the portion of the original sheet between 0 and $X(t)$, with circulation $2\gamma X^{1/2}$, is supposed to roll up into a circle of radius $R = X/\lambda$ with circulation $2\gamma(\lambda R)^{1/2}$ centred at $(\bar{X}(t), \bar{Y}(t))$. From conservation of the y-component of hydrodynamic impulse,

$$2\gamma(\lambda R)^{1/2}\bar{X} = \frac{2}{3}\gamma X^{3/2}, \quad \text{i.e., } \bar{X} = \frac{1}{3}X. \tag{25}$$

The vertical displacement is more difficult to find. Because of the flow at infinity $\int y\omega dS$ is not conserved. We remember that

$$\frac{d\mathbf{I}}{dt} = \oint_\infty \left[\tfrac{1}{2}\mathbf{u}^2\mathbf{n} - \mathbf{u}(\mathbf{u}\cdot\mathbf{n})\right] ds. \tag{26}$$

At infinity, $\mathbf{u} = \nabla\phi$ where $\phi \sim -\gamma r^{1/2} \cos \tfrac{1}{2}\theta$, and the right-hand side of (26) is $-\tfrac{1}{4}\pi\gamma^2\mathbf{i}$.[13] Hence

$$2\gamma(\lambda R)^{1/2}\bar{Y} = -\frac{1}{4}\pi\gamma^2 t, \quad \text{i.e., } X^{\frac{1}{2}}\bar{Y} = -\frac{1}{8}\pi\gamma t. \tag{27}$$

---

[12] Pullin [1978] estimates $\lambda \sim 2$ from a numerical solution of (5).
[13] Compare the leading edge suction calculation §6.2.

A final equation is obtained by equating dX/dt to the x-component of velocity at X induced by the rolled up vortex. This gives

$$\frac{dX}{dt} = -\frac{\gamma(\lambda R)^{1/2}}{\pi} \frac{\overline{Y}}{\overline{Y}^2 + (X - \overline{X})^2}. \tag{28}$$

From (25), (27) and (28), we obtain

$$\overline{X} = \xi(\gamma t)^{2/3}, \quad \overline{Y} = \eta(\gamma t)^{2/3}, \quad \xi = 0.14, \quad \eta = 0.61. \tag{29}$$

Kaden, using a more complicated series of assumptions obtains the values $\xi = 0.57(9/2\pi^2)^{2/3} = 0.34$, $\eta = 0.88(9/2\pi^2)^{2/3} = 0.52$. These are quite close to the values calculated numerically by Pullin [1978]. However, these values do not satisfy (27), which should be a good approximation being based on conservation of impulse.

8.5   General similarity solutions, single and multibranched spirals

The Kaden spiral is just one member of the many families of vortex sheets which are of similarity form. Prandtl [1922] demonstrated the existence of two-dimensional, self-similar, unsteady spiral vortex sheets with the shape of equiangular (logarithmic) spirals (see also Küchemann and Weber [1965], Mangler and Weber [1967]). Assume the form

$$Z = t^m f(\theta) e^{i\theta}, \quad \Gamma = t^n g(\theta), \tag{1}$$

for the parametric equation of the sheet where f and g are real. Substitution into the Birkhoff–Rott equation (1.6) shows that a solution of this form requires that $n = 2m - 1$ (in order that the powers of t cancel) and then that f and g satisfy

$$e^{-i\theta}(mf + (1 - 2m)\frac{g}{g'}(f' - if)) = \frac{-i}{2\pi} \int \frac{\hat{g}' d\hat{\theta}}{fe^{i\theta} - \hat{f}e^{i\hat{\theta}}}, \tag{2}$$

where $'$ denotes derivative and $\hat{\ }$ denotes evaluation at $\theta = \hat{\theta}$. This complex integro-differential equation is two real equations for f and g. We try

$$f = Ae^{\alpha\theta}, \quad g = Be^{\beta\theta}, \quad -\infty < \theta < \infty. \tag{3}$$

Then, on substitution and reduction, we obtain

$$e^{(2\alpha - \beta)\theta} A^2(m + \beta(1 - 2m)(\alpha - i)) = \frac{-iB\beta}{2\pi} \int_{-\infty}^{\infty} \frac{e^{\beta\phi}}{1 - e^{(i+\alpha)\phi}}. \tag{4}$$

This equation cannot be satisfied unless $\beta = 2\alpha$. But the integral in (4) diverges unless α and β are the same sign and $|\beta| < |\alpha|$. The problem is due to unbounded vorticity at infinity, since the strength of the sheet $\kappa \propto e^{(\beta - \alpha)\theta}$.

However, for $0 < \beta < \alpha$,

$$\int_{-\infty}^{\infty} \frac{e^{\beta\phi}d\phi}{1 - e^{(i+\alpha)\phi}} = \frac{\pi i}{i + \alpha} \left\{ \frac{1 + e^u}{1 - e^u} \right\}, \qquad u = \frac{2\pi i \beta}{i + \alpha}, \tag{5}$$

and the integral can be defined for $\beta = 2\alpha$ by analytic continuation. Then

$$\frac{A^2}{B\alpha} \left\{ m(\alpha + i) + \frac{1}{2\alpha}(\alpha^2 + 1)(1 - 2m) \right\} = \frac{1 + e^u}{1 - e^u}. \tag{6}$$

Eliminating A^2/B from the real and imaginary parts gives

$$\frac{1}{m} = \frac{2}{1 + \alpha^2} - \frac{2\alpha}{1 + \alpha^2} \frac{\sinh(4\pi\alpha/(1 + \alpha^2))}{\sin(4\pi\alpha^2/(1 + \alpha^2))}. \tag{7}$$

It follows that $-\infty < m < 1/2$.

Since the solution is $r = At^m e^{\alpha\theta}$, $\Gamma = Bt^{2m-1}e^{2\alpha\theta}$, where $A^2/B = \text{fn}(\alpha)$, then $r^2 = (A^2/B)\Gamma t$ and the vorticity is moving outwards. Also, $(d\theta/dt)_\Gamma = -(2m - 1)/2\alpha t$, so the angular velocity of the vortices slows down as t increases. Any piece of the vortex is stretching, straightening out and slowing down. The circulation Γ inside a circle of radius r is decreasing like $1/t$. The time t can be positive or negative. If $t < 0$ (or the solution runs backwards) we have a flow in which the solution becomes infinite everywhere at the same time.[14]

The complex potential $w(z)$ is given by the integral[15]

$$\frac{dw}{dz} = u - iv = \frac{-i}{2\pi} \frac{B}{A} t^{m-1} \int_{-\infty}^{\infty} \frac{e^{\beta\hat{\theta}}}{\frac{z}{At^m} - e^{i(\alpha+i)\hat{\theta}}} d\hat{\theta} \quad \propto \quad t^{m-1}\left(\frac{z}{t^m}\right)^{\frac{\alpha-i}{\alpha+i}}, \tag{8}$$

so that

$$w \propto t^{2m-1}(z/t^m)^{\frac{2\alpha}{\alpha+i}}. \tag{9}$$

The preceding treatment is based on unpublished notes by D. W. Moore (1972), who introduced the artifice of subtracting the divergent contribution from the vorticity at infinity by using analytic continuation. The divergence arises from the fact that the expressions for **u** as integrals of ω given by

---

[14] Compare the solutions describing the collapse of three point vortices (Aref [1979, 1983]).

[15] Values for $\beta = 2\alpha$ are again found by analytic continuation.

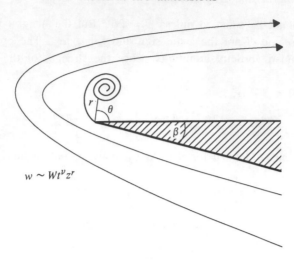

8.5-1 Accelerated flow past a wedge.

the Poisson type integrals of §1.1 may not be a valid representation of the solution of curl $\mathbf{u} = \omega$ if ω does not vanish sufficiently fast at infinity. For example, consider the simple case of a two-dimensional flow for which $\omega = \omega(r) \to \infty$ as $r \to \infty$. The velocity is well defined but the integral representation diverges. Analytic continuation is then appropriate.

Alexander [1971] has generalized the Prandtl solution to multibranched configurations of N sheets, with equations

$$Z_p = t^m f(\theta) e^{i\{\theta + 2\pi(p-1)/N\}}, \qquad 1 \leq p \leq N. \tag{10}$$

The analysis proceeds as before, the only change is in the relation (7) between m and α, where 4π is to be replaced by $4\pi/N$ in the arguments of the last term.

Similarity solutions of the type described by equation (1), but with $m > \frac{1}{2}$, so they are not in the class of Prandtl equiangular spirals, occur for generalizations (Pullin [1978], Pullin and Phillips [1981]) of the Kaden problem (for which $m = 2/3$). Consider accelerated flow past a wedge of angle β as sketched in Figure 8.5-1 (Blendermann [1969]). In the absence of vortex shedding, the complex potential is

$$\mathbf{w} = -Wt^\nu z^\mu, \tag{11}$$

where ν is a given positive number and W is a positive dimensional constant with dimensions $L^{2-\mu} T^{-1-\nu}$. The exponent μ is related to β by

$$\mu = \pi/(2\pi - \beta), \tag{12}$$

so that $\Im w = 0$ on the faces of the wedge. If $\beta < \pi$, the velocity at the tip of the wedge is infinite. A Kutta condition can be imposed by allowing a vortex sheet to separate from the tip. The complex potential then has an extra term w_v, due to the vortex sheet and its image in the wedge, which by dimensional analysis must have the form

$$w_v = W^{\frac{2}{2-\mu}} t^{\frac{2\nu+\mu}{2-\mu}} f\left(\frac{z}{W^{\frac{1}{2-\mu}} t^{\frac{1+\nu}{2-\mu}}}\right). \tag{13}$$

This gives a solution of the form (1), since w and Γ have the same dimensions, with

$$m = \frac{1+\nu}{2-\mu}, \quad n = \frac{2\nu+\mu}{2-\mu}. \tag{14}$$

Note that $2m - 1 = n$. Now the problem is to solve (2), including the image vorticity in the wedge, with appropriate end conditions and the Kutta condition at the wedge tip. This condition is taken in the form of assuming that the streamline approaching the tip separates smoothly. For $\beta > 0$, this implies that there is a stagnation point at the tip on the leeward side of the wedge. This does not violate the continuity of pressure across the separating vortex sheet because the flow is unsteady.

For $0 < \beta < \pi$, $\frac{1}{2} < \mu < 1$, and hence $m > \frac{1}{2}$ if $\nu > 0$. Exact solutions are not known, but the form of the spiral near its centre can be analysed as for the Kaden spiral. The tip of the spiral is at

$$Z_T = W^{\frac{1}{2-\mu}} t^m \zeta_0, \tag{15}$$

where ζ_0 is some unknown constant number determined by the entire solution and not locally by the flow near the centre of the spiral. We take polar co-ordinates r, θ centred on Z_T. Let Γ denote the circulation measured at radius r from the spiral centre. Then from dimensional conditions,

$$\Gamma = k^{1/m} r^{2-1/m}, \quad \text{where } k \propto W^{\frac{1}{2-\mu}}. \tag{16}$$

From $r d\theta/dt = \Gamma/2\pi r$, we obtain $\theta = \Gamma t/2\pi r^2$, and hence

$$r = k(t/2\pi\theta)^m \tag{17}$$

is the equation of the spiral. The parametric equation of the sheet near the centre is

$$Z = W^{\frac{1}{2-\mu}} t^m \zeta_0 + \left(\frac{\Gamma^m}{k}\right)^{\frac{1}{2m-1}} \exp(itk^{2/(2m-1)}/\Gamma^{(1/2m-1)}). \tag{18}$$

Alternatively, we use equation (2) and consider the denominator of the integral, which goes from $\theta = \infty$ at the centre of the spiral to some value θ_0 where it meets the tip of the wedge. We approximate as follows. First note that $f(\theta)$ decreases as θ increases. If $\hat{\theta} < \theta, f e^{i\theta} - \hat{f} e^{i\hat{\theta}} \approx -\hat{f} e^{i\hat{\theta}}$ and the integral oscillates about zero. If $\hat{\theta} > \theta$, the denominator of the integral oscillates about $f e^{i\theta}$. So we replace the integral by

$$\frac{1}{f e^{i\theta}} \int_\theta^\infty \hat{g}' \, d\hat{\theta} = \frac{-g(\theta)}{f(\theta) e^{i\theta}}. \tag{19}$$

(We are essentially using the result $\int_0^{2\pi} (1 - \epsilon e^{i\theta})^{-1} \, d\theta = 0$ if $\epsilon > 1$ and $= 2\pi$ if $\epsilon < 1$.) Then (2) separates into two real equations,

$$mf + (1 - 2m)g \frac{f'}{g'} = 0, \qquad 2\pi(1 - 2m)f^2 = g', \tag{20}$$

which have the solution

$$g = c\theta^{1-2m}, \quad f = (c/2\pi)^{1/2}\theta^{-m}. \tag{21}$$

Some special cases are of particular interest. The case $\beta = 0$, $\mu = \frac{1}{2}$, $\nu = 0$, $m = 2/3$, is the impulsive motion of a semi-infinite flat plate (Anton [1939]). Rott [1956] considers the motion induced by a weak shock incident on a wedge. If u_0 is the amplitude of the velocity jump across the shock and a_0 is the speed of sound, the flow is equivalent to an incompressible flow with an imposed velocity potential (linear in u_0)

$$w = Ku_0 a_0 t \left(\frac{z}{a_0 t} \right)^\mu, \tag{22}$$

where K is a dimensionless constant given by the solution for diffraction of an acoustic pulse by a wedge. Comparing with the general form (13), we see that $\nu = 1 - \mu$ and hence $m = 1$. Also, $W = Ku_0 a_0^{1-\mu}$. Then the position Z_T of the centre of the spiral behaves according to the scaling

$$\frac{Z_T}{a_0 t} \propto \frac{1}{a_0} W^{\frac{1}{2-\mu}} = K^{\frac{1}{2-\mu}} M^{\frac{1}{2-\mu}}, \tag{23}$$

where $M = u_0/a_0$ is the Mach number of the flow.

The similarity approach can be applied to steady three-dimensional flows with conical symmetry, in which all flow quantities are functions of $x/\tilde{z}$ and $y/\tilde{z}$, where $\tilde{z}$ is the axis of symmetry. For example, the velocity potential has the form $\phi = \tilde{z} \, \text{fn}(x/\tilde{z}, y/\tilde{z})$. and the spiral has the equation $r = \sqrt{x^2 + y^2} = \tilde{z}f(\theta)$. Slender body theory, applicable, say, to flow past a slender delta wing with velocity U, allows the equivalence $\tilde{z} \leftrightarrow Ut$, and we are in the class of similarity solutions with $m = 1$. For flow past a

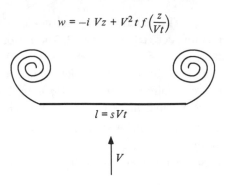

$$w = -i\,Vz + V^2 t\, f\!\left(\frac{z}{Vt}\right)$$

$$l = sVt$$

V

8.5-2 Similarity flow for an extending plate.

slender delta wing, the equivalent two-dimensional problem (Smith [1968], see Figure 8.5-2) is the self-similar flow with velocity V normal to a plate whose length is expanding like sVt. The self-similar velocity potential has the form (corresponding to $m = 1, n = 1$)

$$w = -iVz + V^2tf(z/Vt). \tag{24}$$

Double-branched spirals (i.e., $N = 2$, see (10)) are appropriate for the roll-up of infinite vortex sheets or perhaps the initial stages of further evolution after a sheet develops a Moore singularity (§3). Thus the uniform vortex sheet of strength $\kappa = U$, separating streams of velocity $\pm\frac{1}{2}U$, could roll-up in principle into a double branch spiral with shapes $r = \pm Utf(\theta)$. The Birkhoff–Rott parametrization has the form,

$$Z = \pm Ut\,\zeta(\Gamma/U^2t), \quad 0 < \Gamma < \infty, \tag{25}$$

where $\zeta(\tau) \sim \tau$ as $\tau \to \infty$ and

$$\zeta(\tau) \sim \frac{\tau}{\lambda}\, e^{i(\lambda^2/\pi\tau+\epsilon)} \quad \text{as } \tau \to 0. \tag{26}$$

Application of the Betz approximation implies $\lambda = 1$. However, it has not yet been shown that such solutions exist and the available evidence (see, e.g., Pullin [1989]) suggests non-existence.

More generally, consider the initial distribution

$$Z = \pm A\,\Gamma^q, \quad 0 \le \Gamma < \infty, \tag{27}$$

where $q > 0$, and suppose it rolls up into a double-branched spiral. The parametric equation of the spiral has the form

$$Z = \pm \frac{t^m}{A^n}\,\zeta(\tau), \quad \tau = \Gamma A^2 n/t^n, \tag{28}$$

where $n = 1/(2q - 1)$ and $m = nq$. The function $\zeta(\tau)$ satisfies an ordinary singular integro-differential equation derived from the Birkhoff–Rott equation, with boundary conditions $\zeta(\infty) \sim \tau^q$, $|\zeta(\tau)| \sim \tau^q/\lambda$ as $\tau \to 0$, where λ (the compression or tightening factor) is some to be determined real constant. For an expanding spiral, we require $m > 0$, so that q is in the range $\frac{1}{2} < q < \infty$ and $\infty > m > \frac{1}{2}$. Note that there is a qualitative change in the initial conditions at $q = 1$. The initial sheet strength $\kappa \propto X^{\frac{1}{q}-1}$, and is finite at $x = 0$ if $q < 1$ and infinite if $q > 1$.

The asymptotic shape of the spiral for $r \to 0$, $\Gamma \to 0$, $\tau \to 0$, is

$$r = A\Gamma^q/\lambda. \tag{29}$$

The dependence on θ follows from

$$2\pi r v_\theta = 2\pi r^2 \dot{\theta} = N\Gamma \quad (N = 2), \tag{30}$$

so that

$$\theta = \frac{t\Gamma}{\pi r^2} + \epsilon = \frac{\lambda^2}{\pi\tau^{1/n}} + \epsilon. \tag{31}$$

Thus

$$\zeta(\tau) \sim \frac{\tau^q}{\lambda} e^{i(\lambda^2/\pi\tau^{1/n}+\epsilon)} \quad \text{as } \tau \to 0 \tag{32}$$

It follows again from application of the Betz hypothesis that $\lambda = 1$.

The length of the spiral is infinite if $q \geq 1$ and $m \leq 1$. This follows from (29) and (31), which show on eliminating Γ that

$$r \sim \left(\frac{t}{\theta\pi}\right)^m \left(\frac{\lambda}{A}\right)^n, \quad \text{and} \quad s \sim \int^\infty r \, d\theta. \tag{33}$$

The vorticity at $r = 0$ is always singular, as from (30)

$$\omega = \frac{N}{2\pi r}\frac{d\Gamma}{dr} \sim \frac{1}{\pi q}\left(\frac{\lambda}{A}\right)^{\frac{1}{q}}\frac{1}{r^{1/m}}, \tag{34}$$

but the singularity is integrable.[16]

Pullin [1989] has studied the numerical solutions of the Birkhoff–Rott similarity equation for a range of q. For $\frac{1}{2} < q < 1$, he finds non-spiral solutions and separated single-spiral solutions. The spirals are of finite length. Separated single spirals and double spirals were also found for q

---

[16] We discuss in §13.2 how viscosity can be introduced in order to make the vorticity field smooth in the centre of the spiral, and apply the results to the axial flow in trailing vortices.

very close to, but greater than, 1. The single spirals will differ from the double spirals in having a value of λ greater than 1. The Betz argument can be applied to give $\lambda = (1 + q)/q$ for the single spirals.[17]

[17] Matching of these similarity solutions to the Moore singularity profile, in order to provide a description of vortex sheet evolution after the spiral has formed, has not yet proved possible.

9

DYNAMICS OF TWO-DIMENSIONAL VORTEX PATCHES

9.1 Vortex sheets of finite thickness

In two dimensions, a connected region of finite area containing uniform vorticity, surrounded by irrotational fluid, is called a vortex patch. The boundary of the patch is a vortex jump (§2.1). Sheet patches, which are patches bounded by vortex sheets, are also of considerable interest. Hollow sheet patches containing fluid of zero density bounded by vortex sheets and stagnant sheet patches containing irrotational fluid bounded by vortex sheets are cases in which the vorticity is zero inside the patch. Sheet patches containing non-zero vorticity are also known as Prandtl–Batchelor flows and are important for models of inviscid flow past bluff bodies.[1] Non-homogeneous patches, consisting of a finite number of regions of uniform but different vorticity can model fields of continuously varying vorticity.

Moore [1978] studied the vortex patch formed by a vortex sheet of finite but small thickness.[2] He showed that the effect of small thickness could be described by adding a term to the Birkhoff–Rott equation, which becomes

$$\frac{\partial \overline{Z}}{\partial t} = \frac{-i}{2\pi} \int \frac{d\Gamma'}{Z(\Gamma) - Z(\Gamma')} - \frac{i}{6\omega} \frac{\partial}{\partial \Gamma} \left(\frac{\partial \overline{Z}/\partial \Gamma}{|\partial Z/\partial \Gamma|^4} \right) + O(\epsilon^2). \tag{1}$$

Here, $Z(\Gamma, t)$ is the parametric equation of the centre line of the vortex, and ω is the constant vorticity. The expansion proceeds in powers of $\epsilon = \bar{h}/\rho$,

---

[1] Moore, Saffman and Tanveer [1988] calculated the structure of families of two touching patches of equal and opposite vorticity bounded by vortex sheets known as Sadovskii vortices. Saffman and Tanveer [1984b] calculated asymmetric Prandtl–Batchelor flow past a flat plate with a forward-facing flap in connection with the Kasper wing. See also Smith [1982].

[2] Shelley and Baker [1990] have studied numerically the evolution of the Kelvin–Helmholtz instability for a sheet of finite thickness.

where $\bar{h}$ is the mean thickness of the sheet, and ρ is a characteristic value of the radius of curvature, with $\bar{\kappa} = \omega\bar{h}$ kept constant. The order of the correction term is $\bar{\kappa}\,\bar{h}/\rho$. Note that ω is constant by the dynamics, and $\bar{h}$ is constant by definition, but the thickness h and the local sheet strength $\kappa = \omega h$ will in general vary.

Moore derived (1) by the formal method of matched asymptotic expressions. We shall rederive the equation here by heuristic arguments. The effect of finite thickness can be broken down into two distinct actions. First, the curvature of the sheet will produce an altered flux of vorticity along the sheet so that the speed of the Lagrangian point moving with the circulation is changed; and, second, there is a change in the motion of the sheet normal to itself caused by variation in the sheet thickness along the sheet. Let s denote arc length along the centreline, and θ the angle between the tangent to the centreline and the x-axis. Then

$$\frac{\partial Z}{\partial \Gamma} = \frac{e^{i\theta}}{\kappa}, \quad dZ = e^{i\theta}ds, \quad \frac{\partial \Gamma}{\partial s} = \kappa, \quad \frac{\partial \theta}{\partial s} = \frac{1}{\rho}, \quad \frac{\partial \kappa}{\partial s} = \omega\frac{\partial h}{\partial s}. \quad (2)$$

Moore's correction term to $\partial\bar{Z}/\partial t$ can now be written

$$-\frac{\kappa^2}{6\rho\omega}e^{-i\theta} - \frac{i}{2\omega}\kappa\frac{\partial\kappa}{\partial s}e^{-i\theta}, \quad (3)$$

where the first contribution is parallel to the sheet and is a curvature effect (it vanishes as $\rho \to \infty$) and the second contribution is normal to the sheet and is a variation of thickness effect (it vanishes when $\partial h/\partial s = 0$).

To calculate the curvature effect, it is sufficient to consider a circular vortex sheet, see Figure 9.1-1a. The tangential velocity $v(r)$ in the sheet is found from $\omega = (1/r)d(rv)/dr$, $v(\rho - h/2) = 0$ and is

$$v = \tfrac{1}{2}\omega\left(r - \frac{(\rho - \tfrac{1}{2}h)^2}{r}\right), \quad \rho - \tfrac{1}{2}h < r < \rho + \tfrac{1}{2}h. \quad (4)$$

The velocity defined by the vorticity flux is

$$\frac{1}{\omega h}\int_{\rho-\frac{1}{2}h}^{\rho+\frac{1}{2}h} \omega v\, dr = \frac{\kappa}{2} - \frac{\kappa h}{6\rho} + O\left(\frac{h^2}{\rho^2}\right). \quad (5)$$

The extra velocity in (5) is exactly the first term in (3).

To calculate the thickness effect on the normal velocity, consider the difference $\widehat{V}$ in the induced normal velocity V at a point O in the centreline due to a sheet of varying width $h(x)$ and a sheet of zero thickness and the

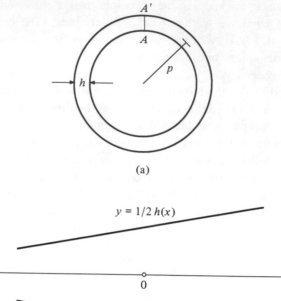

(a)

(b)

9.1-1 Canonical thickness effects. (a) Curvature. (b) Variation of thickness.

same strength, see Figure 9.1-1b. We have

$$
\widehat{V} = \fint_{-\infty}^{\infty} dx \left\{ \left[\int_{-\frac{1}{2}h(x)}^{\frac{1}{2}h(x)} \frac{-x}{x^2 + y^2} \frac{\omega\, dy}{2\pi} \right] + \frac{\omega h}{2\pi x} \right\}
$$
$$
= \frac{\omega}{2\pi} \fint_{-\infty}^{\infty} \left\{ \frac{h}{x} - 2\tan^{-1}\frac{h}{2x} \right\} dx
$$

(6)

on integrating with respect to y. Now integrating by parts, we obtain

$$
\widehat{V} = \frac{\omega}{2\pi} \fint_{-\infty}^{\infty} \frac{h^2}{4} \left(\frac{h}{x} - h' \right) \frac{dx}{x^2 + h^2/4}.
$$

(7)

To find the limit as $h \to 0$, put $h = h_0 + \epsilon x$ and let $\epsilon \to 0$. After some analysis, it is found that

$$
\fint_{-\infty}^{\infty} \frac{h^2}{4} \left(\frac{h}{x} - h' \right) \frac{dx}{x^2 + h^2/4} \quad \sim \pi h_0 \epsilon = \pi h\, dh/dx.
$$

(8)

Hence

$$\widehat{V} = \tfrac{1}{2} \omega h \, dh/dx \ (1 + O(dh/dx)), \tag{9}$$

which agrees with the second term in Moore's formula (3).

Moore demonstrates that the modified equation conserves linear and angular momentum and also energy correct to $O(\epsilon)$. He applies it to the Kelvin–Helmholtz instability (§8.2) of a sheet of thickness h and works out the behaviour of a disturbance of the form

$$Z = \frac{\Gamma}{U} + a(t)e^{ik\Gamma/U} + b(t)e^{-ik\Gamma/U}, \tag{10}$$

where $a(t)$ and $b(t)$ are complex valued functions of time. Linearising in a and b, he finds that

$$\frac{d\bar{a}}{dt} = \frac{iUkb}{2} - ik^2 Uh(2b + \bar{a}), \tag{11}$$

and the same equation with a and b interchanged. These equations imply that disturbances grow like $e^{\sigma t}$, where

$$\sigma = \tfrac{1}{2} Uk \left[\left(1 - \frac{kh}{3}\right)\left(1 - kh\right)\right]^{\frac{1}{2}}. \tag{12}$$

The approximations leading to (12) are only valid if $kh \ll 1$, and then

$$\sigma \approx \pm \tfrac{1}{2} Uk[1 - \tfrac{2}{3} kh]. \tag{13}$$

This agrees with Rayleigh's [1880, 1945 §367] analysis of the infinitesimal stability of a vortex sheet of finite thickness for small kh. The exact dispersion relation for infinitesimal disturbances on a layer of width h is

$$\sigma^2 = \frac{U^2}{H^2} \left[e^{-2kh} - (1 - kh)^2\right]. \tag{14}$$

The instability should therefore disappear when $kh = 1.28$. Moore's approximation shows a stabilization when $kh = 1$ but there is then a spurious instability for $kh > 3$. The well-posedness of Moore's modified Birkhoff–Rott equation is an open question, but the instability for large kh, at a greater rate than for the Birkhoff–Rott, suggests that the modified equation is even more singular.

9.2 Contour dynamics and Schwarz functions

The evolution of vortex patches is governed by the Euler equations, which imply that both the vorticity inside the patch and its area are constant; and also that vortex jumps are material surfaces. Thus in order to determine the motion of a patch it suffices to know the velocity field induced by the patch on its boundary, plus, of course, the external velocity field produced by other vorticity and images. This idea was exploited by Deem and Zabusky [1978], who showed, following a similar technique in plasma physics known as the waterbag model, that an integro-differential equation could be written for the boundary so that the evolution of the boundary is reducible to a 'marching' problem, similar to the Birkhoff–Rott equation for a vortex sheet.

Consider the complex velocity field $u - iv$ produced at a point P with complex co-ordinate $z = x + iy$ by a single patch occupying an area A. It is

$$u - iv = -\frac{i\omega}{2\pi} \iint_A \frac{dx' dy'}{z - z'}. \tag{1}$$

Suppose that z is outside the patch. Then by Green's theorem

$$\iint_A \frac{dx' dy'}{z - z'} = \oint_{\partial A} \log(z - z') \, dy'. \tag{2}$$

By Cauchy's theorem,

$$\oint_{\partial A} \log(z - z')(dx' + idy') = 0. \tag{3}$$

Breaking up (2) and (3) into real and imaginary parts, we have

$$\iint_A \frac{dx' dy'}{z - z'} = i \oint_{\partial A} \log|z - z'|(dx' - idy')$$

and hence[3]

$$u + iv = \frac{\omega}{2\pi} \oint_{\partial A} \log|z - z'| \, dz'. \tag{4}$$

A further transformation of this formula was given by Pullin [1981]. Writing

$$\log|z - z'| = \tfrac{1}{2} \log|z - z'| + \tfrac{1}{2} \log|\bar{z} - \bar{z}'|,$$

we use Cauchy's theorem and integrate by parts to obtain

$$u + iv = -\frac{\omega}{4\pi} \oint_{\partial A} \frac{z - z'}{\bar{z} - \bar{z}'} \, d\bar{z}'. \tag{5}$$

---

[3] Note the change in signs.

This form is advantageous for numerical evaluation, since as $z \to z'$ on the contour, the integrand takes the finite value $dz'/d\bar{z}' = e^{2i\theta}$, where θ is the angle made by the tangent with the x-axis at z.

Let $Z(s,t)$ denote a parametric representation of the boundary. Then the condition that the boundary is a material surface requires that the difference between $\partial Z/\partial t$ and the induced velocity is parallel to the surface. That is,

$$\Im \left[\left\{ \frac{\partial Z}{\partial t} - (U + iV) - (U_E + iV_E) \right\} \frac{\partial \bar{Z}}{\partial s} \right] = 0,$$

or

$$\frac{\partial Z}{\partial s} \, /\!/ \, \frac{\partial Z}{\partial t} - (U + iV) - (U_E + iV_E), \tag{6}$$

where

$$U + iV = \frac{\omega}{2\pi} \oint_{\partial A} \frac{Z - Z'}{\bar{Z} - \bar{Z}'} \frac{\partial \bar{Z}}{\partial s'} \, ds' \tag{7}$$

and (U_E, V_E) are the components of the externally generated velocity field. The integral in (7) is singular, but the singularity is weak enough for the integral to exist as an improper Riemann integral.

If s is chosen to be a Lagrangian variable, such that s is constant for a point which moves parallel to the boundary with the tangential component of $(U + U_E, V + V_E)$, (6) takes the simple form

$$\frac{\partial Z}{\partial t} = (U + U_E) + i(V + V_E). \tag{8}$$

The approach to patch evolution by solution of (6) and (7), or (7) and (8), is called the method of contour dynamics.[4]

An alternative formulation employs the concept of Schwarz functions.[5] The idea is to construct an analytic function $\Phi(z)$ which is equal to $\bar{z}$ on some simple closed contour C. Let $z = A(\zeta)$ be a conformal map of the outside of C into the outside of the unit circle $|\zeta| = 1$. In principle, $\Phi(z)$ is defined by $A(\zeta)$. Now in general, $\Phi(z)$ has singularities inside and outside C, but we can write

$$\Phi(z) = F(z) + G(z),$$

---

[4] Zabusky, Hughes and Roberts [1979]. For a review, see Pullin [1991]. Baker [1990] has studied the numerical stability of the method.

[5] Jimenez (1982 private communication, [1988]). For a discussion of the theory, see Gaier [1964].

where $F(z)$ is analytic inside C and $G(z)$ is analytic outside C. If $\Phi(z)$ has a Laurent expansion in an annulus containing C,

$$\Phi(z) = \sum_0^\infty f_n z^n + \sum_1^\infty g_n z^{-n}, \tag{9}$$

and F and G are the analytic continuations of the series. Remember that

$$\Phi(z) = \bar{z} \quad \text{on } C. \tag{10}$$

Let (u, v) denote the components of velocity induced by the patch. Then $u - iv$ is analytic outside C and $u - iv + \frac{1}{2} i \omega \bar{z}$ is analytic inside C, so that $\partial v / \partial x - \partial u / \partial y = \omega$ inside C. Consider now the velocity field such that

$$u - iv = -\tfrac{1}{2} i \omega G, \quad \text{outside } C, \qquad u - iv = \tfrac{1}{2} i \omega F - \tfrac{1}{2} i \omega \bar{z} \quad \text{inside } C. \tag{11}$$

The velocity is continuous on C, gives constant vorticity inside C and is analytic outside and is therefore the velocity induced by the vortex patch. The evolution is now given by (6) or (8) and leads to an equation for the conformal map $A(\zeta)$.

For vortex patches that are not too deformed from circular shapes, it is expected that the Laurent expansion exists, and if

$$A(\zeta) = \zeta \left(a_0 + \frac{a_1}{\zeta} + \frac{a_2}{\zeta^2} + \cdots \right), \tag{12}$$

then since $\bar{\zeta} = 1/\zeta$ on C,

$$\bar{z} = \frac{1}{\zeta} \left(\bar{a}_0 + \bar{a}_1 \zeta + \bar{a}_2 \zeta^2 + \cdots \right), \tag{13}$$

on C. It follows that

$$g_n = \frac{1}{2\pi i} \oint \bar{z} z^{n-1} \, dz = \text{coefficient of } \zeta^{-1} \text{ in the product } \bar{z} A^{n-1} \frac{dA}{d\zeta}. \tag{14}$$

The evolution equation is now an infinite set of coupled non-linear ordinary differential equations for the coefficients $a_n(t)$.[6]

The existence for all time of smooth solutions for given smooth initial conditions has been demonstrated by Chemin [1991]. Thus corners and cusps cannot form in a finite time from smooth initial data. There is, however, numerical evidence for the formation of highly convoluted surfaces with many points of high curvature by a process known as filamentation

---

[6] An alternative formulation employs the Grunski polynomials (Burbea [1982]). Hamiltonian methods also show promise, especially for nearly circular vortices.

(see §5). It is a consequence of the reversibility of the Euler equations that if the initial boundary is singular, it cannot become smooth.

9.3   The Kirchhoff vortex and elliptical patches in uniform strain

The simplest vortex patch is the circle of radius R in unbounded fluid (Rankine vortex). For this flow, the Schwarz functions are

$$G = \tfrac{1}{2} R^2/z, \quad F = 0, \quad \text{and} \quad u - iv = -\frac{i\omega R^2}{2z} \tag{1}$$

outside the patch.

Kirchhoff (see Lamb [1932 §159]) showed that the solution could be generalized to a rotating ellipse of semi-axes a and b which rotates steadily with angular velocity

$$\Omega = \omega \frac{ab}{(a+b)^2}. \tag{2}$$

There are many ways to derive this result. Here we use the Jimenez–Schwarz function approach, in which the unknown is the conformal map of the exterior of the unit circle $|\zeta| = 1$ into the exterior of the ellipse The map is

$$z = \alpha(t)\zeta + \frac{\beta(t)}{\zeta}, \qquad \zeta = \frac{z + (z^2 - 4\alpha\beta)^{1/2}}{2\alpha}. \tag{3}$$

Without loss of generality, we map $\zeta = 1$ into the end of the major axis. Then

$$\alpha + \beta = ae^{i\theta}, \quad \alpha - \beta = be^{i\theta}, \quad \alpha = \tfrac{1}{2}(a+b)e^{i\theta}, \quad \beta = \tfrac{1}{2}(a-b)e^{i\theta}, \tag{4}$$

where θ is the angle between the major axis and the x-axis. Note that $\alpha\bar{\alpha} - \beta\bar{\beta} = ab = A/\pi$, where A is the area of the ellipse. The equations of vortex motion require that ω and A are both constant.

On the ellipse, $\zeta\bar{\zeta} = 1$, and hence from (3),

$$\bar{z} = \frac{\bar{\alpha}}{\zeta} + \bar{\beta}\zeta = \frac{\bar{\beta}}{\alpha}z + \frac{ab}{\alpha\zeta} \tag{5}$$

on substituting for ζ. Now $1/\zeta$ is analytic outside the ellipse and hence, in the notation of §2,

$$G(z) = \frac{ab}{\alpha\zeta}, \tag{6}$$

and the induced velocity field outside and on the boundary is

$$u - iv = -\frac{i\omega ab}{2\alpha\zeta}.$$ (7)

The parametric equation of the ellipse $Z(s,t)$ is given by (3) with $\zeta = e^{is}$, $0 \le s \le 2\pi$, and hence

$$\frac{\partial Z}{\partial t} = \dot\alpha\zeta + \frac{\dot\beta}{\zeta}, \qquad \frac{\partial Z}{\partial s} = i\alpha\zeta - \frac{i\beta}{\zeta}.$$ (8)

The boundary condition (2.6) gives

$$\dot\alpha\zeta + \frac{\dot\beta}{\zeta} - \frac{i\omega ab}{2\overline\alpha}\zeta \;\; /\!/ \;\; i\alpha\zeta - \frac{i\beta}{\zeta},$$ (9)

from which follows

$$\frac{\dot\alpha - \dfrac{i\omega ab}{2\overline\alpha}}{\dot\beta} = \frac{i\alpha}{-i\beta},$$ (10)

that is,

$$\frac{\dot\alpha}{\alpha} - \frac{i\omega ab}{2\alpha\overline\alpha} = -\frac{\dot\beta}{\beta}.$$ (11)

This equation (combined with the conservation of area) has the unique solution

$$\dot a = 0, \quad \dot b = 0, \quad \dot\theta = \frac{\omega ab}{4\alpha\overline\alpha} = \Omega,$$ (12)

verifying (2).

In the limit $a/b \to 1$, the rotating ellipse behaves like an infinitesimal perturbation to a circular vortex patch of the form $r = a + \epsilon\cos 2(\theta - \Omega_2 t)$. Kelvin [1880], and Lamb [1932 §158] (see §4) showed that $\Omega_2 = \omega/4$. This agrees with the limit of (2) as $a \to b$. In this limit, fluid particles rotate with angular velocity $\omega/2 = 2\Omega$, so that the motion of the boundary is retrograde relative to the fluid in the patch.

As noted by Batchelor [1967 §7.3], the limit $b \to 0$, $\omega \to \infty$, $2\Omega b \to \kappa$ is a vortex sheet of length $2a$ and strength $\kappa(1 - x^2/a^2)^{1/2}$ which rotates steadily with angular velocity κ/a. Note that the strength is zero at the ends, so there is no force unbalance to start roll-up. This example shows that unsteady vortex sheets need not roll up.

Moore and Saffman [1971] generalized the Kirchhoff solution to a vortex patch in a uniform straining field

$$U_E = -\epsilon y, \quad V_E = -\epsilon x, \quad U_E + iV_E = -i\epsilon\overline z.$$ (13)

Then (9) is replaced by

$$\dot{\alpha}\zeta + \frac{\dot{\beta}}{\zeta} - \frac{i\omega ab}{2\bar{\alpha}}\,\zeta + i\epsilon\left(\frac{\bar{\alpha}}{\zeta} + \bar{\beta}\zeta\right) \quad \Big/\!\!\Big/ \quad i\alpha\zeta - \frac{i\beta}{\zeta}, \tag{14}$$

from which follows

$$a\dot{a} - b\dot{b} = -\epsilon(a^2 + b^2)\,\sin 2\theta, \tag{15}$$

$$\dot{\theta} = \frac{\omega ab}{(a+b)^2} - \epsilon\,\frac{(a^2 + b^2)}{a^2 - b^2}\,\cos 2\theta. \tag{16}$$

Steady-state solutions exist if $\theta = 0$, so that the major axis makes an angle $\frac{1}{4}\pi$ with the direction of the principal rate of strain, and

$$\frac{\epsilon}{\omega} = \frac{ab(a-b)}{(a+b)(a^2+b^2)}. \tag{17}$$

The interesting feature of this relation is that the right-hand side as a function of a/b (≥ 1) starts at zero, rises to a maximum of about 0.15 for $a/b = (a/b)_c \doteq 2.9$ and then decreases to zero. Thus steady elliptical patches in a uniform external straining field cannot exist if the strain is too large or the vorticity too small. If the strain is not too large, two solutions exist, one relatively circular and the other relatively linear.

Moore and Saffman [1971] used a representation in elliptical curvilinear co-ordinates and confined the analysis to a study of steady states and their stability to two-dimensional infinitesimal disturbances. The extension to unsteady flow which gives evolution equations (15) and (16) for elliptical patches in a constant uniform straining field was given by Kida [1981a] (see also Neu [1984], Jimenez [1988]). Note that ϵ need not be constant but can be an arbitrary function of time. If the initial strain does not satisfy (17) for some a/b, the vortex is pulled out into a long thin ellipse along the principal axis of extension.

The solution can also be generalized to include changes in the direction of the principal rates of strain, by replacing (13) with

$$U_E + iV_E = -i\epsilon\bar{z} + \gamma\bar{z}, \tag{18}$$

where γ is real and may be an arbitrary function of time. It follows as before that

$$a\dot{a} - b\dot{b} = -\epsilon(a^2 + b^2)\,\sin 2\theta + \gamma(a^2 + b^2)\,\cos 2\theta, \tag{19}$$

$$\dot{\theta} = \frac{\omega ab}{(a+b)^2} - \epsilon\,\frac{(a^2 + b^2)}{a^2 - b^2}\,\cos 2\theta - \gamma\,\frac{a^2 + b^2}{a^2 - b^2}\,\sin 2\theta. \tag{20}$$

Since $\pi ab = A = $ constant, these equations are sufficient to describe the shape and orientation of the ellipse.

Adding a term $i\Omega_0 z$ to the right-hand side of (18) includes a solid-body rotation Ω_0 at infinity. Taking $\gamma = 0$, $\epsilon = -\Omega_0 = -\kappa/2$ corresponds to a patch in a simple shear $u = \kappa y$. Equilibrium configurations and their stability to general two-dimensional disturbances were calculated by Moore and Saffman [1971], (see §4). The modifications to (19) and (20) to describe unsteady motion of a patch in a simple shear were given by Kida; they can be found directly or by moving to axes rotating with angular velocity Ω_0 and changing ω to $\omega - 2\Omega_0$ and then using (19) and (20).

Equations (19) and (20), together with (7) for the external velocity induced by a patch, form the basis of the elliptical vortex approximation for studying the interactions of vortex patches. Using (7), we expand the external velocity field produced by other vortex patches, and images if appropriate, as a Taylor series about the centre of the patch. The linear term is of the form (18) and determines the rates of strain for equations (19) and (20), which describe the evolution of the axis ratio and the orientation of the ellipse. The convection of the centre of the patch is given by the Betz results discussed in §3.2, according to which the velocity of the centre is the average of the external velocity through the patch. This follows from the constant and the quadratic term in the Taylor series expansion of the complex velocity

$$\frac{dW}{dz} = \left(\frac{dW}{dz}\right)_0 + z\left(\frac{d^2W}{dz^2}\right)_0 + \frac{1}{2}z^2\left(\frac{d^3W}{dz^3}\right)_0, \tag{21}$$

where W is the complex potential produced by the other patches and suffix 0 means evaluation at the centre of the patch. It follows that

$$\frac{d}{dt}(X - iY) = \left(\frac{dW}{dz}\right)_0 + \frac{1}{8}(a^2 - b^2)e^{2i\theta}\left(\frac{d^3W}{dz^3}\right)_0. \tag{22}$$

Moore and Saffman [1975a] used this idea to build a model of the turbulent mixing layer, in which vortices are destroyed by 'fission' when they grow too large by entrainment to exist in the straining field induced by the other vortices (coherent structures) in the layer. A simplified quasi-steady version of these equations, with the left-hand sides of (19) and (20) neglected along with the higher-order term in (22), was employed by Saffman [1979] in a study of a vortex pair approaching a plane interface, to study the rebound phenomenon that had been reported experimentally (Harvey and Perry [1971]) and demonstrate that it is not due to the finite size of the vortices.[7]

---

[7] Rebound is due to secondary vorticity produced in a boundary layer on a non-slip wall (e.g., see Peace and Riley [1983]). If viscosity is included but the no-slip boundary

Melander, Zabusky and Styczek [1986] have shown that the system (19), (20) and (22), with (7) used to calculate the contribution to W from the patches, is a set of equations of Hamiltonian form.[8]

9.4 Equilibrium configurations for single patches

The Kirchhoff vortex and the Moore–Saffman generalizations are the only known exact steady solutions for uniform vortex patches.[9]

Numerical solutions using the contour dynamics representation or the Schwarz function approach have been employed to investigate equilibrium configurations for various arrangements.

Consider first the single patch rotating in fluid otherwise at rest. As noted by Deem and Zabusky [1978], the Kirchhoff vortices are just one member ($n = 2$) of an infinite family of rotating vortices with n-polygonal symmetry. The existence of these solutions is easily demonstrated by a bifurcation analysis. Consider infinitesimal (two-dimensional) disturbances of n-fold symmetry to a uniform circular patch of radius a and vorticity ω (Kelvin [1880], Lamb [1932, §158], Batchelor [1967 §7.3]) associated with deformation of the boundary keeping the vorticity (and area) constant. The boundary takes the shape ($\delta \ll 1$, $n \neq 0$)

$$r = a(1 + \delta e^{i(n\theta - \sigma t)}). \tag{1}$$

The stream function inside and outside the patch is

$$\psi = \frac{1}{4}\omega(a^2 - r^2) + A\delta \frac{r^{|n|}}{a^{|n|}} e^{i(n\theta - \sigma t)}, \quad r < a, \tag{2}$$

$$\psi = -\frac{\omega a^2}{2} \log (r/a) + B\delta \left(\frac{a}{r}\right)^{|n|} e^{i(n\theta - \sigma t)}, \quad r > a. \tag{3}$$

The functional forms are determined by the requirement that the disturbances are irrotational. The quantities σ, A, B follow from the conditions

condition is relaxed, to model interaction with a free surface at small Froude number, then there is a slow drift of the vortices away from the wall (see also Saffman [1991]).

[8] This also follows from more general considerations, see §14.3.

[9] There are some exact solutions for non-uniform vortex patches; e.g., Hill's [1975b] hollow vortex in a uniform strain; Lamb's [1932 §165] vortex pair; Pocklington's [1895a] hollow vortex pair, Baker, Saffman and Sheffield's [1976] linear array of hollow vortices. It is expected that a multitude of closed form exact solutions exist for particular distributions of vorticity and that algorithms exist which construct families of such solutions, (e.g., Abrashkin and Yakubovich [1984]).

that the deformed boundary (1) is a material surface across which ψ and $\nabla\psi$ (and hence p) are continuous. Neglecting terms of order δ^2, we obtain

$$A = B = -(\sigma + \tfrac{1}{2}|n|\omega),\tag{4}$$

$$\sigma = \tfrac{1}{2}\omega(|n| - 1)\,\mathrm{sgn}\,n.\tag{5}$$

$n = \pm 1$ is the trivial case in which the patch is displaced without change of shape.

The angular velocity Ω_n of the disturbance is σ/n; that is,

$$\Omega_n = \frac{1}{2}\omega\left(1 - \frac{1}{|n|}\right).\tag{6}$$

Since the fluid in the undisturbed patch is in solid-body rotation with angular velocity $\tfrac{1}{2}\omega$, the disturbance moves backwards relative to the fluid.

In a frame of reference rotating with angular velocity Ω_n, the disturbance is stationary. Consider now the non-linear problem of determining the shapes of a patch of given area and vorticity that is at rest in a frame rotating with angular Ω. This can be expressed formally as an infinite set of equations

$$N(a_n; \Omega) = 0,\tag{7}$$

where the unknown is an infinite set $\{a_n\}$ of coefficients of the Fourier series for the boundary,

$$r = a + \sum_{-\infty}^{\infty} a_n e^{in\theta},\tag{8}$$

$(a_n = \bar{a}_{-n})$, and Ω is regarded as a control parameter.

For all Ω, the trivial solution $\{a_n\} = 0$ exists. The existence of infinitesimal solutions for $\Omega = \Omega_n$ suggests the existence of bifurcations when Ω passes through the values Ω_n (see Figure 9.4-1). This can be verified by straightforward analysis.[10] Su [1979] gives the results for bifurcation into the m-fold symmetric family

$$r = a(1 - \tfrac{1}{2}\delta^2) + \delta\,\cos m\theta + \frac{2m - 1}{4}\delta^2\cos 2m\theta + \cdots\tag{9}$$

and

$$\Omega = \Omega_m\left(1 - \frac{\pi\delta^2}{4}(m - 1) + \cdots\right).\tag{10}$$

For $m = 2$, these results agree with the nearly circular Kirchhoff vortex formulae.

[10] The Schwarz function approach is relatively easy.

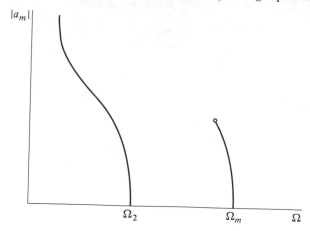

9.4-1 Sketch of bifurcation diagram for a single circular vortex patch. ○ denotes an *m*-sided curvilinear polygon whose corners are right angles.

Deem and Zabusky's numerical studies suggest that each branch (for $m > 2$) ends in a curvilinear polygon. Saffman and Szeto's [1980] analysis of the shape of corners of vortex patches shows that the included angle is $90°$ and that the curvature is infinite. The tangent at the vertex has the equation in local polar co-ordinates, with $\theta = 0$ along the axis of symmetry at the vertex,

$$\theta' = \pm \left(\frac{1}{4}\pi + \frac{\pi}{8} \log r' + O(1/\log r') \right). \tag{11}$$

The patches are initially linearly stable, but the possibility of further bifurcation, to shapes with less symmetry, cannot be ruled out. Explicit calculations have been done for the $m = 2$ case of the Kirchhoff vortex by Moore and Saffman [1971]. They showed by a calculation using elliptical curvilinear co-ordinates that the frequencies σ of disturbances to the Kirchhoff vortex of m-fold symmetry are given by

$$\sigma^2 = \frac{1}{4} \omega^2 \frac{(a^2 + b^2)^2}{(a+b)^4} \left\{ \left[\frac{2mab}{a^2 + b^2} - \frac{(a+b)^2}{a^2 + b^2} \right]^2 - \frac{(a+b)^4}{(a^2 + b^2)^2} \left(\frac{a-b}{a+b} \right)^{2m} \right\},$$

$$\tag{12}$$

where a and b are the major and minor axes of the ellipse. It can be shown that $\sigma^2 > 0$ for $a/b < 3$. The $m = 2$ mode becomes unstable by an exchange of stability when $a/b = 3$ (Love [1893]). There is then a bifurcation to a family of patches without symmetry, which are, however, unstable (at least in the vicinity of the bifurcation point) to disturbances

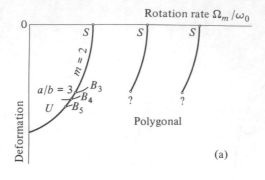

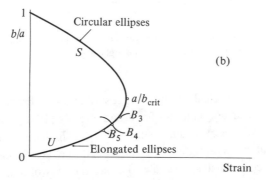

9.4-2 Sketch of families of uniform patches. (a) Kirchhoff vortices.
(b) Moore–Saffman vortices.

with two-fold symmetry. Further bifurcations occur at the values of a/b for which σ vanishes for other values of m. Kamm [1987] has investigated these properties in detail.[11]

Moore and Saffman [1971] also studied the stability properties of two-dimensional infinitesimal disturbances of the strained elliptical vortices described in §3. In this case,

$$\sigma^2 = \frac{\omega^2}{4} \left\{ \left(\frac{2mab}{a^2 + b^2} - 1 \right)^2 - \left(\frac{a - b}{a + b} \right)^{2m} \right\}. \tag{13}$$

It follows that the strained elliptical vortex is linearly stable if $a/b < (a/b)_c$, and unstable otherwise. Bifurcations into families of non-elliptical vortices occur at the values of a/b for which $\sigma = 0$ with $m > 2$. These have also been studied by Kamm [1987]. Figure 9.4-2 sketches some of the families of solutions that can exist.

[11] Kamm's results are summarised by Saffman [1988].

The instabilities discussed here are exchanges of stability and correspond to changes in sign of σ^2, the square of the eigenvalue. Instability can also occur because of collision of eigenvalues. If σ_{m_1} and σ_{m_2} are two frequencies corresponding to different modes or eigenfunctions, and $\sigma_{m_1} = \sigma_{m_2}$ for some value of a control parameter, a/b say, then there is a possibility of a bubble of instability, that is, a range of the control parameter, for which the frequencies are complex and the steady state is unstable. This happens if the modes have opposite signatures, which can be interpreted as the excess energy of the perturbation (see MacKay and Saffman [1986].)

For a vortex patch of vorticity ω in a simple shear $u = \kappa y$, Moore and Saffman [1971] found that there is one steady solution for a co-rotating configuration in which κ and ω have opposite signs when $|\kappa/\omega| < 1$. There are no solutions of elliptical form if this condition is violated. The vorticity outside the patch is $-\kappa$ and inside the patch is ω. The axis ratio is given by

$$|\kappa/\omega| \, (a^2 + b^2) = a^2 - ab. \tag{14}$$

The patch is orientated with its major axis parallel to the shear streamlines, and it can be shown to be stable to infinitesimal two-dimensional disturbances. There are no steady solutions of elliptical form if $|\kappa/\omega| > 1$. For the counter-rotating configuration in which κ and ω have the same sign, there exist two elliptical solutions if $\kappa/\omega < \frac{\sqrt{2}-1}{2}$, the less deformed being stable, and none if κ/ω is greater than this critical value. The counter-rotating patches are orientated with the major axis perpendicular to the direction of the shear flow. The axis ratio is given by

$$\kappa/\omega \, (a^2 + b^2) = ab - b^2. \tag{15}$$

Generally, depending on the details of the external flow, there may be 0, 1, 2 or 3 steady solutions of elliptical form, which may be stable or unstable to linear two-dimensional disturbances.

9.5 Filamentation

Numerical calculations (Deem and Zabusky [1978], Dritschel [1988]) by contour dynamics of isolated vortex patches of uniform vorticity show that small bumps steepen and form points of high curvature from which spring thin filaments of vortical fluid. This could be expected if the patch were linearly unstable (see Polvani, Flierl and Zabusky[1989]), but it occurs if the patch is circular or elliptical and linearly stable. Also, the height of the bumps does not increase.[12] The steepening and filamentation is also observed for layers of uniform vorticity adjacent to a rigid wall; for instance, the vorticity $\omega = \omega_0$, $0 < y < h$ and $\omega = 0$, $h < y < \infty$, with $y = 0$ a rigid wall (Pullin [1981]), for conditions such that the layer is linearly stable. However, note that in this case, there is also for layers that are not too thick a non-linear secondary instability mechanism which can lead to filamentation at a faster rate (Pullin et al. [1990]).

The filamentation occurs even though the patch is non-linearly stable in an L^1-norm.[13] This can be shown simply for a circular patch of initial radius a where the norm is the area swept out by the boundary by the use of Schwarz's inequality (Saffman [1985]). Suppose $r(\theta, t)$ is the radius of the patch at time t. The L^1 norm is

$$L_1 = \frac{1}{2} \int |r^2 - a^2| \, d\theta. \tag{1}$$

By Schwarz's inequality,

$$L_1 \leq \sqrt{\int d\theta} \sqrt{\int (r^4 - 2a^2 r^2 + a^4) \, d\theta}$$

$$= \sqrt{2\pi \int (r^4 - a^4) \, d\theta} = \sqrt{8\pi \, \delta(H)}, \tag{2}$$

since $\int r^2 \, d\theta = \int a^2 \, d\theta$ by conservation of area, where $\delta(H)$ is the difference in angular momentum between the initially disturbed patch and the circle.[14] For an alternative approach, see Wan and Pulvirenti [1985].

This result does not prevent vorticity from escaping to infinity (i.e., does not imply convergence in an L^∞ maximum displacement norm), provided the amount that goes a long distance is sufficiently small. Chemin's [1991] analysis shows that non-linear steepening cannot lead to the formation of cusps or corners.

For a bump of height h, width l, on a circular patch of radius R containing vorticity ω, which steepens and forms a filament in a time t_B, it follows from dimensional analysis that

$$t_B = \frac{1}{\omega} \, \text{fn} \left(\frac{l}{R}, \frac{h}{R} \right). \tag{3}$$

---

[12] The behaviour resembles shock-wave formation in this respect.
[13] Stability in an L^2-norm is automatically ensured by conservation of angular momentum.
[14] It is a consequence of Kelvin's variational principle for steady flow (see §14.2) that the uniform circular patch has the least angular momentum for given vorticity and area.

Dritschel's [1988] calculations, supported to some extent by those of Pullin and Moore [1990],[15] suggest the formula

$$t_B \approx \frac{l^2}{2\pi \omega h^2} \left(15 + \frac{100l}{\pi R} \right). \tag{4}$$

Filamentation can occur at a much faster rate for elliptical patches with $a/b > 3$, owing to the linear instability.

The steepening filamentation is observed for layers of uniform vorticity adjacent to a rigid wall; for instance, the vorticity distribution $\omega = \omega_0$, $0 < y < h$, and $\omega = 0$, $y > h$, with $y = 0$ a rigid wall (Pullin [1981]). But in this case, the above-mentioned non-linear secondary instability mechanism can lead to filamentation at a faster rate provided the layer is not too thick.

There is an interesting difference between the patch and layer filamentation. In the former case, the filamentation is extrusive, with thin filaments of vorticity invading the irrotational fluid, whereas in the latter case it is intrusive, with filaments of irrotational fluid invading the vortex layer. Pullin et al. [1990] argue, however, that the difference may be more apparent than real, since in the case of the patch the phenomenon should be observed in a frame of reference in which the bump is quasi-stationary, in which case the external fluid is not irrotational.

Marsden and Weinstein [1983] and Dritschel [1988] have given a weakly non-linear equation for small disturbances to the boundary of a uniform circular vortex patch. For the dependent variable

$$\phi(\theta, t) = \tfrac{1}{2} (r(\theta, t)^2 - R^2), \tag{5}$$

the equation is

$$\phi_t + \tfrac{1}{2} \omega_0 (\phi_\theta + \mathcal{H}(\phi)) \tag{6}$$

$$= \omega_0 \frac{\partial}{\partial \theta} \left(\frac{1}{4R^2} \phi^2 - \frac{1}{3R^4} \phi^3 + \frac{1}{24\pi R^4} \int_0^{2\pi} \frac{(\phi(\theta) - \phi(\theta'))^3}{1 - \cos(\theta - \theta')} d\theta' \right),$$

where

$$\mathcal{H}(\phi) = \frac{1}{2\pi} \int_0^{2\pi} \phi(\theta') \cot \frac{\theta - \theta'}{2} d\theta'$$

is the Hilbert transform on the circle. The error is $O(\phi^4)$.[16]

---

[15] These calculations did not vary l/R.

[16] Marsden and Weinstein gave this equation with the quadratic non-linearity. The cubic terms are due to Dritschel.

This equation is Hamiltonian[17] with Poisson structure $(1/\omega_0)\,\partial/\partial\theta$; that is, it can be written

$$\frac{\partial \phi}{\partial t} = \frac{1}{\omega_0} \frac{\partial}{\partial \theta} \frac{\delta H}{\delta \phi}, \tag{7}$$

where $H[\phi]$ is the Hamiltonian,

$$
\begin{aligned}
H = & -\frac{\omega_0^2}{4} \int_0^{2\pi} \left(\phi^2(\theta) - \frac{1}{3R^2} \phi^3(\theta) + \frac{1}{3R^4} \phi^4(\theta) \right) d\theta \\
& + \frac{\omega_0^2}{4\pi} \int_0^{2\pi} \int_0^{2\pi} \phi_\theta(\theta) \phi_{\theta'}(\theta') \log \left| \sin \frac{\theta - \theta'}{2} \right| d\theta\, d\theta' \\
& + \frac{\omega_0^2}{192\pi R^4} \int_0^{2\pi} \int_0^{2\pi} \frac{(\phi(\theta) - \phi(\theta'))^4}{1 - \cos(\theta - \theta')} d\theta\, d\theta'.
\end{aligned}
\tag{8}
$$

A. Rouhi (private communication) has derived the corresponding equations for a sheet of uniform vorticity ω_0 bounded by an infinite plane wall, in motion with a period L in the x-direction parallel to the wall. If the equation of the interface is

$$y = \eta(x, t) + h_0, \tag{9}$$

where h_0 is the mean thickness, The Hamiltonian is

$$H = \frac{\omega_0}{2} \int_{-\frac{L}{2}}^{\frac{L}{2}} \int_0^{h_0 + \eta(x,t)} \psi(x, y)\, dy\, dx, \tag{10}$$

where ψ is the stream function inside the patch, given by the integral (3.10.5) over the vorticity and its image in the wall, or as the solution of the equations $\nabla^2\psi = -\omega_0$ for $y < h_0 + \eta$, $\nabla^2\psi = 0$ for $y > h_0 + \eta$, $\psi(x, 0) = 0$, ψ and its derivatives continuous on the interface and vanishing as $y \to \infty$, and $\psi(x, y) \equiv \psi(x+L, y)$. The equation of motion in Hamiltonian form is

$$\frac{\partial \eta}{\partial t} = -\frac{1}{\omega_0} \frac{\partial}{\partial x} \frac{\delta H}{\delta \eta} \quad \left(= -\frac{\partial}{\partial x} \psi(x, h_0 + \eta(x, t)) \right). \tag{11}$$

The Hamiltonian computed by Rouhi to quartic order in η is ($\alpha = 2\pi/L$, $p = e^{-2\alpha h_0}$),

$$H[\eta] = -\frac{\omega_0^2}{2} \int_{-\frac{L}{2}}^{\frac{L}{2}} \left(h_0 \eta^2(x, t) + \frac{1}{6} \eta^3(x, t) \right) dx \tag{12}$$

$$-\frac{\omega_0^2}{8\pi} \int_{-\frac{L}{2}}^{\frac{L}{2}} \int_{-\frac{L}{2}}^{\frac{L}{2}} \log \left[\frac{1 - \cos \alpha(x - x')}{1 - 2p \cos \alpha(x - x') + p^2} \right] dx\, dx'$$

[17] See also §14.3.

$$+ \frac{\alpha p \omega_0^2}{24\pi} \int_{-\frac{L}{2}}^{\frac{L}{2}} \int_{-\frac{L}{2}}^{\frac{L}{2}} (\eta(x,t) + \eta(x',t))^3 \frac{\cos\alpha(x-x') - p}{1 - 2p\cos\alpha(x-x') + p^2} \, dx \, dx'$$

$$+ \frac{\alpha^2 \omega_0^2}{192\pi} \int_{-\frac{L}{2}}^{\frac{L}{2}} \int_{-\frac{L}{2}}^{\frac{L}{2}} \frac{(\eta(x,t) - \eta(x',t))^4}{1 - \cos\alpha(x-x')} \, dx \, dx'$$

$$- \frac{\alpha^2 \omega_0^2 p}{96\pi} \int_{-\frac{L}{2}}^{\frac{L}{2}} \int_{-\frac{L}{2}}^{\frac{L}{2}} (\eta(x,t) + \eta(x',t))^4 \frac{(1+p^2)\cos\alpha(x-x') - 2p}{(1 - 2p\cos\alpha(x-x') + p^2)^2} \, dx \, dx'.$$

9.6 Vortex pairs

The structure of a pair of equal and opposite counter-rotating vortices with circulations $\pm\Gamma$, moving steadily with speed V, was obtained by Deem and Zabusky [1978]. Pierrehumbert [1980] calculated the family of solutions as a function of $\theta = a/x_c$, where a is the equivalent radius of each vortex, and $2x_c$ is the distance between the centroids. For $\theta \ll 1$, the vortices are nearly circular and translate without change of shape at a speed $V_0 = \Gamma/4\pi x_c$. As θ increases and the gap between the vortices decreases, V/V_0 decreases from 1 to about 0.6. The aspect ratio (length/width) increases from 1 to 3.34. The limiting value of θ is 2.16.

The limit flow of two touching vortex patches exists according to numerical calculations. It was calculated by Sadovskii [1971], and later by Saffman and Tanveer [1982], who corrected an error in Pierrehumbert's formulation.[18] The speed of the pair is $V_c = 0.16\sqrt{S\omega^2}$, where S is the area of each vortex; the limiting value of $S/(\text{length})^2 = 0.22$.

Exact solutions in closed form for a pair of hollow or stagnant vortices were given by Pocklington [1895a] (see also Tanveer [1986]). These have the property that the aspect ratio goes to infinity as $\theta \to \infty$, there being in this case no finite limit with the vortices touching.

The stability of counter-rotating vortex patches to infinitesimal two-dimensional disturbances does not seem to have been studied in detail, but it is believed that they are stable. Keady's [1985] existence proof based on a variational principle shows that the speed of a pair of vortex patches is less than that of a pair of point vortices with the same circulation at the centroids of the patches.

The steady co-rotating pair of vortex patches was investigated by Saffman and Szeto [1980], who solved the equations of contour dynamics numerically. When far apart, the vortices are approximately circles. As the

---

[18] Pierrehumbert assumed the vortices were cusped at the ends of the axis of symmetry. In fact, the boundary meets the axis at right angles, but with infinite curvature (see §4).

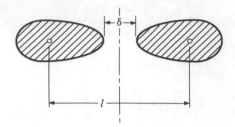

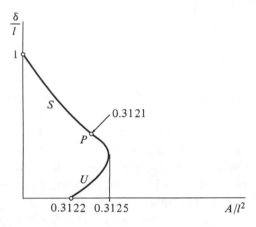

9.6-1 Properties of a counter-rotating pair.

distance between them decreases, the vortices become more deformed, until they finally touch. As A/l^2 increases from zero, where A is the area of each vortex and l is the distance between the centroids, the distance δ between the vortices decreases until the vortices touch when $A/l^2 = 0.3122$.[19] Figure 9.6-1 shows δ/l as a function of A/l^2. Note the existence of a maximum value of the area, greater than the critical value at which the vortices touch, which is not, however, associated in this case with a change of stability. The stability changes at a point P on the locus, with $A/l^2 = 0.3121$, for which the angular momentum is a minimum and the excess kinetic energy is a maximum.

[19] The quoted numbers are from the calculations of Kamm [1987] who employed the Schwarz function approach. The existence of a minimum separation of the centroids for a co-rotating pair to exist was demonstrated numerically from initial value calculations by Roberts and Christiansen [1972]. See also Christiansen and Zabusky [1973]. Fine et al. [1991] describe vortex merging experiments using the electron plasma analogue.

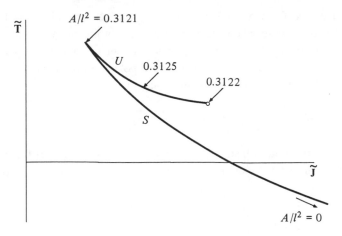

9.6-2 Energy and angular momentum for a co-rotating pair.

To see this, let H denote the angular impulse of the vortex pair. Define a dimensionless angular momentum by

$$\tilde{J} = -H/\Gamma A. \tag{1}$$

The actual kinetic energy is infinite, but we can define an excess kinetic energy T by

$$T = \iint \omega\psi\,dx\,dy, \tag{2}$$

where the stream function ψ is normalised by the requirement that $\psi \sim -\Gamma/\pi \log r + o(1)$ as $r \to \infty$. A dimensionless excess energy is defined by

$$\tilde{T} = T/\Gamma^2. \tag{3}$$

The dimensionless angular momentum is infinite when $A/l^2 = 0$, and decreases to a minimum when $A/l^2 = 0.3121$, and then increases until the vortices touch. A similar behaviour holds for $\tilde{T}$, except that it initially increases and then decreases. The minimum of $\tilde{J}$ and maximum of $\tilde{T}$ (see Figure 9.6-2) occur at the same value of A/l^2. This is a consequence of Kelvin's variational principle, and it also implies that there is a change of stability to two-dimensional infinitesimal disturbances at this value, since if the lower branch is a minimum, the upper branch is not, and hence is unstable. This is consistent with Kamm's calculations of linear stability of counter-rotating vortex pairs. It is expected that after the vortices touch, the shapes connect with a single dumb-bell form of simply connected vortices which arise from bifurcations of the Kirchhoff ellipses.

An exact closed form solution for touching vortices was given by Lamb [1932 §165], for a case when the vorticity is not uniform. Consider two-dimensional steady flow of speed U past a cylinder of radius a containing a vorticity distribution $\omega(r, \theta)$, where r, θ are polar co-ordinates centred in the cylinder. For $r > a$, the stream function ψ is

$$\psi = -U \left(r - \frac{a^2}{r} \right) \sin \theta. \tag{4}$$

If we have steady motion inside the cylinder, $\omega = -\nabla^2 \psi = f(\psi)$ for $r < a$, where f is an arbitrary function. Put $f(\psi) = -k^2 \psi$, where k is a constant. Then

$$\frac{\partial^2 \psi}{\partial r^2} + \frac{1}{r} \frac{\partial \psi}{\partial \theta^2} + \frac{1}{r^2} \frac{\partial^2 \psi}{\partial \theta^2} = -k^2 \psi. \tag{5}$$

This equation has the solution

$$\psi = C J_1(kr) \sin \theta, \qquad \omega = -C k^2 J_1(kr) \sin \theta. \tag{6}$$

Now choose C and k so that ψ and $\partial \psi / \partial r$ are continuous across $r = a$. This can be done if

$$J_1(ka) = 0, \qquad C = -2U/k J_0(ka). \tag{7}$$

Superposing a velocity U, we have a vortex propagating steadily through unbounded fluid. The impulse of the vortex is

$$\iint y \omega r \, dr \, d\theta = -C k^2 \pi \int_0^a r^2 J_1(kr) \, dr = 2\pi a^2 U. \tag{8}$$

D. Bliss (1970, private communication) has pointed out that Lamb's solution can be generalised to describe a touching vortex pair moving in a circle. Take, for $r < a$,

$$\psi = Ur \sin \theta + C J_1(kr) \sin \theta + B J_0(kr) - \frac{Dr^2}{4},$$
$$\omega = C k^2 J_1(kr) \sin \theta + B k^2 J_0(kr) + D. \tag{9}$$

Clearly, $\omega = -\nabla^2 \psi$. For $r > a$,

$$\omega = 0, \qquad \psi = \frac{Ua^2}{r} \sin \theta - \frac{Da^2}{2} \log r. \tag{10}$$

We now verify that this describes motion in which the centre of the circular boundary $r = a$ describes a circle of radius b with angular velocity Ω and speed $U = b\Omega$. First we note that from (9), ω is constant on $r = a$. Next, the velocity components $\partial \psi / \partial \theta$ and $\partial \psi / \partial r$ are continuous on $r = a$.

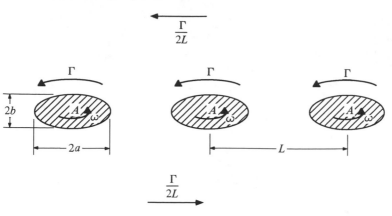

9.7-1 Geometry of the linear array.

Finally, we need to show that $D\omega/Dt = 0$. Now,

$$\frac{\partial r}{\partial t} = -U\cos\theta, \qquad \frac{\partial\theta}{\partial t} = \frac{U\sin\theta}{r} - \Omega, \tag{11}$$

for polar co-ordinates centred in the circle. Direct substitution shows that $D = \Omega$ satisfies the vorticity conservation equation. Note the curious feature that the solution is independent of B.

9.7   Arrays of vortex patches

Saffman and Szeto [1981] and Pierrehumbert and Widnall [1981] have investigated the properties of a linear array of patches, of equal areas A and circulations Γ lying in a straight line with centres distance L apart (see Figure 9.7-1). They carried out numerical integrations of the equations of contour dynamics. As A/L^2 increases from zero to a limiting value, the gap between the vortices decreases. Small patches are nearly circular, while large ones are like ellipses. The configuration is determined by $\alpha = A/L^2$, and the properties are sketched in Figure 9.7-2. These results assume that the patches have elliptical symmetry. Note that solutions with the assumed symmetry only exist for $\alpha < 0.2377$, but this value is a fold or limit point and is not the critical value at which the patches touch.[20] The existence of the fold implies the existence of neighbouring solutions with the same area, and hence a neutral disturbance of the equations for infinitesimal perturbations. Because the flow is inviscid and time reversible, the neutral

[20] The quoted numbers were obtained by Kamm [1987].

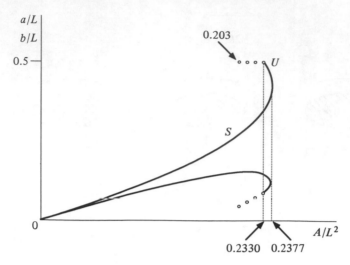

9.7-2 Sketch of properties of the linear array. Dotted line is family of connected
    vortices.

disturbance can be expected to be a collision of eigenvalues and therefore likely to be a transition of stability to instability for two-dimensional
disturbances with wavelength L. This is the superharmonic tearing-mode
instability identified by Moore and Saffman [1975a].

On the other hand, the Kelvin variational principle predicts a change of
stability when the excess kinetic energy is stationary. Thus we expect the
fold to be a point for which the kinetic energy as a function of α is a
maximum or a minimum. Saffman and Szeto in their numerical calculations found that the excess energy was a minimum where the area was a
maximum. Baker [1980] checked the agreement analytically for the linear
array of hollow vortices studied by Baker, Saffman and Sheffield [1976].

When the patches touch at the critical value, the family of solutions
continues into a family of connected vortices or finite-amplitude waves on
a vortex sheet of finite thickness. The critical wavelength for which a
sheet of uniform thickness $2b$ bifurcates into a family of waves is $9.83b$
(Rayleigh [1945 vol. II p. 396]). The shapes and properties of the families
are independent of Γ, which just fixes the time scale of the flow.

We now consider the stability of the configuration to two-dimensional
infinitesimal disturbances using a normal-mode analysis. This is the extension to vortices of finite area of the analysis of the stability of a row
of point vortices described in §7.5. The stream function $\Psi(x, y, t)$ of the

disturbed flow can be written

$$\Psi(x, y, t) = \Psi_0(x, y) + \epsilon e^{\sigma t} \psi(x, y) e^{ip2\pi x/L}, \tag{1}$$

by virtue of Floquet theory, where $\Psi_0(x, y)$ is the undisturbed stream function and is periodic in x; that is, $\Psi_0(x, y) \equiv \Psi_0(x + L, y)$, $\epsilon \ll 1$, $\psi(x, y)$ is the eigenfunction, σ is the eigenvalue and p is an arbitrary real number which can be called the disturbance wavenumber. The eigenfunction $\psi(x, y)$ is also of period L in x, and has the expansion

$$\psi(x, y) = \sum_n \psi_n(y) e^{i2\pi nx/L}. \tag{2}$$

If p is an integer, it can be taken to be zero without loss of generality, as this just corresponds to a relabelling of the eigenfunction components ψ_n. In this case, the disturbance has the same wavelength as the undisturbed flow, and is said to be superharmonic; all the vortices are deformed in the same way. Note that the superharmonic disturbance case is trivial in the limit of zero area.

If p is not an integer, we can without loss of generality take $0 < p < 1$. If p is rational, $p = M/N$ say, then every Nth vortex will repeat. The disturbance will now be referred to as subharmonic. A special case of particular interest is $p = \frac{1}{2}$. Every second vortex is deformed in the same way, the wavelength being $2L$, and this is referred to as the pairing instability. For vortices of small area, it is the most unstable disturbance.

Substituting (1) and (2) into the Euler equations and linearising in ϵ leads to a generalised eigenvalue problem for the value of σ, the solutions of which have the form

$$\sigma = \frac{\Gamma}{L^2} \, \mathrm{fn}\left(\frac{A}{L^2}, p\right), \tag{3}$$

where fn is a multivalued function with an infinity of branches when $A/L^2 > 0$ and the infinite sets of $\{\psi_n(y)\}$ describe the corresponding eigenfunctions.

When the undisturbed vorticity distribution is smooth so that Ψ_0 is also a smooth function, the generalised eigenvalue problem can be converted into matrix form by using, for example, a finite difference truncation of the Euler equations. Pierrehumbert and Widnall [1982] used this method to study the stability of the Stuart vortices (see below),[21] but it is not suitable for vortex patches. The method of contour dynamics can be used for

---

[21] Actually their main interest was the stability to three-dimensional disturbances for which this is the only practical method unless $A/L^2 \ll 1$.

superharmonic disturbances but is awkward for subharmonic disturbances unless p is the reciprocal of a small integer.[22]

The Schwarz function technique can be used for subharmonic perturbations of general subharmonic wave number p. The boundary of a patch is taken to be

$$z = a_0 \zeta \left(1 + \frac{a_1}{\zeta} + \frac{a_2}{\zeta^2} + \cdots \right), \qquad z^* = a_0^* \zeta \left(1 + a_1^* \zeta + a_2^* \zeta^2 + \cdots \right), \qquad (4)$$

with ζ on the unit circle. Now a_n and a_n^* are complex conjugates – that is, $\bar{a}_n = a_n^*$ – but we proceed as if they are independent variables. Then the Schwarz factorisation of $\bar{z}$ leads to expressions for the components u and v of the velocity induced by the patch

$$u - iv = -\tfrac{1}{2} i\omega \sum_0^\infty \frac{g_n}{z^n}, \qquad u + iv = \tfrac{1}{2} i\omega \sum_0^\infty \frac{g_n^*}{z^{*n}}, \qquad (5)$$

where the g_n and g_n^* are analytic functions of the a_n and a_n^*. The steady equilibrium shapes in a suitably translating frame will be described by sets of coefficients $\{A_n\}$ and $\{\bar{A}_n\}$.

An infinitesimal perturbation of the mth patch will have the form

$$a_n^{(m)}(t) = A_n + \epsilon b_n^{(m)} e^{\sigma t}, \qquad a_n^{*(m)}(t) = \bar{A}_n + \epsilon b_n^{*(m)} e^{\sigma t}. \qquad (6)$$

It is a consequence of the symmetry of the configuration that the perturbations of the different vortex patches are of the form

$$b_n^{(m)} = B_n e^{2\pi ipm}, \qquad b_n^{*(m)} = B_n^* e^{2\pi ipm}, \qquad (7)$$

where p is the subharmonic wave number. Substitution into the boundary condition (2.6), expressed as analytic constraints for the $\{a_n\}$ and $\{a_n^*\}$, for one of the patches, $m = 0$ say, leads to an infinite linear algebraic eigenvalue problem for the eigenvalue σ and eigenvector $\{B_n\}$, $\{B_n^*\}$. (For further details of the formulation, see Meiron, Saffman and Schatzman [1984] and Kamm [1987].) There is no problem with the fact that B_n and B_n^* are not complex conjugates of each other; this just corresponds to the eigenvalues of a real eigenvalue problem being complex numbers.

When the area A is zero, only subharmonic disturbances are relevant, and σ is given by equation (7.5.8). As the area increases, the dependence of σ on the area can be found by solving a truncation of the algebraic system numerically or by perturbation methods. See Jimenez [1987, 1988].

---

[22] Contour dynamics can be employed for numerical simulations of finite amplitude disturbances (see, e.g., Christiansen and Zabusky [1973]).

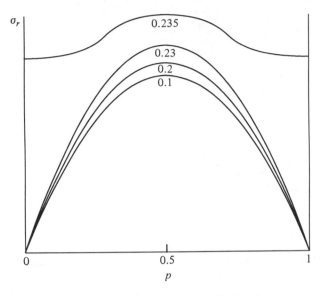

9.7-3 Instability of subharmonic disturbances. Sketch of growth rate of subharmonic disturbances for single rows versus subharmonic wavenumber for various areas (Kamm [1987]).

Figure 9.7-3 shows the dependence of the leading subharmonic growth rate on area. Curves are plotted for $\alpha = 0.10$, 0.20, 0.23 and for the top branch past the fold for $\alpha = 0.235$. Note that past the fold, the superharmonic perturbations ($p = 0$ or $p = 1$) are unstable. For this class of subharmonic perturbation, the $p = \frac{1}{2}$ pairing instability is most unstable. The pairing instability is stationary; that is, $\Im \sigma = 0$ and the disturbance does not propagate.

The effect of area on the oscillation frequency of superharmonic or shape oscillations is shown in Figure 9.7-4. For $A = 0$, these are the frequencies for an isolated circular vortex. As the area increases, there is a collision of eigenvalues for the lowest mode at the value corresponding to the fold. One can also consider superharmonic-subharmonic perturbations – for instance, shape oscillations in which every other vortex moves in the same way. The modifications are slight, however. The frequencies for $p = \frac{1}{2}$ are shown in the figure.

A similar analysis can be carried out for the Karman vortex street. Of particular interest is the effect of finite area on the subharmonic point vortex instability. It is not necessary to assume that the vortices in each row have the same area, but they must have equal and opposite circulations.

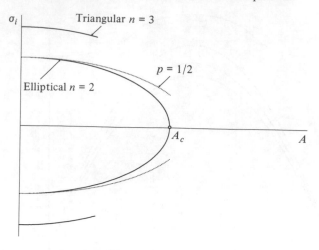

9.7-4 Dependence of oscillation frequency of vortex patches on area (Kamm [1987]).

Also, we restrict attention to the symmetrically staggered case, in which the centroid of each vortex patch lies midway between the centroids of the patches in the other row, and moreover each patch has fore-aft symmetry.[23] There are extra geometrical parameters and now the eigenfrequency will have the form

$$\sigma = \frac{\Gamma}{L^2} \, \text{fn} \left(p, \frac{A}{L^2}, \frac{H}{L}, \Delta \right), \tag{8}$$

where H is the distance between the rows and the vortex areas are $1 \pm \Delta A$. We call $k = H/L$ the aspect ratio. Again, p is the subharmonic wavenumber. The analysis for point vortices $(A = 0)$ was carried out in §7.6. The neutral boundary for point vortices, that is, the locus in the p-k plane for which $\Re\sigma = 0$, is shown in Figure 7.6-1. Note the cross, from which follows the conclusion that for any k, there exists a range of p for which the perturbation is unstable, except for the critical value $k_c = \sinh^{-1} 1 = 0.28055$, for which there is no unstable linear mode. Note also that the eigenvalue is symmetric about $p = \frac{1}{2}$; that is, $\sigma(p) = \sigma(1 - p)$. Meiron, Saffman and Schatzman [1984] were the first to show numerically that the cross-structure persists for finite area, except that there are two crosses and the symmetry about $p = \frac{1}{2}$ changes. Jimenez [1988] showed analytically

---

[23] The symmetric stagger is not necessary, although then the vortices do not move parallel to the rows. It is not known if rows exist without the fore-aft symmetry.

that the centre of the cross moves to

$$k_c = 0.28055 + 0.55536(1 + \Delta^2)A^2 + O(A^4), \tag{9}$$

$$p_c = 0.5 \pm 4.63089(1 + \Delta^2)A^2 + O(A^4). \tag{10}$$

These values are consistent with Meiron, Saffman and Schatzman's and Kamm's [1987] numerical calculations. Jimenez [1987] and MacKay [1987] have proved from the Hamiltonian structure of the motion that the point vortex cross unfolds in this manner, so that finite area neither stabilizes nor destabilizes the array.

Figure 9.7-5a shows the locus of the stability boundaries. Notice that for each k there is always a p for which the mode is unstable, except at the special value k_c, which is a function of A. Note further that the $p = \frac{1}{2}$ pairing mode is stabilized by finite area for k near k_c. This was found by Saffman and Schatzman [1982a]. The pairing mode is not the most unstable. Figure 9.7-5b shows the dependence of the growth rate on p for a given k.[24] Christiansen and Zabusky found by numerical integration of the contour dynamics equations that subharmonic finite-amplitude disturbances are stabilized by finite area.

Dritschel [1985] has calculated the equilibrium shapes and stability of equal vortex patches at the vertices of regular polygons. Dhanak [1992] has obtained asymptotic results using the elliptical vortex model.

Although not vortex patches, it is appropriate to mention here the Stuart [1967] vortices as a closed form solution for a smooth vorticity distribution which can model a linear array of finite cored vortices. In steady (two-dimensional, inviscid) flow, the stream function ψ and vorticity ω satisfy

$$\nabla^2 \psi = -\omega = F(\psi). \tag{11}$$

Stuart took the special case $F(\psi) = e^{-2\psi}$, for which (11) has the exact solution

$$\psi = \log(C \cosh y + \sqrt{C^2 - 1} \cos x). \tag{12}$$

The limit $C = 1$ gives $u = \psi_y = \tanh y$, which is a model of a homogeneous shear layer profile. The other limit $C \to \infty$ gives

$$\psi \sim \log C + \log(\cosh y + \cos x), \tag{13}$$

---

[24] Saffman and Schatzman claimed incorrectly that finite area stabilized the street because they assumed that the symmetry about $p = \frac{1}{2}$ of the point vortex case still held for $A > 0$, in which case the pairing mode would be the most unstable. Kida [1982] developed a perturbation expansion in area and found that the pairing mode was not the most unstable, but algebraic errors led him still to conclude that finite area stabilized.

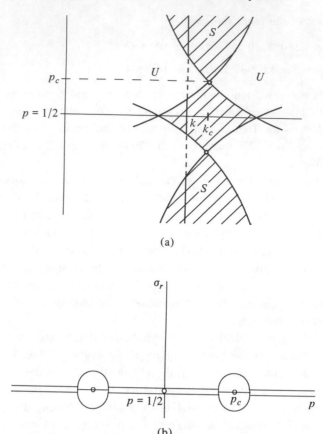

(a)

(b)

9.7-5 Stability properties of the vortex street. (a) Locus of stability boundaries for
Karman vortex street of finite area vortex patches. Stable areas are hatched.
(b) Variation of σ_r with p for a given k marked in (a).

which is the stream function for an array of point vortices with circulation
$\Gamma = -4\pi$ distance 2π apart. For C between ∞ and 1, the streamlines have
the cat's-eye pattern characteristic of periodic disturbances of a shear layer.
Pierrehumbert and Widnall [1982] calculated the linear stability of these
Stuart vortices to two- and three-dimensional disturbances. The results
for two-dimensional subharmonic perturbations are in qualitative agreement
with the results for patches. The pairing mode is always unstable and has
the highest growth rate. The behaviour of two-dimensional superharmonic

modes differs in that there is no superharmonic instability for large vortices.[25] That is, there is pairing but no tearing in the Stuart vortex model of the mixing layer.

---

[25] The absence of a transition is associated with the non-existence of a fold or limit point in the properties of the family of Stuart vortices as C decreases from ∞ to 1.

10

AXISYMMETRIC VORTEX RINGS

10.1 Formulation

The usual manifestation of an axisymmetric vortex is the vortex ring, commonly called a smoke ring. General conditions for the existence of steady axisymmetric flows were discussed in §3.13, and an exact solution (the Hill's spherical vortex) was described in §2.1. Here we give a different formulation aimed at the calculation of thin cored rings. We introduce cylindrical polars (y, ϕ, x), with y radial, x axial and ϕ azimuthal. The corresponding components of the velocity $\mathbf{q}$ are (v, w, u). See Figure 10.1-1. Axisymmetry requires that $\partial/\partial\phi = 0$. The fluid is supposed to be incompressible, and there exists a (Stokes) stream function $\psi(x, y, t)$ such that

$$u = \frac{1}{y}\frac{\partial\psi}{\partial y}, \qquad v = -\frac{1}{y}\frac{\partial\psi}{\partial x} \tag{1}$$

(see §3.13, where, however, the notation is different). Continuity or kinematics impose no constraints on the swirl velocity w.

The components of the vorticity $\boldsymbol{\omega}$ are

$$\left(-\frac{\partial w}{\partial x}, \frac{\partial v}{\partial x} - \frac{\partial u}{\partial y}, \frac{1}{y}\frac{\partial(wy)}{\partial y}\right) = (\omega_y, \omega_\phi, \omega_x). \tag{2}$$

In terms of the stream function, the azimuthal vorticity is

$$\omega_\phi = -\frac{1}{y}\frac{\partial^2\psi}{\partial x^2} - \frac{\partial}{\partial y}\left(\frac{1}{y}\frac{\partial\psi}{\partial y}\right). \tag{3}$$

Note that if $w = 0$, (i.e., there is no swirl), then the vector potential $\mathbf{A}$ is related to the stream function by

$$\mathbf{A} = \frac{\psi}{y}\,\hat{\phi}, \tag{4}$$

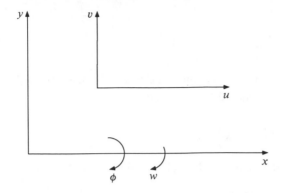

10.1-1 Co-ordinate system for axisymmetric vortex flows.

where $\hat{\phi}$ is a unit vector in the azimuthal direction.

Now equation (3) can be integrated to give the velocity in terms of the vorticity (see Equation (1.1.18)). Substituting into this equation

$$r = [(x - x')^2 + y^2 + y'^2 - 2yy' \cos(\phi - \phi')]^{\frac{1}{2}},$$
$$d\mathbf{x}' = y' \, dx' \, dy' \, d\phi', \tag{5}$$
$$\omega' = \omega'_{\phi} \hat{\phi}' \cos(\phi - \phi') + \omega'_{\phi} \hat{\mathbf{y}}' \sin(\phi - \phi')$$

and putting $\theta = \phi - \phi'$, we obtain for the stream function at the field point (x, y)

$$\psi(x, y) = \frac{1}{4\pi} \iint yy' \omega(x', y') dx' dy' \int_0^{2\pi} \frac{\cos \theta d\theta}{[(x - x')^2 + y^2 + y'^2 - 2yy' \cos \theta]^{\frac{1}{2}}}. \tag{6}$$

We can interpret equation (6) in terms of the stream function $G(x, y, \sigma)$ at (x, y) due to an infinitesimal circular vortex filament of radius σ and strength $dx \, dy$ lying in the plane $x = 0$. Then

$$G = \frac{y\sigma}{4\pi} \int_0^{2\pi} \frac{\cos \theta \, d\theta}{[x^2 + y^2 + \sigma^2 - 2y\sigma \cos \theta]^{\frac{1}{2}}}. \tag{7}$$

This integral can be expressed in terms of the complete elliptic integrals of the first and second kind of modulus k defined by

$$K(k) = \int_0^{\frac{1}{2}\pi} (1 - k^2 \sin^2 u)^{-\frac{1}{2}} \, du, \qquad E(k) = \int_0^{\frac{1}{2}\pi} (1 - k^2 \sin^2 u)^{\frac{1}{2}} \, du,$$

where

$$k^2 = \frac{4y\sigma}{[x^2 + (y+\sigma)^2]} = 1 - \frac{r_1^2}{r_2^2}.$$

Here r_1 and r_2 are the minimum and maximum distances from the field point to the circle. By standard transformations,

$$G = \frac{(y\sigma)^{\frac{1}{2}}}{2\pi} \left[\left(\frac{2}{k} - k \right) K(k) - \frac{2}{k} E(k) \right]. \tag{8}$$

Landen's transformation (Lamb [1932 §161]) gives an alternative form

$$G = \frac{\kappa}{2\pi} (r_1 + r_2) [K(\lambda) - E(\lambda)], \quad \text{where } \lambda = \frac{r_2 - r_1}{r_2 + r_1}. \tag{9}$$

Since a circular filament of circulation κ is equivalent to a surface distribution of dipoles of uniform strength κ filling the circle, separation of variables can be used to obtain the velocity potential and hence the stream function as integrals of Bessel functions (Lamb [1932 §161]), giving

$$G = \tfrac{1}{2} y\sigma \int_0^\infty e^{-\lambda x} J_1(\lambda y) J_1(\lambda \sigma) \, d\lambda, \tag{10}$$

for $x > 0$. This formulation does not appear to confer any advantages.

A question of interest is the existence of steady vortex rings in inviscid fluid. In the absence of swirl, the problem can be expressed as follows. Does there exist an area A, boundary ∂A, with $\omega_\phi = 0$ outside A, and a velocity U and vorticity distribution $\omega_\phi = yF(\Psi)$ inside A, such that the solution Ψ of the integral equation

$$\Psi(x, y) = -\tfrac{1}{2} Uy^2 + \iint\limits_A y' F(\Psi(x', y')) \, G(x - x', y, y') \, dx' \, dy' \tag{11}$$

is constant on ∂A? In general, ∂A is a vortex jump, and the automatic continuity of velocity implies continuity of pressure. The function F is arbitrary, and two limiting cases are of interest. The case $F \equiv$ constant is called the uniform vortex ring. If $F = 0$ inside A, the ring is hollow or stagnant; in this case the boundary will be a vortex sheet and F will be a delta function.

This mathematical problem was studied first by Lichtenstein [1925], but his treatment was incomplete and dynamically inconsistent as he assumed that $\omega_\phi =$ constant inside A was possible. Fraenkel [1970] proved existence for rings of sufficiently small core size, that is, for $A/\overline{R} \ll 1$, where $\overline{R}$ is the distance of the centroid of the area from the x-axis. Existence results for finite $A/\overline{R}$ have been obtained by Fraenkel and Berger [1974].

If there is swirl, then in addition

$$yw = C(\Psi), \qquad \omega_\phi = y F(\Psi) + \frac{C}{y} \frac{dC}{d\Psi} \qquad (12)$$

must also be satisfied for steady motion (see §3.13). No existence proofs appear to have been given yet for this case.

For rings in which the cross-section is convex, with a well defined boundary, Fraenkel [1972] and Norbury [1973] take the radius of the ring as $R = \frac{1}{2}(y_{min} + y_{max})$ and define a non-dimensional core size by $\epsilon = \sqrt{A/\pi R^2}$. For $A \to 0$, $\epsilon \sim a/R$ and $R \sim \bar{R}$, where a is the radius of the nearly circular core. With this normalisation, the Hill spherical vortex has $\epsilon = \sqrt{2}$. Fraenkel and Norbury have discussed by approximate analytical and numerical methods uniform vortex rings without swirl for $0 < \epsilon \leq \sqrt{2}$.

Norbury [1972] has considered the existence of solutions for $\sqrt{2} - \epsilon \ll 1$. These are Hill spherical vortices with a small hole.

10.2 Thin cored rings

Kelvin [1867b] (in a note added to Tait's [1867] translation of Helmholtz's [1858] paper) gave the formula without explanation for the speed U of a thin cored vortex ring of circulation Γ, ring radius R and core radius a ($a/R \ll 1$),

$$U = \frac{\Gamma}{4\pi R} \left[\log \frac{8R}{a} - \frac{1}{4} \right]. \qquad (1)$$

Hicks [1885] confirmed Kelvin's result for uniform thin vortex rings.[1] Hicks also calculated the speed of a hollow or stagnant thin vortex ring;

$$U = \frac{\Gamma}{4\pi R} \left[\log \frac{8R}{a} - \frac{1}{2} \right]. \qquad (2)$$

Dyson [1893] worked out the velocity of a uniform ring with error $O(\Gamma a^4/R^5 \log 8R/a)$ and the boundary deformation with error

$$O(a^4/R^4 \log R/a).$$

Fraenkel [1970] describes a systematic expansion scheme in ϵ for arbitrary $F(\Psi)$, but no swirl. The essence of the technique is as follows. In the frame of reference in which the motion is steady, let O be the stagnation point in A. This exists because Ψ is constant on ∂A and has at least one

---

[1] There was some controversy as other workers obtained formulae with the $-\frac{1}{4}$ replaced by -1.

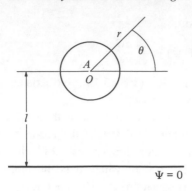

10.2-1 Co-ordinates for finite area vortex ring.

minimum which is a stagnation point. Take polar co-ordinates centred on O; $x = r \cos \theta$, $y = r \sin \theta + l$ (see Figure 10.2-1). Consider the unknown mapping

$$r = s + q(s, \theta, a), \quad 0 \le s \le a, \tag{3}$$

with the property that curves $s = $ constant map into $\Psi = $ constant. The boundary ∂A is $s = a$. Now $\omega = (y/l)\,\Omega(s)$ is supposed given; that is, $\Omega(s)$ is a known function, zero for $s > a$, but $\Omega(a) \ne 0$ in general, which is equivalent to the function $F(\Psi)$, which specifies the value of the vorticity on the streamlines in steady flow. Substitution into (1.11) gives the integro-differential equation

$$\Psi(s) = -\tfrac{1}{2} U y^2 + \int_0^a \int_0^{2\pi} (s' + q') \left(1 + \frac{\partial q'}{\partial s'} \right) \frac{y' \Omega'}{l} G(x - x', y, y') \, ds' \, d\theta',$$

$$\tag{4}$$

where primes denote evaluation at r' and θ', for $0 \le s \le a$, and $0 \le \theta \le 2\pi$. This equation can be solved as an expansion in $\epsilon = a/l$, since $q/s \to 0$ as $\epsilon \to 0$.

Using the expansion (9) below, Fraenkel obtained from (4) the result

$$U = \frac{\Gamma}{4\pi\overline{R}} \left[\log \frac{8\overline{R}}{a} - \frac{1}{2} + 2\pi^2 a^2 \overline{V^2} \right] \left\{ 1 + O\left[\left(\frac{a}{\overline{R}} \right)^2 \log \frac{\overline{R}}{a} \right] \right\}, \tag{5}$$

where V and $\overline{V^2}$ are defined by

$$\frac{1}{s} \frac{d(Vs)}{ds} = \Omega(s), \qquad \overline{V^2} = \frac{2}{a^2} \int_0^a s V^2 \, ds. \tag{6}$$

V is the tangential velocity around the core for a single rectilinear circular vortex. $\overline{R}$ and Γ are the radius of the centroid and the total circulation about the core. This result contains those of Kelvin, for which

$$\Omega = \frac{\Gamma}{\pi a^2}, \qquad V = \frac{\Gamma s}{2\pi a^2}, \qquad \overline{V^2} = \frac{\Gamma^2}{8\pi^2 a^2}, \tag{7}$$

and that of Hicks, for which $V = 0$ and the integral in equation (6) is zero.

For a uniform core (i.e., $\Omega = \text{constant}$), one can proceed directly. From the asymptotic properties of the complete elliptic integrals, we have

$$\left(\frac{2}{k} - k\right) K(k) - \frac{2}{k} E(k) \sim \log \frac{4}{k'} - 2 + O(k'^2 \log k') \tag{8}$$

as $k \to 1$, where $k'^2 = 1 - k^2$. k' is called the complementary modulus. Putting $x = r\cos\theta$, $y = \overline{R} + r\sin\theta$, one obtains, after some algebra,

$$G = \frac{\overline{R}}{2\pi} \left\{ \log\left(\frac{8\overline{R}}{[r^2 + r'^2 - 2rr'\cos(\theta - \theta')]^{\frac{1}{2}}}\right) \left[1 + \frac{r\sin\theta + r'\sin\theta'}{2\overline{R}}\right] \right.$$
$$\left. - 2 - \frac{1}{2\overline{R}}(r\sin\theta + r'\sin\theta') + O\left(\frac{r^2}{\overline{R}^2}\log\frac{\overline{R}}{r}\right) \right\}. \tag{9}$$

Now in order to evaluate the stream function, we need the distribution of vorticity. For a general core structure, this is difficult and Fraenkel's procedure or equivalent is needed. However, for a uniform core we know that

$$\omega = \omega_0 y/\overline{R} = \omega_0(1 + r/\overline{R} \sin\theta). \tag{10}$$

The question now is the shape. Properly, the radius should be expanded as a Fourier series in θ, but for brevity the result is anticipated and the boundary is assumed to be circular to leading order; that is, $r = a(1 + O(a^2/\overline{R}^2 \log \overline{R}/a))$. Remembering that the element of area $dx\,dy$ is $(1 + s/\overline{R} \sin\theta)\,r\,dr\,d\theta$, we obtain to leading order

$$\psi(r, \theta) = \frac{\overline{R}\omega_0}{2\pi} \int_0^a \int_0^{2\pi} \left\{ \log \frac{8\overline{R}}{[r^2 + r'^2 - 2rr'\cos(\theta - \theta')]^{\frac{1}{2}}} \right. \tag{11}$$
$$\left. \times \left[1 + \frac{r\sin\theta + 3r'\sin\theta'}{2\overline{R}}\right] - 2 - \frac{1}{2\overline{R}}(r\sin\theta + 5r'\sin\theta') \right\} r'\,dr'\,d\theta'.$$

To evaluate the integrals, we have

$$\int_0^{2\pi} \log\sqrt{r^2 + r'^2 - 2rr'\cos(\theta - \theta')}\,d\theta' = \begin{cases} 2\pi\log r' & \text{if } r' > r, \\ 2\pi\log r & \text{if } r > r', \end{cases}$$

and

$$\int_0^{2\pi} \cos(\theta - \theta') \log \sqrt{r^2 + r'^2 - 2rr' \cos(\theta - \theta')}d\theta'$$

$$= \begin{cases} -\pi r/r' & \text{if } r' > r, \\ -\pi r'/r & \text{if } r > r'. \end{cases}$$

This enables us to integrate (11), and defining $\Gamma = \pi a^2 \omega_0$ (it is the flux of vorticity through the core), we find that for $0 \le r \le a$ (the core being assumed circular),

$$\psi = \frac{\Gamma \overline{R}}{2\pi} \left\{ \log \frac{8\overline{R}}{a} - \frac{3}{2} - \frac{r^2}{2a^2} + \frac{r}{2\overline{R}} \sin \theta \left[\log \frac{8\overline{R}}{a} + 1 - \frac{5r^2}{4a^2} \right] \right.$$

$$\left. + O\left(\frac{a^2}{\overline{R}^2} \log \frac{\overline{R}}{a}\right) \right\}. \tag{12}$$

The leading-order approximation for G is insufficient to obtain (12) correct to $O(a/R)$.

In vortex fixed co-ordinates,

$$\Psi = -\frac{1}{2} U y^2 + \psi, \quad \text{where } y^2 = \overline{R}^2 \left(1 + \frac{2r}{\overline{R}} \sin \theta + O\left(\frac{a^2}{\overline{R}^2}\right) \right). \tag{13}$$

Thus $\Psi = $ constant on $r = a(1 + O(a^2/\overline{R}^2 \log(\overline{R}/a))$ (verifying that the core is circular to the order assumed) if the coefficient of $\sin \theta$ in (12) equals $Ua\overline{R}$, giving the Kelvin formula (1) with error $O(\Gamma a^2/\overline{R}^3 \log \overline{R}/a)$.

The interior streamlines in the core, in the vortex fixed frame, are given by $r = r_0 + \eta(\theta)$, where $\eta \ll r_0$ for $0 \le r_0 \le a$. Inserting the value of U into the expression for Ψ, we obtain for $r \le a$,

$$\Psi = \frac{\Gamma \overline{R}}{2\pi} \left\{ \frac{3}{4} \log \frac{\overline{R}}{a} - \frac{23}{16} - \frac{r^2}{2a^2} \right\} + \frac{\Gamma r \sin \theta}{4\pi} \left\{ \frac{5}{4} - \frac{5r^2}{4a^2} \right\} + o\left(\frac{a}{\overline{R}}\right). \tag{14}$$

Thus the equations of the streamlines (on which $\Psi = $ constant) are

$$\frac{r^2}{a^2} = \frac{5r}{4\overline{R}} \sin \theta \left(1 - \frac{r^2}{a^2} \right) + \frac{r_0^2}{a^2},$$

that is, to leading order,

$$r = r_0 + \frac{5}{8} \left(\frac{a^2 - r_0^2}{\overline{R}} \right) \sin \theta \quad \text{for } 0 \le r_0 \le a. \tag{15}$$

These curves constitute a family of non-concentric circles whose centres are a distance $5(a^2 - r_0^2)/8\overline{R}$ above the centre of the boundary. The stagnation point ($r_0 = 0$) is at a distance $\overline{R} + 5a^2/8\overline{R}$ from the axis.

The formulae for higher-order approximations worked out by Dyson (see also Fraenkel [1972]) can be expressed in terms of R and ϵ as defined at the end of the preceding section. The velocity is

$$U = \frac{\Gamma}{4\pi R} \left\{ \log \frac{8}{\epsilon} - \frac{1}{4} + \epsilon^2 \left[-\frac{3}{8} \log \frac{8}{\epsilon} + \frac{15}{32} \right] + O\left(\epsilon^4 \log \frac{8}{\epsilon} \right) \right\}.$$

(16)

The shape of the core in polar co-ordinates centred on $(R, 0)$ is

$$r = \epsilon R \left[1 - \epsilon^2 \left\{ \frac{3}{8} \log \frac{8}{\epsilon} - \frac{17}{32} \right\} \sin 2\theta \right.$$

$$\left. - \epsilon^3 \left\{ \frac{21}{128} \log \frac{8}{\epsilon} - \frac{273}{1024} \right\} (\sin \theta + \sin 3\theta) + \cdots \right].$$

(17)

The vortex is deformed to leading order into an ellipse with axis ratio

$$1 + \epsilon^2 \left\{ \frac{3}{8} \log \frac{8}{\epsilon} - \frac{17}{32} \right\},$$

(18)

flattened parallel to the axis. The stagnation point in the core is at $\theta = 0$ (which is immediate from the fore and aft symmetry) and

$$r = \frac{5}{8} \epsilon^2 + \epsilon^3 \left[-\frac{99}{128} \log \frac{8}{\epsilon} + \frac{739}{1024} \right].$$

(19)

The kinetic energy E and hydrodynamic impulse I (only the component parallel to the axis is non-zero) can be calculated from 3.11.7 and 3.2.8, respectively, which give

$$E = \pi \int \omega \psi \, dx \, dy = \frac{1}{2} \Gamma^2 R \left\{ \log \frac{8}{\epsilon} - \frac{7}{4} + \frac{3}{16} \epsilon^2 \log \frac{8}{\epsilon} + \cdots \right\},$$

(20)

and

$$I = \pi \int \omega y^2 \, dx, \, dy = \pi R^2 \Gamma \left\{ 1 + \frac{3}{4} \epsilon^2 + \cdots \right\}.$$

(21)

For the Hill spherical vortex, $R = \frac{1}{2} a$, $U = \Gamma/10R$, $I = 2\pi a^3 U$, $E = (10/7) \pi a^3 U^2$. The dimensionless ratios have values $UR/\Gamma = 0.1$, $E/\Gamma^2 R = 0.359$, $I/R^2\Gamma = 5.027$. The above formulae with $\epsilon = \sqrt{2}$ give the values 0.089, 0.316, 7.85, respectively. Norbury [1973] has numerically calculated the properties of fat uniform vortex rings as functions of ϵ.

Of some interest is the amount of irrotational fluid carried along by the ring. This is the volume bounded by the stream surface which contains

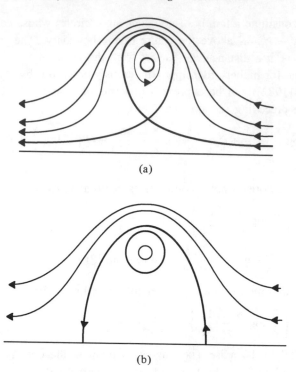

(a)

(b)

10.2-2 Sketch of relative streamlines for (a) thin and (b) fat vortex rings.

the external stagnation points (i.e., those outside the core) in vortex fixed co-ordinates. There are either two on the axis, for the case when the fluid on the axis is faster than the core and the shape of the fluid moving with the vortex is a closed oval, or just one between the core and the axis, in which case the co-travelling fluid is ring shaped. See Figure 10.2-2. An estimate for the core radius at which the changeover occurs can be found by the Biot–Savart formula. On $y = 0$, the velocity of the fluid obtained from the Biot–Savart formula which replaces the ring by a circular filament is $\Gamma R^2/2(x^2 + R^2)^{3/2}$. There is therefore a stagnation point in vortex fixed co-ordinates for a thin cored ring when x satisfies the equation

$$\frac{\Gamma R^2}{2(x^2 + R^2)^{3/2}} = \frac{\Gamma}{4\pi R} \left(\log \frac{8R}{a_e} - \frac{1}{4} \right), \tag{22}$$

where $a_e = a$ for a uniform vortex ring, $a_e = ae^{1/4}$ for a hollow or stagnant ring and a_e can be expressed in terms of the core structure for a general distribution using the formulae of the next section. This equation has no solution if $a_e/R < 1/86$, in which case fluid at the centre of the ring is not moving with it.

10.3 Lamb's transformation and general core structure

An alternative method of calculating the speed of thin cored vortex rings is suggested by the treatment given by Lamb [1932 §162]. Not only does this avoid the heavy algebra associated with the direct methods, but it allows relatively easy extension to deal with viscosity, swirl, unsteadiness and compressibility. Lamb's method was applied to non-uniform vorticity distributions by Saffman [1970].

Consider, first, a thin cored ring in fluid fixed co-ordinates, in which the velocity at infinity vanishes. The kinetic energy E is given by (3.11.3) and the impulse $\mathbf{I}$ by (3.2.8). Substitute into these expressions

$$\mathbf{u} = \mathbf{U} + \tilde{\mathbf{u}}, \tag{1}$$

where $\mathbf{U}$ is the velocity of the ring which needs to be defined precisely when considering viscosity and unsteadiness. We obtain

$$E = 2\mathbf{U} \cdot \mathbf{I} + \int \tilde{\mathbf{u}} \cdot (\mathbf{x} \times \boldsymbol{\omega}) \, dV. \tag{2}$$

For thin rings, we can calculate E and U with error $O(a/R)$ using only the leading-order approximation to the Green's function (2.9)

$$G(r, \theta, r', \theta') = \frac{R}{2\pi} \left[\log \frac{8R}{[r^2 + r'^2 - 2rr' \cos(\theta - \theta')]^{\frac{1}{2}}} - 2, \right] \tag{3}$$

where r and θ are polar co-ordinates about an origin in the core distance R from the axis. In the expression for the stream function as an integral over the vorticity in the core

$$\psi = \iint_A \omega'_\phi G \, dA', \tag{4}$$

we substitute (3) and

$$\omega_\phi = \omega_0(r), \tag{5}$$

assuming to leading order that the core is circular (with radius a_0) and the particle paths are circles about the centre, which gives inside the core

$$\psi \approx \psi_0(r) = R \int r' \omega_0(r') [\log 8R - 2 - \log r_>] \, dr', \tag{6}$$

where $r_> = r$ if $r > r'$ and is r' if $r < r'$. If we define $\Gamma_0(r)$ to be the circulation about circles of radius r, $\Gamma_0(a) = \Gamma$,

$$\Gamma_0(r) = \int_0^r 2\pi s \omega_0(s)\, ds, \tag{7}$$

we find (after some integration by parts) for $r < a$,

$$\psi_0 = \frac{R\Gamma}{2\pi} \left(\log \frac{8R}{a} - 2 \right) + \frac{R}{2\pi} \int_r^a \frac{\Gamma_0(s)}{s}\, ds. \tag{8}$$

Hence, to leading order

$$\pi \iint_A \omega_\phi \psi\, dA \approx \tfrac{1}{2} R\Gamma^2 \left[\log \frac{8R}{a} - 2 \right] + \tfrac{1}{2} R \int_0^a \frac{\Gamma_0(s)^2}{s}\, ds, \tag{9}$$

and the axial component of the impulse is

$$I = \pi \iint_A y^2 \omega_\phi\, dA \approx \pi \Gamma R^2. \tag{10}$$

Suppose now that in addition to the meridional flow, there is a swirl velocity w around the axis of the ring localized inside the core. Then the kinetic energy E is given exactly by

$$E = \pi \iint_A y(u^2 + v^2 + w^2)\, dx\, dy = \pi \iint_A \omega_\phi \psi\, dA + \pi \iint_A yw^2\, dA, \tag{11}$$

and the last term in (2) is

$$\int \tilde{\mathbf{u}} \cdot \mathbf{x} \times \boldsymbol{\omega}\, dV = 2\pi \iint_A yw_\phi(y\tilde{u} - x\tilde{v})\, dA + 2\pi \iint_A yw(x\omega_y - y\omega_x)\, dA. \tag{12}$$

Substituting the vorticity components given by (1.2) into the last term in (12), we obtain for this expression the value

$$\pi \iint_A yw^2\, dA - \pi \iint_A \left\{ \frac{\partial}{\partial x}(xyw^2) + \frac{\partial}{\partial y}(y^2w^2) \right\}\, dA. \tag{13}$$

The last term in (13) vanishes if w is zero at infinity, and we have, exactly, from (2)

$$\pi \iint_A \omega_\phi \psi\, dA = 2UI + 2\pi \iint_A yw_\phi(y\tilde{u} - x\tilde{v})\, dA. \tag{14}$$

Now the first two terms in (14) are given to leading order, that is, with error $O(a/R)$, by (9) and (10) as functions of $\omega_0(r)$ and the dimensions and strength of the ring. Note also that these terms are $O(R\Gamma^2)$. The problem is the last term. Since $\omega_\phi \sim \Gamma/a^2$, $\tilde{u} \sim \Gamma/a$, $y \sim R$, $dA \sim a^2$, the last term $\sim R^2\Gamma^2/a$, and must therefore vanish to leading order. However, this requires that ω_ϕ and $\tilde{u}$ be calculated correctly to $O(a/R)$, which is a non-trivial task.[2]

Lamb's transformation circumvents this difficulty. Suppose now that the motion is steady. Then U is well defined. The velocity components $\tilde{u}$, $\tilde{v}$ and azimuthal vorticity ω_ϕ satisfy the equations of steady motion relative to the vortex, cf (3.13.4),

$$\frac{\partial \tilde{u}}{\partial x} + \frac{1}{y}\frac{\partial y\tilde{v}}{\partial y} = 0, \qquad \tilde{u}\frac{\partial \omega_\phi}{\partial x} + \tilde{v}\frac{\partial \omega_\phi}{\partial y} = \frac{\tilde{v}\omega_\phi}{y} + \frac{1}{y}\frac{\partial w^2}{\partial x}. \tag{15}$$

It follows that

$$y^2\tilde{u}\omega_\phi - xy\tilde{v}\omega_\phi = yw^2 - 3xy\tilde{v}\omega_\phi - \frac{\partial}{\partial x}(w^2xy - xy^2\tilde{u}\omega_\phi) + \frac{\partial}{\partial y}(xy^2\tilde{v}\omega_\phi). \tag{16}$$

Hence,

$$2\pi \iint_A (y^2\tilde{u}\omega_\phi - xy\tilde{v}\omega_\phi)\, dA = -6\pi \iint_A xy\tilde{v}\omega_\phi\, dA + 2\pi \iint_A yw^2\, dA, \tag{17}$$

exactly for steady flow. The right-hand side is $O(\Gamma^2R)$ and can therefore be evaluated using the leading-order approximations $\omega_\phi = \omega_0(r)$, $\tilde{v} = \Gamma_0(r)/2\pi r \, \cos\theta$, $w = w_0(r)\, dA = 2\pi r\, dr$. This gives

$$2\pi \iint_A (y^2\tilde{u}\omega_\phi - xy\tilde{v}\omega_\phi)\, dA \approx -\frac{3}{4}R\Gamma^2 + 4\pi^2R\int_0^a rw_0^2\, dr. \tag{18}$$

(Shariff and Leonard (1992) note that Helmholtz (1858) gives the equations of Lamb's transformation.)

Substitution into (14) gives

$$\tfrac{1}{2}\Gamma^2R\left(\log\frac{8R}{a} - 2\right) + \tfrac{1}{2}R\int_0^a \frac{\Gamma_0(r)^2}{r}\, dr$$
$$= 2U\pi R^2\Gamma - \frac{3}{4}R\Gamma^2 + 4\pi^2R\int_0^a rw_0^2\, dr, \tag{19}$$

which gives

$$U = \frac{\Gamma}{4\pi R}\left(\log\frac{8R}{a} - \frac{1}{2} + 2\frac{\pi^2 a^2 \overline{v_\theta^2}}{\Gamma^2} - 4\frac{\pi^2 a^2 \overline{w_0^2}}{\Gamma^2}\right), \tag{20}$$

---

[2] It would, for instance, be necessary to extend Fraenkel's method (§2) to swirl.

where $v_\theta = \Gamma_0(r)/2\pi r$, and the overbar denotes the average over the cross-section. To the same order,

$$E = \tfrac{1}{2} R\Gamma^2 \left(\log \frac{8R}{a} - 2 + 2 \frac{\pi^2 a^2 \overline{v_\theta^2}}{\Gamma^2} + 2 \frac{\pi^2 a^2 \overline{w_0^2}}{\Gamma^2} \right). \tag{21}$$

The swirl velocity contribution to the speed of the ring was given by Widnall, Bliss and Zalay [1971], who derived it in a different way. Note that the swirl velocity slows down the ring, and in principle by having a sufficiently large swirl the ring could go backwards.[3] For a hollow or stagnant core, v_θ and w_0 are zero.

If the distribution of vorticity inside the core does not have a sharp boundary at $r = a$, the formulae (20) and (21) are not well defined. In this case, we replace expressions involving $\log a$ and v_θ by the combination

$$-\frac{\Gamma}{4\pi R} \log a + \frac{1}{4\pi\Gamma R} \int_0^a \frac{\Gamma_0(r)^2}{r} \, dr, \tag{22}$$

which is independent of a when a is sufficiently large that $\Gamma(a) = \Gamma$, the circulation of the core. The same combination is used for the energy. If we take the distribution of circulation corresponding to the Oseen–Lamb viscous vortex filament,

$$\Gamma(r,t) = \Gamma(1 - e^{-r^2/4\nu t}), \tag{23}$$

we obtain for the speed of a viscous vortex ring

$$U = \frac{\Gamma}{4\pi R} \left[\log \frac{8R}{\sqrt{4\nu t}} - 0.558 \right]. \tag{24}$$

However, this formula is at present meaningless as the motion is unsteady and U needs to be defined. It is also based on Lamb's transformation which was predicated on the assumption of steady flow. However, Saffman [1970] found that the Lamb transformation holds for unsteady axisymmetric flow, for which $\partial\omega_\phi/\partial t + U\partial\omega_\phi/\partial x$ is added to the left-hand side of the second equation in (15), provided U is defined by

$$U \equiv \frac{dX}{dt}, \quad \text{where } X = \tfrac{1}{2} \int \frac{\mathbf{x} \times \boldsymbol{\omega} \cdot \mathbf{I}}{l^2} \, x \, dV \tag{25}$$

is the position of the vortex centroid (§3.9). In other words, (24) gives the speed of the three-dimensional vortex centroid.

Moore [1980] has used the Lamb transformation method to calculate the speed of a vortex ring whose core is to leading order a Kirchhoff rotating

---

[3] Compare §3.13, where the same effect is described for Hill's spherical vortex with swirl.

ellipse with semi-axes a and b. The basic motion is then unsteady, and Moore finds that the average speed of the ring is

$$\bar{U} = \frac{\Gamma}{4\pi R} \left[\log \frac{16R}{a+b} - \frac{1}{4} \right].$$ (26)

Moore [1985] discusses a uniform vortex ring in compressible fluid and finds to lowest order in the Mach number that the speed is

$$U = \frac{\Gamma}{4\pi R} \left[\log \frac{8R}{a} - \frac{1}{4} - \frac{5M^2}{12} \right],$$ (27)

where $M = \Gamma/2\pi ac_\infty$, c_∞ being the speed of sound at infinity.

A physical explanation for the effect of swirl on the speed of a thin cored ring is provided by the concept of tension of a curved vortex filament.[4] When the filament is curved, the streamlines on the side towards the centre of curvature will be closer together and this will reduce the pressure. Similarly, the pressure on the other side will be increased. Such a force directed along the normal $\mathbf{n}$ to the filament is equivalent to a tension T_0 which produces an inward-directed force $T_0 \, ds \, \mathbf{n}/\rho$ on an element ds, where ρ is the radius of curvature. Suppose the element is moving with speed U relative to surrounding fluid. It has circulation Γ and hence is subject to a Kutta lift $\Gamma \mathbf{s} \times \mathbf{U} \, ds$, where $\mathbf{s}$ is the unit tangent vector. If the core is hollow or stagnant, these forces will balance to leading order, as the inertia and virtual mass effects are of smaller order. Thus

$$\frac{T_0}{R} \mathbf{n} + \Gamma \mathbf{s} \times \mathbf{U} = 0, \qquad \mathbf{U} = \frac{T_0}{\Gamma R} \mathbf{b},$$ (28)

where $\mathbf{b} = \mathbf{s} \times \mathbf{n}$ is the unit binomial. A comparison with Hick's result for the speed of a stagnant ring gives

$$T_0 = \frac{\Gamma^2}{4\pi} \left(\log \frac{8R}{a} - \frac{1}{2} \right).$$ (29)

This result can be confirmed using matched asymptotic expansions to calculate the pressure on the surface of the core (Moore and Saffman [1972]).

Now suppose the filament is not empty or stagnant and contains a rotational velocity $v_\theta(r)$. Then the pressure in the filament is reduced by centrifugal force, equivalent to an additional suction or tension force

$$-2\pi \int_0^a r(p - p(a)) \, dr = \pi \int_0^a r v_\theta^2 \, dr = \frac{1}{2} \pi a^2 \overline{v_\theta^2},$$ (30)

---

[4] S. Crow (1970 private communication). Workers in liquid Helium (e.g., Hall [1958]) have used the concept of tension of a vortex line.

since $\partial p/\partial r = -v_\theta^2(r)/r$ inside the core. If in addition there is an axial flow in the filament $w_0(r)$, this is equivalent to a compression or negative tension equal to the momentum flux $2\pi \int_0^a rw_0^2\,dr = \pi a^2\overline{w_0^2}$. Thus the internal structure is equivalent to replacing T_0 in (28) by

$$T_0 + \frac{1}{2}\pi a^2(\overline{v_\theta^2} - 2\overline{w_0^2}), \tag{31}$$

which gives (20).

10.4 Canonical co-ordinates for thin rings

Several authors have shown that the motion of an incompressible inviscid fluid is a Hamiltonian system and have given the Lie–Poisson brackets describing the motion on an appropriate manifold (see, e.g., Marsden and Weinstein [1983]). However, the problem of finding canonical co-ordinates for the Hamiltonian evolution has not been solved for general systems. For the case of thin rings, with uniform or stagnant core, Roberts and Donnelly [1970] have shown that the hydrodynamic impulse and position are canonical in the sense that

$$U = \frac{\partial E}{\partial I}, \tag{1}$$

where E is the energy. It will now be shown that this result remains true for arbitrary internal structure.

For a thin cored ring, E and I are given by (3.21) and (3.10). We consider variations which conserve volume and circulation in the following sense. Define $s = r/a$ in the core. Then the rotational velocity around the core is $v_\theta = \Gamma(s)/2\pi sa$, and the axial velocity is $w(s)$. Then $\delta\Gamma(s) = 0$ and $\delta(wR) = 0$ express conservation of circulation around the core and around the axis of the ring. In particular, $\delta\Gamma = 0$. Further from conservation of volume, $\delta(Ra^2) = 0$. We write the energy as

$$E = \tfrac{1}{2}R\Gamma^2\left(\log\frac{8R}{a} - 2\right) + \tfrac{1}{2}R\int_0^1 \frac{\Gamma(s)^2}{s}\,ds + 2\pi^2Ra^2\int_0^1 sw^2\,ds. \tag{2}$$

Then

$$\delta E = \tfrac{1}{2}\Gamma^2\left(\log\frac{8R}{a} - 2\right)\delta R + \tfrac{1}{2}R\Gamma^2\left(\frac{\delta R}{R} - \frac{\delta a}{a}\right) + \tfrac{1}{2}\delta R\int_0^1 \frac{\Gamma(s)^2}{s}\,ds$$

$$+ 2\pi^2(a^2\,\delta R + 2aR\,\delta a)\int_0^1 sw^2\,ds + 2\pi Ra^2\int_0^1 2sw\,\delta w\,ds. \tag{3}$$

Now $\delta a/a = -\frac{1}{2}\delta R/R$ and $\delta w/w = -\delta R/R$. Substituting into (2), we find

$$\frac{\delta E}{\delta R} = 2\pi\Gamma R U, \tag{4}$$

which gives (1) on using $\delta I = 2\pi\Gamma R\,\delta R$.

If the core is hollow (as could be caused by cavitation), then $v_\theta = 0$ and $w = 0$ and the pressure p in the core is constant. However, the volume of the core can change. To leading order,

$$p = p_\infty + \frac{\Gamma^2}{8\pi^2 a^2}, \tag{5}$$

where the second term is the hydrostatic pressure reduction due to the circulation about the core. Then $\delta p = 0$ implies that the core radius is constant; that is, $\delta a = 0$. If $V = 2\pi^2 R a^2$ denotes the volume of the core, $\delta V/V = \delta R/R$. For the hollow vortex, we take the energy to be the sum of the kinetic energy of a hollow core plus the potential energy of compression of the core, that is,

$$E = \frac{1}{2}R\Gamma^2\left(\log\frac{8R}{a} - 2\right) + (p_\infty - p)V. \tag{6}$$

Taking the variation, we have

$$\begin{aligned}
\delta E &= \frac{1}{2}\Gamma^2\left(\log\frac{8R}{a} - 2\right)\delta R + \frac{1}{2}R\Gamma^2\frac{\delta R}{R} + \frac{\Gamma^2}{8\pi^2 a^2}\delta V \\
&= \frac{1}{2}\Gamma^2\left(\log\frac{8R}{a} - \frac{1}{2}\right)\delta R = U\,\delta I,
\end{aligned} \tag{7}$$

where U is the speed of the hollow ring.

This Hamiltonian representation is not sufficient to study the general motion of vortex rings interacting with each other or with bodies because it relies on the rings remaining circular, and in general rings will deform as they interact. The formulation can be used to study leap-frogging of rings in which they remain co-axial and axisymmetric. For two rings centres X_1 and X_2, and radii R_1 and R_2, we have the equations

$$\dot{X}_1 = \frac{\partial H}{\partial I_1}, \qquad \dot{X}_2 = \frac{\partial H}{\partial I_2}, \qquad \dot{I}_1 = -\frac{\partial H}{\partial X_1}, \qquad \dot{I}_2 = -\frac{\partial H}{\partial X_2}, \tag{8}$$

where $I_1 = \pi\Gamma_1 R_1^2$, $I_2 = \pi\Gamma_2 R_2^2$. The Hamiltonian is the kinetic energy

$$H = T_1 + T_2 + T_{12}, \tag{9}$$

where T_1 and T_2 are the kinetic energies of the individual rings and T_{12} is the mutual kinetic energy $\pi\int \omega_1\psi_1\,dA_1 + \omega_2\psi_2\,dA_2$ which for thin rings takes the form

$$T_{12} = 2\pi\Gamma_1\Gamma_2 \, G(X_1 - X_2, R_1, R_2). \tag{10}$$

Experimental observations of leap-frogging are described by Oshima [1978] and Yamada and Matsui [1978].

11

DYNAMICS OF VORTEX FILAMENTS

11.1 Local induction approximation

The ring is a special case of a vortex filament of circular shape. If we wish to deal with the general interaction of vortex rings, a more general approach is needed, unless they are axisymmetric and co-axial.[1] For thin cored rings, the approach is via the Biot–Savart law (see §2.3), which gives the velocity induced by the filament as a line integral along the filament (see Equation (2.3.3)). If the filament has the parametric equation $\mathbf{r} = \mathbf{R}(\xi, t)$, where ξ is a Lagrangian co-ordinate, and t is the time, then the evolution of the filament would be given by $\partial \mathbf{R}/\partial t$ equals the induced velocity plus any external velocity evaluated on the filament, except that, as is well known, the Biot–Savart law diverges as the field point approaches the filament (see Equation (2.3.9)). The classical method of overcoming this problem (Thomson [1883]) is to employ a 'cut-off'. Thus at a station s on the filament, where the radius of the core is $a(s, t)$,[2] and $s = s(\xi)$ is the distance along the filament, the fluid velocity induced by the filament is taken to be

$$\mathbf{u}(s) = \frac{\Gamma}{4\pi} \int_{[\delta]} \mathbf{s}' \times \frac{\mathbf{R}(s) - \mathbf{R}(s')}{|\mathbf{R}(s) - \mathbf{R}(s')|^3} \, ds', \tag{1}$$

where $[\delta]$ denotes that the portion $-a\delta < s' - s < a\delta$ is omitted from the integral. It follows from 2.3.8, on putting $x = y = 0$, that the integrand in (1) behaves like

$$\frac{1}{2} \frac{\mathbf{b}}{\rho|s' - s|} + \text{bounded terms} \tag{2}$$

---

[1] Dyson [1893] extends his analysis to consider interacting rings and shows that uniform rings are stable to infinitesimal fluted disturbances

[2] There is no a priori reason why the core radius should be constant.

as $s' \to s$, where $\mathbf{b}$ is the binormal and ρ is the radius of curvature. Then to leading order,

$$\mathbf{u}(s) \sim \frac{\Gamma \mathbf{b}}{8\pi\rho} \int_{[\delta]} \frac{ds'}{|s' - s|} \sim \frac{-\Gamma}{4\pi} \log a\delta \, \frac{\mathbf{b}}{\rho}. \tag{3}$$

The error terms are bounded in the limit $a \to 0$, and δ is supposed to be $O(1)$. Then the leading approximation

$$\frac{\partial \mathbf{R}}{\partial t} = -\frac{\Gamma \log a}{4\pi} \frac{\mathbf{b}}{\rho}, \tag{4}$$

with a kept constant, is called the local (or localised) induction approximation.[3] Now the left-hand side of (4) is a partial derivative with respect to t with the Lagrangian parameter ξ kept constant, whereas the right-hand side is regarded as a function of s. To relate s and ξ, we use the kinematical result that

$$\frac{d}{dt} \log \left(\frac{ds}{d\xi} \right) = \mathbf{s} \cdot \frac{\partial \mathbf{u}}{\partial s}. \tag{5}$$

Since $\mathbf{s} \cdot \mathbf{b}$ and $\mathbf{s} \cdot d\mathbf{b}/ds = 0$ (see 2.3.2), it follows that $ds/d\xi = $ constant, and without loss of generality we can take $s = \xi$. Then we can write the evolution equation, with a rescaling of the time $t \to 4\pi t/(-\Gamma \log a)$, as

$$\frac{\partial \mathbf{R}}{\partial t} = \frac{\partial \mathbf{R}}{\partial s} \times \frac{\partial^2 \mathbf{R}}{\partial s^2}. \tag{6}$$

It follows from the above result for $ds/d\xi$ that the local induction approximation keeps the length of the filament constant.

Hasimoto [1972] elegantly transformed the equations of the local induction approximation into the cubic Schrödinger equation. Define

$$\psi(s, t) = \frac{1}{\rho} \exp \left(i \int^s \tau \, ds \right), \tag{7}$$

where τ is the torsion; then

$$\partial \psi / \partial t = i \left(\partial^2 \psi / \partial s^2 + \tfrac{1}{2} |\psi|^2 \psi \right). \tag{8}$$

This is the so-called cubic or non-linear Schrödinger equation, and, as is well known, its solutions are related to those of a corresponding linear problem and a variety of exact solutions are known. In particular, it possesses soliton solutions. This topic has been pursued by several authors (e.g., Lamb [1976]).

---

[3] The approximation was given by Da Rios [1906], and has been rediscovered several times. For historical reviews and further references, see Ricca [1991], Hama [1988] and Hasimoto [1988].

It can be verified that besides conserving length, the approximation conserves the linear and angular impulses

$$\mathbf{I} = \tfrac{1}{2}\Gamma \oint \mathbf{R} \times \mathbf{s}\, ds, \qquad \mathbf{A} = -\tfrac{1}{2}\Gamma R^2 \oint \mathbf{s}\, ds. \tag{9}$$

Note that conservation of impulse has a geometric interpretation, since $\oint \mathbf{R} \times \mathbf{s}\, ds = 2 \int d\mathbf{S}$; that is, it is twice the projected area of the surface bounded by the filament. A direct proof of the conservation is as follows:

$$
\begin{aligned}
\frac{d}{dt} \oint \mathbf{R} \times \mathbf{s}\, ds &= \oint \frac{\partial \mathbf{R}}{\partial t} \times ds + \oint \mathbf{R} \times \frac{\partial ds}{\partial t} \\
&= \oint \frac{\mathbf{b}}{\rho} \times ds + \oint \mathbf{R} \times \frac{\partial}{\partial s}\left(\frac{\mathbf{b}}{\rho}\right) ds \\
&= \oint \frac{\mathbf{b}}{\rho} \times ds + \oint \frac{\partial}{\partial s}\left(\mathbf{R} \times \frac{\mathbf{b}}{\rho}\right) ds - \oint ds \times \frac{\mathbf{b}}{\rho} \\
&= 2 \oint \frac{\mathbf{b}}{\rho} \times \mathbf{s}\, ds.
\end{aligned}
\tag{10}
$$

But $(\mathbf{b}/\rho) \times \mathbf{s} = \mathbf{n}/\rho = \partial \mathbf{s}/\partial s$ and $\oint (\partial \mathbf{s}/\partial s)\, ds = 0$. Hence $\mathbf{I} = $ constant.

An alternative approach to the change of length of the filament is to use the evolution equation in terms of the distance s, in the form

$$\mathbf{q} \equiv \frac{\partial \mathbf{R}(s,t)}{\partial t} = \mathbf{u}(s) + \alpha(s)\mathbf{s}, \tag{11}$$

where α is chosen to be an additional component of velocity along the filament such that the point moves with its distance from some origin on the filament kept constant. Choose an origin O such that $\mathbf{q}(0) \cdot \mathbf{s} = 0$, that is, $\alpha(0) = -(\mathbf{u} \cdot \mathbf{s})_0$. Now the condition that the length stays constant for the point moving with speed $\mathbf{q}$ is $\mathbf{s} \cdot \partial \mathbf{q}/\partial s = 0$. Further, $\partial \mathbf{q}/\partial s = \partial \mathbf{u}/\partial s + \alpha \mathbf{n}/\rho + \mathbf{s}\partial\alpha/\partial s$. Hence, $\partial\alpha/\partial s = -\mathbf{s} \cdot \partial \mathbf{u}/\partial s = -\partial(\mathbf{s} \cdot \mathbf{u})/\partial s + \mathbf{u} \cdot \mathbf{n}/\rho$. Integrating,

$$\alpha = -(\mathbf{u} \cdot \mathbf{s}) + (\mathbf{u} \cdot \mathbf{s})_0 + \int_0^s \frac{\mathbf{u} \cdot \mathbf{n}}{\rho}\, ds. \tag{12}$$

In particular, $\alpha = 0$ when $\mathbf{u} = \mathbf{b}/\rho$.

In addition to the soliton and cnoidal wave solutions of the Schrödinger equation, there are planar solutions in which the filament is of permanent form and lies in a rotating plane (Hasimoto [1971]). The condition for such solutions to exist with shape $y(x)$ is

$$\frac{\Gamma}{4\pi\rho} \log \frac{1}{a} = -\Omega y, \tag{13}$$

where Ω is the angular velocity of the plane. Notice that the filament rotates in the retrograde sense, that is, the direction of rotation of the filament is in the opposite sense to that of the core. The shapes are the solutions of the equation

$$\frac{y''}{(1+y'^2)^{3/2}} = k^2 y. \tag{14}$$

These are the elastica (e.g., Love [1927 p. 405]). Kida [1981b] gives a complete description of filaments of permanent form under the local induction approximation.

The approximation leads to elegant mathematics and is especially suitable for numerical integration, but it has serious flaws and is of limited physical validity. First, it is a crude approximation, requiring that $\log 1/a \gg 1$, which is unlikely to be satisfied for real vortex filaments.[4] Second, it is a dimensionally unsound approximation; clearly a needs to be scaled with some length characteristic of the entire filament. Third, it allows unphysical behaviour, such as filaments passing through one another.

The further objection that it is an approximation to the equations obtained from the cut-off hypothesis, which is itself an ad hoc approximation, is not valid since the cut-off hypothesis can be formally justified in the limit $a/\rho \to 0$ (Moore and Saffman [1972]) for filaments of arbitrary shape. This problem is examined in §4.

11.2   The cut-off method

We can verify the cut-off method and obtain an estimate for the parameter δ by examining the case of a thin cored vortex ring with arbitrary structure whose speed was calculated by Lamb's transformation in §10.3. The velocity of a ring of radius R is given by the cut-off formula to be

$$U = \frac{\Gamma}{4\pi} \int_{\frac{\delta a}{R}}^{2\pi - \frac{\delta a}{R}} \frac{R \sin \frac{1}{2}\theta}{4R^2 \sin^2 \frac{1}{2}\theta} \, d\theta = \frac{\Gamma}{4\pi R} \log \frac{4R}{\delta a}. \tag{1}$$

Comparison with the analytical result (10.3.20) shows there is agreement if the cut-off parameter is given by

$$\log 2\delta = \frac{1}{2} - \frac{2\pi^2 a^2 \overline{v_\theta^2}}{\Gamma^2} + \frac{4\pi^2 a^2 \overline{w_0^2}}{\Gamma^2}. \tag{2}$$

---

[4] This objection is perhaps least serious for liquid Helium in which classical vortex filaments are used to model the evolution of quantized vortex lines, and a is taken to be of the order of interatomic spacing, but the second and third objections still apply in this case.

The hypothesis of the cut-off method is that this form is valid for all filaments, at least to leading order. Note, however, that $\delta = \delta(a, v_\theta, w)$. The cut-off therefore depends on the local structure, and the analysis is incomplete until the local structure is found.[5] For the uniform core without axial flow or the hollow vortex

$$\delta = \tfrac{1}{2} e^{1/4}, \quad \text{or } \delta = \tfrac{1}{2} e^{1/2}, \tag{3}$$

respectively.

Rosenhead [1930] suggested that the cut-off formula be replaced by

$$\mathbf{u(R)} = \frac{\Gamma}{4\pi} \oint \frac{\mathbf{s'} \times (\mathbf{R} - \mathbf{R'})}{[(\mathbf{R} - \mathbf{R'})^2 + \mu^2 a^2]^{3/2}} \, ds'. \tag{4}$$

This form has advantages for numerical integration (Moore [1972]). If

$$\mu = 2\delta/e, \quad \log \mu = \log 2\delta - 1, \tag{5}$$

then the two forms are equivalent. Thomson [1883] used this form with $\mu = 1$, but this gives the speed of a ring incorrectly; for a uniform ring, it replaces the 1/4 in the Kelvin formula by 1.

An example of the use of the cut-off method is the helical vortex, which can be used as a model of propeller wakes or the wakes of spinning bodies. The parametric equation of a moving left-handed helix with axis in the z-direction is the curve

$$\mathbf{R} = D(\mathbf{i} \cos \theta + \mathbf{j} \sin \theta - \mathbf{k}(\theta + \sigma t)/\gamma). \tag{6}$$

This is a helix of radius D, axis $\mathbf{k}$ and pitch $1/\gamma$, rotating with period $2\pi/\sigma$. (Zero pitch, $\gamma = \infty$, is a circle; infinite pitch, $\gamma = 0$, is a straight line.) The radius of curvature is $\rho = D(1 + \gamma^2)/\gamma^2$. If a vortex filament of radius a is wrapped up into such a helix, the cut-off theory can be applied if

$$a \ll D/\gamma^2. \tag{7}$$

We take ds and $\mathbf{k}$ in the same sense, so that ds increases when $d\theta$ decreases. To calculate the velocity, we substitute into the cut-off formula

$$\mathbf{s} = \frac{\gamma}{\sqrt{1+\gamma^2}} (\sin \theta, -\cos \theta, 1/\gamma), \quad ds = -D \frac{\sqrt{1+\gamma^2}}{\gamma} \, d\theta,$$

$$\frac{\mathbf{n}}{\rho} = -\frac{\gamma^2}{D(1+\gamma^2)} (\cos \theta, \sin \theta, 0).$$

---

[5] Widnall, Bliss and Zalay [1971] obtained the result for nearly straight vortices using matched asymptotic expansions. An alternative method is given below in §4.

Without loss of generality, we can put $t = 0$ and evaluate the velocity at $\theta = 0$. From symmetry, $\mathbf{u} \cdot \mathbf{i} = 0$. The y and z components are given by

$$\mathbf{u} = -\frac{\Gamma\gamma^2}{4\pi D} \left\{ \mathbf{j} \int_{[]} \frac{(\theta \sin\theta - 1 + \cos\theta)}{[\theta^2 + 2\gamma^2(1 - \cos\theta)]^{3/2}} \, d\theta \right.$$
$$\left. + \mathbf{k}\gamma \int_{[]} \frac{(1 - \cos\theta)}{[\theta^2 + 2\gamma^2(1 - \cos\theta)]^{3/2}} \, d\theta \right\}. \tag{8}$$

The integrals run from $-\infty$ to ∞ and $[]$ implies that the contribution for $|\theta| < (a\delta/D)\gamma/\sqrt{1 + \gamma^2}$ is excluded.

We now specialise to the case of large pitch, that is, $\gamma \ll 1$. Neglecting terms $O(\gamma^2)$, we obtain

$$\mathbf{u} = -\frac{\Gamma\gamma^2}{2\pi D} \left\{ \mathbf{j} \int_{\frac{a\gamma\delta}{D}}^{\infty} \frac{\theta \sin\theta - 1 + \cos\theta)}{\theta^3} \, d\theta + \mathbf{k}\gamma \int_{\frac{a\gamma\delta}{D}}^{\infty} \frac{1 - \cos\theta)}{\theta^3} \, d\theta. \right\}$$

$$\tag{9}$$

These integrals can be expressed in terms of the cosine integral function Ci,

$$Ci(\theta) = -\int_{\theta}^{\infty} \frac{\cos\theta}{\theta} \, d\theta \sim C + \log\theta + O(\theta^2)$$

as $\theta \to 0$. C denotes Euler's constant 0.5772. Thus for

$$\frac{a\gamma\delta}{D} \ll 1, \tag{10}$$

$$\mathbf{u} = -\frac{\Gamma\gamma^2}{4\pi D} \left\{ \mathbf{j} \left(\frac{1}{2} - C - \log\left(\frac{a\gamma\delta}{D} \right) \right) \right.$$
$$\left. + \mathbf{k}\gamma \left(\frac{3}{2} - C - \log\left(\frac{a\gamma\delta}{D} \right) \right) \right\}. \tag{11}$$

We can now find σ. Any point on the curve moves normal to the curve with the normal component of $\mathbf{u}$; that is, $(\partial\mathbf{R}/\partial t - \mathbf{u}) \times \mathbf{s} = 0$. From (6), $\partial\mathbf{R}/\partial t = -D\sigma/\gamma \, \mathbf{k}$ and $\mathbf{s}$ is parallel to $-\mathbf{j} + \mathbf{k}/\gamma$ at $\theta = 0$. Hence

$$-(D\sigma/\gamma)\mathbf{k} - \mathbf{u} = \lambda(-\mathbf{j} + \mathbf{k}/\gamma) \tag{12}$$

for some λ. We take scalar products with $\mathbf{k}$ and $\mathbf{j}$ and eliminate λ, to give $D\sigma = -\gamma(\mathbf{u} \cdot \mathbf{k}) - (\mathbf{u} \cdot \mathbf{j})$. It follows from (11) that

$$\sigma = \frac{\Gamma\gamma^2}{4\pi D^2} \left[\log\frac{2D}{a\gamma} + \frac{1}{2} - C - \log 2\delta \right]. \tag{13}$$

Note that the condition (10) for validity implies also that $a \ll D/\gamma^2$, that is, $a \ll \rho$.

This result for σ with the cut-off parameter given by (2) can be checked against Kelvin's theory for the helical oscillations of a columnar vortex, which is discussed in the next section.

Moore [1972] has used the cut-off approximation to calculate the finite-amplitude oscillations of a counter-rotating pair of vortex filaments, which is a model of aircraft trailing vortices. Dhanak and de Bernardinis [1981] have studied the evolution of an elliptical ring. In both calculations, the filaments cross and the calculation then becomes invalid, and an ad hoc filament breaking and reconnection hypothesis is required to continue the calculation.[6]

11.3 Kelvin waves on a filament

Consider a straight circular vortex filament of radius a parallel to the z-axis. In cylindrical polars (r, θ, z), the velocity components are $(0, V(r), W(r))$. We consider infinitesimal perturbations (u_r, u_θ, u_z), which satisfy the Euler equations linearised about the basic flow and the equation of continuity with homogeneous boundary conditions. To proceed, we use the method of normal modes, searching for well-behaved (discrete proper) solutions of the form

$$u_r = u(r)e^{i(kz+m\theta+\sigma t)}, \quad u_\theta = v(r)e^{i(kz+m\theta+\sigma t)},$$
$$u_z = w(r)e^{i(kz+m\theta+\sigma t)}, \quad p = p(r)e^{i(kz+m\theta+\sigma t)}, \tag{1}$$

where p is the pressure. The axial wavenumber is $2\pi/k$, and m is the azimuthal wavenumber, which must be an integer. The possible σ are the eigenvalues (which constitute the discrete spectrum) and the corresponding u, v, w, p are the components of the eigenfunction. The boundary conditions are that these variables should tend to zero as $r \to \infty$, and should give analytic expressions (i.e., should be expressible as Taylor series in x, y, z) for $u_r \cos\theta - u_\theta \sin\theta$, $u_r \sin\theta + u_\theta \cos\theta$, u_z, p as $r \to 0$. Some calculation shows that this requires

$$u \sim \alpha r^{|m|-1}, \quad v \sim i\alpha r^{|m|-1}\operatorname{sgn} m, \quad w \sim \beta r^{|m|}, \quad p \sim \gamma r^{|m|}, \tag{2}$$

where α, β, γ are complex constants. If $m = 0$ (axisymmetric disturbances), then $\alpha = 0$. If V or W or their derivatives are discontinuous at $r = a$, where there is a vortex sheet or vortex jump, the disturbed flow satisfies

---

[6] Experimentally, this appears to happen. An attempt at a simple theory has been made by Saffman [1990].

the appropriate boundary condition on the displaced surface $(D \ll a)$

$$r = a + De^{i(kz+m\theta+\sigma t)}, \tag{3}$$

of continuity of normal velocity and pressure $(D \ll a)$. For a vortex jump, continuity of the three components of velocity suffices. These conditions will be expressed in the usual way as homogeneous equations relating the variables and their derivatives at $r = a$.

The question of determining if a general solution can be expressed as an infinite sum of such solutions is the completeness problem, which is in general difficult. If the discrete spectrum is not complete, then the general solution will contain integrals over the continuous spectrum of improper eigenfunctions (see, e.g., Drazin and Reid [1981]). If complex σ are found, then the vortex is unstable. If σ is real, the vortex is believed to be stable, at least to infinitesimal disturbances, and the behaviour of the contributions from the continuous spectrum is generally not regarded as significant as it is believed that they decay algebraically with time. This is, however, not necessarily the case. Rosenbluth and Smith [1990] present an example of a vortex filament whose continuous spectrum for two-dimensional disturbances gives perturbations which grow with time.

We will not pursue here the question of stability of filaments of arbitrary profiles, although this is an important problem.[7] Instead the discussion is confined to the relatively simple but important case of the uniform vortex with uniform axial flow for which $V = \Omega r$, $W = $ constant for $r < a$, and $V = \Omega a^2/r$, $W = 0$ for $r > a$. The analysis is lengthy but follows the treatments in Chandrasekhar [1961] and Drazin and Reid and will not be described here. The end result is the transcendental equation

$$\frac{(\sigma + m\Omega)^2}{4\Omega^2 - g^2} \left[\frac{\beta a J'_{|m|}(\beta a)}{J_{|m|}(\beta a)} + \frac{2\Omega m}{g} \right] = -a|k| \frac{K'_{|m|}(|k|a)}{K_{|m|}(|k|a)}, \tag{4}$$

where

$$g = \sigma + m\Omega + kW, \qquad \beta^2 = k^2(4\Omega^2 - g^2)/g^2. \tag{5}$$

The nature of the disturbance depends upon the value of m. If $m = 0$, the disturbance is axisymmetric, and we use the term 'sausaging mode' for the shape deformation. If $|m| \geq 2$, we refer to the disturbance as a fluted mode. Thus for $m = 2$, the cross-section is deformed into an ellipse which

---

[7] As remarked by Kelvin [1880], 'crowds of interesting cases present themselves.' For work on this problem, see Chandrasekhar [1961]. Lessen, Singh and Paillet [1974] have studied the stability of exponential profiles $V = \Gamma(1 - e^{-r^2})/2\pi r$, $W = 1 - \mu e^{-r^2}$.

rotates with z. In all these cases, $u = v = 0$ at $r = 0$, and the axis of the vortex is undisturbed.

The modes with $m = \pm 1$ are called the bending modes. In these, the axis of the vortex is deformed and the velocity perturbation does not vanish at $r = 0$. The deformed core is given by (3). Supposing that D is real and taking the real part, we have that it is a circle with centre at $r = D$ on the radius with $\theta = -(kz + \sigma t)/m$. For $m = 1$, this is a left-handed spiral with inverse pitch $\gamma = kD$, which has period $2\pi/\sigma$. Now if $\gamma \lll 1$, it is possible to satisfy $D \ll a$ (which is the condition for the Kelvin analysis to hold) and the condition $a\gamma/D \ll 1$, which is the condition for the cut-off analysis of §3 to hold. Thus, the value of σ given by (4) in the limit $ka \to 0$ should agree with that given by (2.13). Following Moore and Saffman [1972], we examine (4) in the long wave limit $ka \to 0$. We take $m = 1$ and $k > 0$. The right-hand side of (4) is $1 + k^2 a^2 (K - \frac{1}{4}) + O(k^4 a^4)$, where $K = \log(2/ka) - C + \frac{1}{4}$. There are still in this limit infinitely many modes, for as $g \to 0$, the quantity $\beta a \to \infty$, and passes through the infinity of roots of $J_1(\beta a)$, and a solution of the dispersion relation (4) can always be found with βa near one of these roots. However, these roots have frequencies $O(\Omega)$, and highly oscillatory radial dependence, and our interest is in a root for which $\sigma \ll \Omega$ and there is little radial structure. For this case, $\beta a \ll 1$ and we can use the expansion

$$\beta a\, J_1'(\beta a)/J_1(\beta a) = 1 - \frac{1}{4}\beta^2 a^2 - \frac{1}{96}\beta^4 a^4.$$

The dispersion relation can then be put into the form after some algebra

$$\frac{(\tilde{\sigma} + 1)^2}{1 - (\tilde{\sigma} + kaR)^2} = 1 + Kk^2 \tilde{a}^2 - k^3 a^3 R \frac{2\tilde{\sigma} + 2 + kaR}{4(\tilde{\sigma} + 1 + kaR)^2}, \tag{6}$$

where $\tilde{\sigma} = \sigma/\Omega$ and $R = W/\Omega a$ can be regarded as an internal Rossby number. If this number is of order unity, and this is the condition that swirl around the core and axial flow contribute equally to the motion of the filament, then the consistent root of (6) is

$$\tilde{\sigma} = \frac{1}{2} k^2 a^2 (K - R^2) - \frac{1}{2} k^3 a^3 R \left(K + \frac{1}{2}\right) + \frac{1}{2} k^3 a^3 R^3, \tag{7}$$

with error $O(k^4 a^4)$.

Comparison with (2.13) shows that there is agreement with the cut-off calculation if $\log 2\delta = R^2 + \frac{1}{4}$, which is the value predicted by (2.2).

We note for future reference (see §14.4) that the speed c of long axisymmetric waves ($ka \to 0$, $m = 0$) with zero axial velocity ($W = 0$) is given by

$$c = 2\Omega a/\xi, \quad \text{where } J_0(\xi) = 0. \tag{8}$$

11.4 Justification of the cut-off and higher-order approximations

We now consider another approach, based essentially on the local conservation of momentum, to calculate the equations of motion of a filament, and also justify the tension concept for a general filament using the matched asymptotic expansion technique. The treatment is based on the work of Moore and Saffman [1972]. Fukumoto and Miyazaki [1991] have treated the entire problem from the viewpoint of matched asymptotic expansions and confirmed the results.

The equation of the filament is taken to be $\mathbf{r} = \mathbf{R}(\xi, t)$, where the point $\xi = $ constant moves with the velocity of a fluid particle made precise as follows. The 'self-induced' velocity $\mathbf{V}_I$ is defined by

$$\mathbf{V}_I(\xi) = \frac{\Gamma}{4\pi} \int \left\{ \mathbf{s}' \times \frac{\mathbf{R}(\xi) - \mathbf{R}(\xi')}{|\mathbf{R}(\xi) - \mathbf{R}(\xi')|^3} \, ds' - \mathbf{s}_\odot \times \frac{\mathbf{R}(\xi) - \mathbf{R}_\odot}{|\mathbf{R}(\xi) - \mathbf{R}_\odot|^3} \, ds_\odot \right\}.$$

$$\tag{1}$$

The suffix $\odot$ refers to the osculating circle to the filament at station ξ which has equation $\mathbf{r} = \mathbf{R}_\odot(s_\odot)$. The explicit time dependence is suppressed. The velocity $\mathbf{V}_I$ is finite and well defined as the singularities in the integrands at ξ cancel. We suppose in the first instance that there are no other velocity fields and all motion is due to the self-induced velocity field. We can then define ξ unambiguously by the requirement that

$$\mathbf{V}_I(\xi) \cdot \mathbf{s} = \partial \mathbf{R}/\partial t \cdot \mathbf{s}. \tag{2}$$

The advantage of using momentum balance considerations is that less detailed knowledge of the flow field is required than in a direct attack on the problem through the vorticity equation. We have a relative motion of fluid and filament, which we denote by

$$\mathbf{V}_I - \partial \mathbf{R}/\partial t = \mathbf{Q}(\xi). \tag{3}$$

By construction, $\mathbf{Q}$ is perpendicular to $\mathbf{s}$. Consider an element $d\xi$ of the filament. We argue that the relative velocity gives rise to a force per unit length on the filament equivalent to the Kutta lift $\Gamma(\mathbf{V}_I - \partial \mathbf{R}/\partial t) \times \mathbf{s}$, which is $O(\Gamma^2/\rho)$.[8] In addition, there are apparent mass acceleration terms of order $\partial/\partial t\, \pi a^2(\mathbf{V}_I - \partial\mathbf{R}/\partial t) = O(\Gamma^2 a^2/\rho^3)$. It is anticipated here that all motion of the filament is due to its self-induced velocity and $\partial/\partial t = O(\Gamma/\rho^2)$. In addition, there is force due to the curvature of the filament (cf §10.3) which we write as $T_0\,\mathbf{n}/\rho$. Then we assert that this force is comparable to the Kutta lift, this being the basic balance of forces which controls the speed of the filament, and that the force per unit length on the surface of the filament is

$$\Gamma\left(\mathbf{V}_I - \frac{\partial\mathbf{R}}{\partial t}\right) \times \mathbf{s} + T_0\,\frac{\mathbf{n}}{\rho}. \tag{4}$$

We now provide formal justification of this result and calculate T_0 from the Euler equations, showing that it has the value (10.3.29) deduced heuristically.

Introduce a local co-ordinate system with unit vectors $\mathbf{i}, \mathbf{j}$ in the plane of $\mathbf{n}$ and $\mathbf{b}$. For a point P, choose station ξ as the foot of the perpendicular from P to the axis of the filament. The position vector of P is $\mathbf{R}(\xi) + x\mathbf{i} + y\mathbf{j}$. Let χ be the angle between $\mathbf{i}$ and $\mathbf{n}$. Then if χ varies so that $\partial\psi/\partial s = \tau$ (the torsion), the co-ordinate system is orthogonal with metric $h_s = 1 - (x\cos\chi - y\sin\chi)/\rho = 1 - r\cos(\theta + \chi)/\rho$, $h_x = 1$, $h_y = 1$. Here, $r = \sqrt{(x^2 + y^2)}$ and $\theta = \tan^{-1}(y/x)$ are local polar co-ordinates in the plane $s = $ constant.

Outside the core, there is no vorticity and a velocity potential ϕ exists. Expand $\phi = \phi_0 + \phi_1 + \cdots$ in orders of a/ρ, where $\phi_0 = \Gamma\theta/2\pi$. Substitute into Laplace's equation, and scale $r \sim a$, $s \sim \rho$, and obtain the equation for ϕ_1,

$$\frac{\partial^2\phi_1}{\partial r^2} + \frac{1}{r}\frac{\partial\phi_1}{\partial r} + \frac{1}{r^2}\frac{\partial^2\phi_1}{\partial\theta^2} = -\frac{\Gamma}{2\pi\rho r}\sin(\theta + \chi). \tag{5}$$

The surface of the core is circular to leading order and is expanded $r = a_0(s) + a_1(s, \theta) + \cdots$. The velocity of the centre is $\partial\mathbf{R}/\partial t$. Then we have the inner boundary condition on $r = a_0$,

$$\frac{\partial\phi_1}{\partial r} = \frac{\partial\mathbf{R}}{\partial t} \cdot (\mathbf{n}\cos(\theta + \chi) + \mathbf{b}\sin(\theta + \chi)) + \frac{\Gamma}{2\pi a_0^2}\frac{\partial a_1}{\partial\theta}. \tag{6}$$

We need an outer boundary condition on ϕ_1 as $r \to \infty$. For this purpose, we employ a matching condition that the expansions match as $r \to \infty$ the

---

[8] There are logarithmic contributions to the orders of magnitude, but these will not be given explicitly. In practice, they are never large, except possibly in the quantized vortex line model of liquid Helium II.

expression resulting from the Biot–Savart integral as $r/\rho \to 0$. It follows from (1) that the outer limit asymptotes

$$\mathbf{V}_I + \frac{\Gamma}{2\pi} \oint \mathbf{s}_\odot \times \frac{\mathbf{r} - \mathbf{R}_\odot}{|\mathbf{r} - \mathbf{R}_\odot|^3} \, ds_\odot = \mathbf{V}_I + \frac{\Gamma}{2\pi r^2} (X\mathbf{b} - Y\mathbf{n}) \tag{7}$$

$$+ \frac{\Gamma}{4\pi\rho} \left[\frac{-XY}{r^2} \mathbf{n} - \frac{Y^2}{r^2} \mathbf{b} + \mathbf{b} \log \left(\frac{8\rho}{r} \right) \right] + O \left(\frac{\Gamma r}{\rho^2} \right),$$

where $\mathbf{r} = x\mathbf{i} + y\mathbf{j} = X\mathbf{n} + Y\mathbf{b}$, on using the result of (2.3.9). Hence as $r \to \infty$,

$$\nabla\phi_1 \sim \mathbf{V}_I + \frac{\Gamma}{4\pi\rho} \left[\frac{-XY}{r^2} \mathbf{n} - \frac{Y^2}{r^2} \mathbf{b} + \mathbf{b} \log \left(\frac{8\rho}{r} \right) \right]. \tag{8}$$

ϕ_1 is now determinate and is

$$\phi_1 = \mathbf{V}_I \cdot \mathbf{r} - \frac{a_0}{r^2} \left(\frac{\partial \mathbf{R}}{\partial t} - \mathbf{V}_I \right) \cdot \mathbf{r} + \frac{\Gamma Y}{4\pi\rho} \log \frac{8\rho}{r}$$

$$+ \frac{\Gamma Y a_0^2}{4\pi\rho r^2} \left(\log \frac{8\rho}{a_0} - 1 \right) + \int \mathbf{V}_I \cdot d\mathbf{s} + \phi_1^{\text{def}}. \tag{9}$$

The last term arises from $a_1(s, \theta)$.

The matched asymptotic expansion method now solves inside the core to the same accuracy and matches the pressure or the other components of velocity. This gives $\partial\mathbf{R}/\partial t$. However, it is laborious because the interior flow is not irrotational and the modifications to the vorticity field must be found. The force balance method avoids this work (cf. Lamb's transformation for the ring §10.3).

We now calculate the pressure from Bernoulli's equation, keeping in mind that the co-ordinate system is not inertial and is both translating with a speed $O(V_I)$ and rotating with an angular velocity $O(V_I/\rho)$. But to leading order in a/ρ, only the translation with speed $\partial\mathbf{R}/\partial t$ need be retained, and we have for the pressure p,

$$p + \frac{1}{2} (\nabla\phi_0)^2 + \nabla\phi_0 \cdot \nabla\phi_1 - \frac{\partial \mathbf{R}}{\partial t} \cdot \nabla\phi_0 = O \left(\frac{\Gamma^2}{\rho^2} \right). \tag{10}$$

There is no net force from the deformation $a_1(s, \theta)$, and we can ignore it and ϕ_1^{def}. Then on $r = a$ (henceforth, we drop the suffix $_0$),

$$p = \frac{\Gamma}{\pi a^2} \left(\mathbf{V}_I - \frac{\partial \mathbf{R}}{\partial t} \right) \cdot (Y\mathbf{n} - X\mathbf{b})$$

$$- \frac{\Gamma^2 X}{4\pi^2 \rho a^2} \left(\log \frac{8\rho}{a} - \frac{1}{2} \right) - \frac{\Gamma^2}{8\pi^2 a^2} + O \left(\frac{\Gamma^2}{\rho^2} \right). \tag{11}$$

Integrating this pressure over the curved surface gives a force/unit length

$$
\mathbf{F}_E = \Gamma \left(\mathbf{V}_I - \frac{\partial \mathbf{R}}{\partial t} \right) \times \mathbf{s} + \frac{\Gamma^2}{4\pi\rho} \mathbf{n} \left(\log \frac{8\rho}{a} - \frac{1}{2} \right) - \frac{\Gamma^2}{8\pi a^2} \frac{\partial a^2}{\partial s} \mathbf{s} - \frac{\Gamma^2}{8\pi\rho} \mathbf{n}.
$$

(12)

Comparison with (4) shows that the hypothesis of a Kutta force and a vortex tension is verified with the tension T_0 given by the value for the hollow ring (10.3.29) on writing $R = \rho$. The last two terms in (12) are the contributions from the uniform pressure reduction $-\Gamma^2/8\pi^2 a^2$ caused by the swirl over the surface of the vortex and are intuitively obvious. (These terms will be partly cancelled by the forces transmitted in the internal structure.) The error in (12) is of interest. By the method of derivation, it appears that it is $O(\Gamma^2 a/\rho^2)$, but actually it is smaller and is $O(\Gamma^2 a^2/\rho^3)$. This can be verified by direct calculation, but can be seen more easily by a symmetry argument, since the force must be invariant under the change $\mathbf{n} \to -\mathbf{n}$, $\rho \to -\rho$.

Hollow vortices

Consider now the case of a hollow but incompressible vortex or a vortex with a massless core. Then the pressure inside the core will to leading order be independent of s, and it follows that $a = a(t)$. The core size is then determined by

$$
a^2 L = \text{constant},
$$

(13)

where L is the length of the filament.[9]   The force balance equation determining the motion of the filament is

$$
\mathbf{F}_E + \mathbf{F}_I = 0,
$$

(14)

where $\mathbf{F}_E$ is given by (12) and $\mathbf{F}_I$ is the momentum transmitted through the ends of the element $d\xi$ of the filament. We will suppose additionally that there is an interfacial tension γ at the surface. Then the pressure inside the core is to leading order

$$
p_I = -\frac{\Gamma^2}{8\pi^2 a^2} + \frac{\gamma}{a}.
$$

(15)

---

[9]   If the core contained vapor, the radius would be determined by a condition that the internal pressure equals the vapor pressure and volume would not be conserved.

Then

$$\mathbf{F}_I = \frac{\partial}{\partial s}(2\pi a\gamma - \pi a^2 p_I)\mathbf{s} = \left(\frac{\Gamma^2}{8\pi} + \pi a\gamma\right)\frac{\mathbf{n}}{\rho}. \tag{16}$$

Combining (12), (14) and (16), we have

$$\frac{\partial \mathbf{R}}{\partial t} = \mathbf{V}_I + \frac{1}{\Gamma}(T_0 + \pi a\gamma)\frac{\mathbf{b}}{\rho}, \tag{17}$$

which gives the equation of motion of the hollow filament. Note that (2) removes the arbitrary component parallel to **s** allowed by (12).

This is the cut-off approximation when $\gamma = 0$. For by construction,

$$\left.\frac{\partial \mathbf{R}}{\partial t}\right|_{\text{cut off}} = \mathbf{V}_I + \frac{\Gamma}{4\pi}\int_{[\delta]} ds_{\odot} \times \frac{\mathbf{R} - \mathbf{R}_{\odot}}{|\mathbf{R} - \mathbf{R}_{\odot}|^3}$$

$$= \mathbf{V}_I + \frac{\Gamma}{4\pi\rho}\left(\log\frac{8\rho}{a} - \log 2\delta\right)\mathbf{b}. \tag{18}$$

Taking the hollow vortex cut-off value $\delta = \frac{1}{2}e^{1/2}$ given by (2.2) shows the equivalence. Further, interfacial tension gives

$$\log 2\delta = \frac{1}{2} - \frac{4\pi^2 a\gamma}{\Gamma^2}, \tag{19}$$

which is equivalent to replacing T_0 by $T_0 + \pi a\gamma$.[10]

The length L of the vortex required to determine the core radius a can be found as follows. We have

$$L = \oint ds = \oint \frac{\partial s}{\partial \xi}\, d\xi.$$

Then

$$\frac{dL}{dt} = \oint \frac{\partial}{\partial t}\left(\log\frac{\partial s}{\partial \xi}\right) ds = -\oint \frac{\partial \mathbf{R}}{\partial t}\cdot\frac{\mathbf{n}}{\rho}\, ds, \tag{20}$$

because

$$\frac{\partial}{\partial s}\frac{\partial \mathbf{R}}{\partial t} = \frac{\partial \xi}{\partial s}\frac{\partial}{\partial \xi}\left(\frac{\partial \mathbf{R}}{\partial t}\right) = \frac{\partial \xi}{\partial s}\frac{\partial}{\partial t}\left(\frac{\partial \mathbf{R}}{\partial \xi}\right)$$

$$= \frac{\partial \xi}{\partial s}\frac{\partial}{\partial t}\left(\mathbf{s}\frac{\partial s}{\partial \xi}\right) = \frac{\partial \mathbf{s}}{\partial t} + \mathbf{s}\frac{\partial}{\partial t}\left(\log\frac{\partial s}{\partial \xi}\right). \tag{21}$$

---

[10] The additional term is not $2\pi a\gamma$, which is γ times circumference, because there is also a contribution from the pressure reduction γ/a inside the core acting over the cross-section area πa^2.

Take the scalar product with **s** (remembering that $\mathbf{s} \cdot \partial \mathbf{s}/\partial t = 0$ because $s^2 = 1$) to give

$$\frac{\partial}{\partial t} \log \frac{\partial s}{\partial \xi} = \mathbf{s} \cdot \frac{\partial}{\partial s} \frac{\partial \mathbf{R}}{\partial t} = \frac{\partial}{\partial s} \left(\mathbf{s} \cdot \frac{\partial \mathbf{R}}{\partial t} \right) - \frac{\mathbf{n}}{\rho} \cdot \frac{\partial \mathbf{R}}{\partial t}. \tag{22}$$

The expression (20) follows on noting that $\oint \partial/\partial s(\mathbf{s} \cdot \partial \mathbf{R}/\partial t)\,ds = 0$.

If the vortex is in an external irrotational velocity field $\mathbf{U}_E(\mathbf{r}, t)$, we define $\mathbf{V}_E(\xi) = \mathbf{U}_E(\mathbf{R}, t)$, and simply add $\mathbf{V}_E$ to $\mathbf{V}_I$ wherever it occurs in the formulae.

Vortices with internal structure

We proceed in the same way but calculate in addition the effect of internal structure on the pressure and momentum fluxes that determine $\mathbf{F}_I$. Consider pressure first. To leading order, $\partial p/\partial r = v_\theta^2/r$ in the core and $p = -\Gamma^2/8\pi^2 a^2$ on $r = a$. Then the pressure force is

$$-\int_0^a 2\pi r p \, dr = \pi \int_0^a r v_\theta^2 \, dr + \frac{\Gamma^2}{8\pi}. \tag{23}$$

Thus the tension is increased in effect by $\frac{1}{2}\pi a^2 \overline{v_\theta^2}$, a term whose order of magnitude is Γ^2.

Now consider the momentum flux or Reynolds stress. The velocity in the core is

$$w\mathbf{s} + \mathbf{u}_\perp + \partial \mathbf{R}/\partial t, \tag{24}$$

where w is defined to be the axial velocity relative to a point moving with constant ξ and $\mathbf{u}_\perp$ is the velocity in the plane of a cross-section. These terms can be of order $O(\Gamma/a)$. The last term is smaller and is $O(\Gamma/\rho)$. The Reynolds stress contribution is

$$-\int_{\text{core}} w \left(w\mathbf{s} + \mathbf{u}_\perp + \frac{\partial \mathbf{R}}{\partial t} \right) dA = -\pi a^2 \overline{w^2} \, \mathbf{s} - \pi a^2 \overline{w} \frac{\partial \mathbf{R}}{\partial t} - \int w \mathbf{u}_\perp \, dA. \tag{25}$$

The first and second terms on the right-hand side of (25) are of order $O(\Gamma^2)$ and $O(\Gamma^2 a/\rho)$, respectively. The last term is zero if the vortex filament is straight, and because of curvature effects is expected to be of order $O(\Gamma^2 a/\rho)$; its calculation therefore requires the $O(a/\rho)$ corrections to the internal velocity field. However, it can be shown, using a method like Lamb's transformation (see Moore and Saffman [1972, Appx. A]) that to leading order

$$\int w\mathbf{u}_\perp \, dA = \frac{2\pi \mathbf{b}}{\rho} \int_0^a r^2 v_\theta w_0 \, dr = \lambda \Gamma \overline{w_0} a^2 \frac{\mathbf{b}}{\rho}, \tag{26}$$

where $\lambda = \frac{1}{4}$ for the uniform vortex.

Combining (23), (25) and (26), we have

$$
\begin{aligned}
\mathbf{F}_I &= \frac{\partial}{\partial s}\left[\frac{\Gamma^2}{8\pi}\mathbf{s} + \left(\frac{1}{2}\pi a^2 \overline{v_\theta^2} - \pi a^2 \overline{w_0^2}\right)\mathbf{s} - \pi a^2 \overline{w} \frac{\partial \mathbf{R}}{\partial t} - \lambda \Gamma \overline{w} a^2 \frac{\mathbf{b}}{\rho}\right] \\
&= \left(\frac{\Gamma^2}{8\pi} + \frac{1}{2}\pi a^2 \overline{v_\theta^2} - \pi a^2 \overline{w_0^2}\right)\frac{\mathbf{n}}{\rho} + \frac{\partial}{\partial s}\left(\frac{1}{2}\pi a^2 \overline{v_\theta^2} - \pi a^2 \overline{w_0^2}\right)\mathbf{s} \\
&\quad - \pi a^2 \overline{w_0} \frac{\partial \mathbf{s}}{\partial t} - \pi a^2 \overline{w_0}\mathbf{s}\frac{\partial}{\partial t}\left(\log \frac{\partial \mathbf{s}}{\partial \xi}\right) \\
&\quad + \Gamma \lambda \tau \overline{w_0} \frac{\mathbf{n}}{\rho} - \Gamma \overline{w_0} a^2 \frac{\partial}{\partial s}\left(\frac{\lambda}{\rho}\right)\mathbf{b},
\end{aligned}
\tag{27}
$$

where we have used (22) and anticipated (see (32) below) that $\partial(a^2 \overline{w})/\partial s = O(\gamma a^2/\rho^2)$, the terms neglected in (27) being $O(\Gamma^2 a^2/\rho^3)$.

Now momentum conservation implies that the sum of the external and internal forces must balance the rate of change of momentum of fluid inside the core; that is,

$$ds\,(\mathbf{F}_E + \mathbf{F}_I) = d\xi\,\frac{\partial}{\partial t}\left[\left(\pi a^2 \frac{\partial \mathbf{R}}{\partial t} + \pi a^2 \overline{w}\mathbf{s}\right)\frac{\partial \mathbf{s}}{\partial \xi}\right]. \tag{28}$$

Hence, neglecting the $\partial \mathbf{R}/\partial t$ term as being of smaller order, since $\partial/\partial t = O(\Gamma/\rho^2)$,

$$\mathbf{F}_E + \mathbf{F}_I = \pi a^2 \overline{w}\frac{\partial \mathbf{s}}{\partial t} + \mathbf{s}\frac{\partial}{\partial t}(\pi a^2 \overline{w}) + \pi a^2 \overline{w}\mathbf{s}\frac{\partial}{\partial t}\log\frac{\partial \mathbf{s}}{\partial \xi}. \tag{29}$$

From (12), (27) and (29), we have from the components perpendicular to $\mathbf{s}$,

$$
\begin{aligned}
\Gamma\left(\mathbf{V}_I - \frac{\partial \mathbf{R}}{\partial t}\right) \times \mathbf{s} + T\frac{\mathbf{n}}{\rho} &= 2\pi a^2 \overline{w_0}\frac{\partial \mathbf{s}}{\partial t} + \Gamma \overline{w_0}\, a^2 \frac{\partial}{\partial s}\left(\frac{\lambda}{\rho}\right)\mathbf{n} \\
&\quad - \Gamma \lambda \tau \overline{w_0}\, a^2 \frac{\mathbf{n}}{\rho},
\end{aligned}
\tag{30}
$$

where

$$T = T_0 + \tfrac{1}{2}\pi a^2 \overline{v_\theta^2} - \pi a^2 \overline{w_0^2}. \tag{31}$$

The terms on the left-hand side of (30) are of order $O(\Gamma^2/\rho)$ and those on the right-hand side are $O(\Gamma^2 a/\rho^2)$. As demonstrated above for the hollow vortex (see (18)), neglect of the right-hand side gives the cut-off approximation.

However, the velocity given by (30) is a higher approximation. A check is provided by calculating the rotation rate ω of a large pitch helix with a uniform core and comparing it with the result obtained from Kelvin's small perturbation theory and given in (3.7). For a uniform core of radius a, rotating with angular velocity Ω, and with constant axial velocity W,

$$\lambda = \frac{1}{4}, \quad \Gamma = 2\pi\Omega a^2, \quad \overline{v_\theta^2} = \frac{1}{2}\Omega^2 a^2, \quad , \overline{w_0} = W, \quad \overline{w_0^2} = W^2.$$

If the radius of the spiral is D, and the radius of curvature ρ is $(Dk^2)^{-1}$, the torsion is $-k$ and $\partial\mathbf{R}/\partial t = \omega D\mathbf{b}$, $\partial\mathbf{s}/\partial t = k\omega D\mathbf{n}$. Direct calculation (§2) gives

$$\mathbf{V}_I = \frac{\Gamma}{4\pi} k^2 D \left(K - \log\frac{8\rho}{a} + \frac{1}{4} \right) \mathbf{b},$$

where K is given in §3. Substitution into (30) gives a value of ω/Ω which agrees exactly with (3.7).

However for general motion, equation (30) is not complete since the internal structure needs to be determined. There are two remaining equations, namely the equation of continuity and the equation for the longitudinal momentum balance, that is, the $\mathbf{s}$ component of (28).[11] The former gives

$$\frac{\partial}{\partial t}\left(a^2 \frac{\partial s}{\partial \xi} \right) + \frac{\partial}{\partial \xi}(\overline{w_0}a^2) = 0, \tag{32}$$

and the latter is

$$\frac{\partial}{\partial t}(\pi a^2 \overline{w_0}) = -\frac{\Gamma^2}{8\pi a^2}\frac{\partial u^2}{\partial s} + \frac{\partial}{\partial s}\left(\frac{1}{2}\pi a^2 \overline{v_\theta^2} - \pi a^2 \overline{w_0^2} \right)$$
$$- 2\pi a^2 \overline{w_0}\frac{\partial}{\partial t}\left(\log\frac{\partial s}{\partial \xi} \right). \tag{33}$$

Moore and Saffman [1972] argued on the basis of invariance under changes in the sign of the curvature that spatial variations of the core radius are of relative order $O(a^2/\rho^2)$, and that to leading order changes in the core radius are negligible; that is,

$$a = a(t). \tag{34}$$

---

[11] The following two equations with assumptions about the velocity profiles can be used to study the propagation of axisynimetric disturbances to a straight filament. However, these propagate with a fast time scale a^2/Γ (see §12.1). Our concern here is variations on the slow time scale ρ^2/Γ caused by variations of the filament's axis.

The following discussion based on this result is consistent with equations (32) and (33). Conservation of volume of the core then leads to the conclusion

$$La^2 = \text{constant},\tag{35}$$

where L is the total length of the filament. It is a consequence that the leading-order swirl velocity v_θ is independent of s and depends only on r and t, and by conservation of circulation

$$v_\theta = \frac{\Gamma}{2\pi r}\tilde{\Gamma}\left(\frac{r}{a}\right),\qquad \tilde{\Gamma}(1) = 1,\tag{36}$$

where $\tilde{\Gamma}$ is determined by the initial structure of the filament. Note that

$$a^2\overline{v_\theta^2} = \frac{\Gamma^2}{8\pi^2}\mu \quad\text{where } \mu = 4\int_0^1 \frac{1}{\eta}\tilde{\Gamma}^2(\eta)\,d\eta.\tag{37}$$

The quantity μ is constant throughout the motion and has the value 1 for uniform rotation.

The axial velocity w_0 is more difficult. Uniform stretching of the vortex preserves the axial velocity profile. On the other hand, external pressure gradients give axial accelerations uniform through the core. In addition, the average axial velocity will vary with position to order $O(\Gamma a/\rho)$ because of variations of $V_\| = \mathbf{s}\cdot\partial\mathbf{R}/\partial t$. Thus we have

$$w_0 = W(t) + q(\xi,t) + \frac{\Gamma}{b}\chi\left(\frac{r}{a}\right),\tag{38}$$

where we suppose without loss of generality that

$$\int_0^1 \eta\chi(\eta)\,d\eta = 0.\tag{39}$$

The length b is a constant $O(a)$ determined by the initial conditions. Then

$$\overline{w_0} = W + q,\qquad a^2\overline{w_0^2} = a^2W^2 + \frac{a^2\Gamma^2}{b^2}\nu + 2a^2Wq + O\left(\frac{\Gamma^2a^2}{\rho^2}\right),\tag{40}$$

where $\nu = 2\int_0^1 \eta\chi^2(\eta)\,d\eta$ is a constant determined by the initial conditions.

The conservation of core volume gives the variable velocity q. Substituting (38) into the continuity equation (32) and using (35), we obtain

$$\begin{aligned}\frac{\partial q}{\partial s} &= -\frac{\partial}{\partial t}\left(\log\frac{\partial s}{\partial \xi}\right) + \frac{1}{L}\frac{dL}{dt}\\[2mm] &= -\frac{\partial V_\|}{\partial s} + \frac{\partial\mathbf{R}}{\partial t}\cdot\frac{\mathbf{n}}{\rho} + \frac{1}{L}\frac{dL}{dt}\quad\text{on using (22),}\end{aligned}\tag{41}$$

where it is to be noted that

$$L = \oint ds, \quad \frac{dL}{dt} = \oint \frac{\partial}{\partial t} \left(\log \frac{\partial s}{\partial \xi} \right) ds = - \oint \frac{\partial \mathbf{R}}{\partial t} \cdot \frac{\mathbf{n}}{\rho} \, ds.$$

Equation (40) determines q in terms of the filament velocity but is arbitrary to the extent of a function of t. This can be fixed by redefining W so that

$$\oint q \, ds = 0. \tag{42}$$

As is seen from (45) below, this requirement is consistent with the conservation of circulation around the contour defined by the filament axis.

It remains to determine the mean axial velocity $W(t)$. Substituting into the axial momentum balance equation (33), we obtain

$$-2\pi a^2 W \frac{\partial q}{\partial s} - 2\pi a^2 W \frac{\partial}{\partial t} \left(\log \frac{\partial s}{\partial \xi} \right) - \pi \frac{d}{dt}(a^2 W) = 0, \tag{43}$$

and substituting for $\partial q / \partial s$ from (40), we obtain

$$\frac{2a^2 W}{L} \frac{dL}{dt} + \frac{d}{dt}(a^2 W) = 0. \tag{44}$$

Hence,

$$Wa^2 L^2 = \text{constant}, \quad \text{or} \quad WL = \text{constant}, \tag{45}$$

which can be interpreted as either conservation of angular impulse or conservation of circulation.

The remaining parameter of internal structure is λ, see (26). We find that

$$\lambda = \int_0^1 \left(\eta \tilde{\Gamma}(\eta) + \eta \frac{\Gamma}{Wb} \chi(\eta) \tilde{\Gamma}(\eta) \right) d\eta + O\left(\frac{a}{\rho} \right). \tag{46}$$

Thus λ is constant along the filament, but varies with time because of the dependence on W.

We conclude that the pertinent properties of the internal structure are determined by the initial conditions and L. The length L is itself determined by the equation of motion which describes how the vortex moves, so the equations are closed. They are, however, implicit in their present form, but they can be made explicit as follows to the same order of accuracy.

Using the results for the structure, we can write the equation of motion (30) as

$$\Gamma \left(\mathbf{V}_I - \frac{\partial \mathbf{R}}{\partial t} \right) \times \mathbf{s} + T \frac{\mathbf{n}}{\rho} = 2\pi a^2 W \frac{\partial \mathbf{s}}{\partial t} + 2\pi a^2 W q \frac{\mathbf{n}}{\rho}$$

$$+ \lambda \Gamma W a^2 \frac{\partial}{\partial s} \left(\frac{\mathbf{b}}{\rho} \right) + O \left(\frac{\Gamma^2 a^2}{\rho^3} \right), \tag{47}$$

where

$$T = T_0 + \frac{\Gamma^2 \mu}{16\pi} - \pi a^2 W^2 - \frac{\pi a^2 \Gamma^2 \nu}{b^2}. \tag{48}$$

Taking the vector product with **s**, we obtain

$$\frac{\partial \mathbf{R}}{\partial t} - \mathbf{V}_I - \frac{T}{\Gamma \rho} \mathbf{s} = -2\pi a^2 W \mathbf{s} \times \frac{\partial \mathbf{s}}{\partial t} - 2\pi a^2 W q \frac{\mathbf{b}}{\rho} - \lambda \Gamma W a^2 \, \mathbf{s} \times \frac{\partial}{\partial s} \left(\frac{\mathbf{b}}{\rho} \right). \tag{49}$$

Neglecting the right-hand side of (49) gives the leading approximation

$$\frac{\partial \mathbf{R}_0}{\partial t} = \mathbf{V}_I + \frac{T}{\rho} \mathbf{b}, \tag{50}$$

which is the cut-off approximation with

$$\log 2\delta = \frac{1}{2} + \left[\frac{\mu}{4} - \frac{4\pi^2 a^2 W^2}{\Gamma^2} - \frac{4\pi^2 a^2 \nu}{b^2} \right]. \tag{51}$$

Note that the cut off parameter is a function of t.

To a higher approximation, we evaluate $\partial \mathbf{s}/\partial t$ and q from (21) and (43) using the leading approximation (50) for $\partial \mathbf{R}/\partial t$. From (21) we obtain

$$\mathbf{s} \times \frac{\partial \mathbf{s}}{\partial t} = \mathbf{s} \times \frac{\partial}{\partial s} \left(\mathbf{V}_I + \frac{T}{\Gamma \rho} \mathbf{b} \right). \tag{52}$$

From (43), we have

$$q = \int^s \left(\frac{\partial}{\partial s} (\mathbf{V}_I \cdot \mathbf{s}) - \frac{\mathbf{V}_I \cdot \mathbf{n}}{\rho} \right) ds + \frac{s}{L} \frac{dL}{dt} = \int^s \mathbf{s} \cdot \frac{\partial \mathbf{V}_I}{\partial s} ds + \frac{s}{L} \frac{dL}{dt}. \tag{53}$$

The absence of a lower limit of integration indicates that an arbitrary function of t is to be added to ensure that (42) is satisfied.

The velocity $\mathbf{V}_I$ is well defined and (49) with the subsidiary equations provides closed equations for the evolution of the filament. In the existence of an external velocity field, one simply adds the external field $\mathbf{V}_E(\xi)$ to $\mathbf{V}_I$.

Moore and Saffman [1972] show that the linear and angular impulses, **I** and **A**, are given by

$$I = \tfrac{1}{2}\Gamma \oint R \times s \, ds + O(\Gamma a^2)$$

$$A = -\tfrac{1}{2}\Gamma \oint R^2 s \, ds + \pi W a^2 \oint R \times s \, ds + O(\Gamma a^2 \rho). \tag{54}$$

They verify that when $\partial R/\partial t$ is given by (49) with the external velocity added, then

$$d\mathbf{I}/dt = \Gamma \oint V_E \times s \, ds, \qquad d\mathbf{A}/dt = \Gamma \oint R \times (V_E \times s) \, ds. \tag{55}$$

An expression can also be given for the contribution to the kinetic energy of an isolated filament as a line integral around the filament. The energy density per unit length of filament is

$$E = T_0 - \frac{3\Gamma^2}{8\pi} - \Gamma(R \times (V_I + V_E) \cdot s) + \frac{\Gamma^2}{4\pi} \frac{s \cdot R}{\rho} \frac{\partial \rho}{\partial s}$$

$$+ \tfrac{1}{2}\pi a^2 (\overline{v_\theta^2} + \overline{w_0^2}) + O\left(\frac{\Gamma^2 a}{\rho}\right). \tag{56}$$

Conservation of impulse and energy can provide a useful test of the accuracy of numerical investigation.

12

THREE-DIMENSIONAL VORTEX INSTABILITY

12.1 Oscillations of a vortex column

We now describe some properties of the infinitesimal perturbations (Kelvin waves) of a uniform rectilinear vortex filament. The dispersion relation (11.3.4) for disturbances of axial wavenumber k and azimuthal wavenumber m to a filament of radius a in solid body rotation Ω with uniform axial velocity W was examined in §11.3 for the case $m = 1$ and $ka \ll 1$ in order to compare with the predictions of the cut-off approximation for long waves. The interest here is for $ka \approx 1$, when the axial wavelength of the disturbance is comparable with or small compared with the core radius.[1] We shall also restrict attention to $W = 0$, a case which is conveniently called a vortex column.

Consider first the sausaging mode, $m = 0$. The dispersion relation is

$$\frac{1}{\beta a}\frac{J_0'(\beta a)}{J_0(\beta a)} = -\frac{K_0'(ka)}{kaK_0(ka)},\tag{1}$$

where

$$\beta^2 = k^2\frac{4\Omega^2 - \sigma^2}{\sigma^2}, \qquad \sigma = \pm\frac{2\Omega k}{\sqrt{k^2 + \beta^2}},\tag{2}$$

σ being the angular frequency of the disturbance. From the properties of the Bessel function J_0, the left-hand side has the shape sketched in Figure 12.1-1. From the properties of the modified Bessel function K_0, the right-hand side is a positive monotonically decreasing function of ka, decreasing from ∞ to 0 as ka increases from 0 to ∞. It is clear from Figure 12.1-1 that for each value of ka there are infinitely many values of βa, call them β_1, β_2, ... say, which can be found graphically, and the frequencies ω_n

---

[1] The treatment in this section is based on unpublished notes of D. W. Moore.

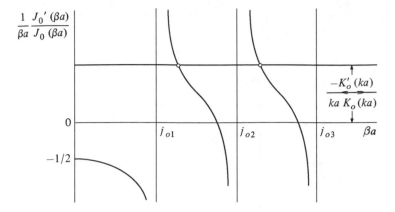

12.1-1 Left-hand side of (1). $J_0(j_{0n}) = 0$.

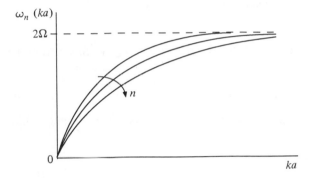

12.1-2 Sketch of frequency of the nth sausaging mode versus axial wavenumber.

then follow from (2),

$$\omega_n = \frac{2\Omega}{\sqrt{1 + \beta_n^2(ka)/(ka)^2}} .$$

(3)

As $ka \to \infty$, β_n is bounded and $\omega_n \to 0$. The behaviour of ω_n is sketched in Figure 12.1-2. The group velocity of the nth mode is $\partial \omega_n/\partial k$. The maximum group velocity is the slope at $k = 0$ and is $2\Omega a/j_{01}$. The overall maximum speed occurs for $n = 1$, which is the mode with least internal structure and is $0.835\Omega a$.[2]

[2] This implies that changes of cross-section in a filament are propagated with speed $O(\Gamma/a)$, large compared with the speed of the filament, and supports the argument in §11.4 that the general filament has a uniform cross-section.

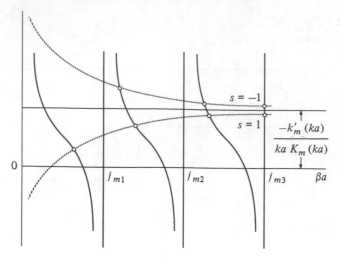

12.1-3 Equation (6) ———— left-hand side, ······ right-hand side.

We now consider bending and fluted modes with $m \geq 1$. Since from the symmetry,

$$\sigma(k,m) = \sigma(-k,m) = -\sigma^*(-k,-m), \tag{4}$$

there is no loss of generality in taking both k and m positive. From (11.3.5),

$$\sigma + m\Omega = \frac{2\Omega k s}{\sqrt{\beta^2 + k^2}}, \tag{5}$$

where $s = \pm 1$. The angular velocity of the disturbance as seen by an observer moving with the fluid is $-\sigma/m - \Omega < 0$ if $s = 1$, in which case the disturbance is retrograde; it is co-grade if $s = -1$. The dispersion relation which determines β is

$$\frac{1}{\beta a} \frac{J_m'(\beta a)}{J_m(\beta a)} = -\frac{K_m'(ka)}{ka K_m(ka)} - \frac{sm\sqrt{\beta^2 + k^2}}{ka^2 \beta^2}. \tag{6}$$

The left- and right-hand sides are sketched in Figure 12.1-3, which shows that there are an infinite number of roots, both retrograde and co-grade. It follows from the expansions of (6) for small ka and βa that there is no co-grade root for small βa. For $ka \to \infty$, βa is found to be bounded and from (5) $\sigma \to (2s - m)\Omega$.

As $ka \to 0$, $\beta a \to j_{mn}$ (the nth root of $J_m(x) = 0$) and $\sigma \to -\Omega$, except for the smallest retrograde root. This has a different behaviour. For $m > 1$, it can be shown that $\sigma/\Omega \sim 1 - m$. In this limit, the z-dependence is weak

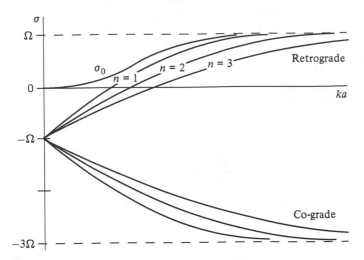

12.1-4 Bending mode oscillation frequencies.

and the frequencies are those for the oscillations of a circular patch (cf. §9.4). The case $m = 1$ is special; this was examined in §11.3 for $W \neq 0$; when $W = 0$, the root σ_0 behaves like

$$\sigma_0 \sim \tfrac{1}{2}\Omega a^2 k^2 \left(\log \frac{2}{ka} - C + \tfrac{1}{4} \right), \tag{7}$$

in agreement with the predictions of the cut-off approximation as is appropriate for the lowest-frequency bending mode in the absence of axial flow.

The dependence of σ on ka is sketched in Figure 12.1-4 for $m = 1$. It is important to notice the existence of axial wavelengths for which $\sigma = 0$. For these, $\beta a = \sqrt{3}ka$, and the values of ka are the roots of the equation

$$\frac{1}{\sqrt{3}ka} \frac{J_1'(\sqrt{3}ka)}{J_1(\sqrt{3}ka)} = -\frac{2s}{3k^2a^2} - \frac{K_1'(ka)}{kaK_1(ka)}. \tag{8}$$

Roots exist only if $s = 1$, and the first three are 2.5, 4.4, 6.2. The eigenfunctions have internal structure. Detailed analysis shows that the radial dependence $p(r)$ of the pressure perturbation can be taken as $J_1(\beta r)$, and then the radial dependence $u(r)$ of the radial velocity perturbation is given by

$$-5k\Omega u = \frac{dp}{dr} + \frac{2p}{r} = \beta J_1'(\beta r) + \frac{2}{r}J_1(\beta r), \tag{9}$$

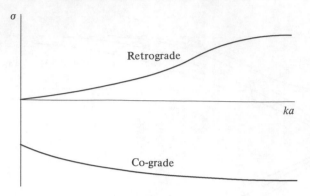

12.1-5 Dependence of frequencies on wavelength for a hollow vortex.

where the right-hand side changes sign for $0 < r < a$. Thus the inner and outer parts of the core are moving in opposite directions.[3] These solutions are of fundamental importance for the parametric instability of vortex filaments (§3). Because of (4), we can superpose perturbations of the boundary of the form

$$r = a + \delta e^{i(kz+\theta+\sigma t)} + \delta e^{i(-kz+\theta+\sigma t)} + \text{c.c.} \tag{10}$$

to give a rotating stationary plane wave

$$r = a + 4\delta \cos kz \cos(\theta + \sigma t), \tag{11}$$

which is stationary at the special values of ka satisfying (8).

Analysis of the dispersion relation for hollow vortices shows the existence of only two modes, as sketched in Figure 12.1-5. The absence of stationary waves implies also that a stagnant cored vortex column (whose boundary is a vortex sheet) also has no stationary waves.

It is obvious that filaments with arbitrary profiles of axial and azimuthal velocity may well be unstable to infinitesimal disturbances, both viscous and inviscid. However, this topic falls more naturally into the field of hydrodynamic stability than that of vortex dynamics and will not be considered here.

---

[3] The expression (7) for σ_0 has a zero when $ka = 2e^{\frac{1}{4}-C}$, but this is of course spurious and its use has led to incorrect predictions of instability (see Moore and Saffman [1974]).

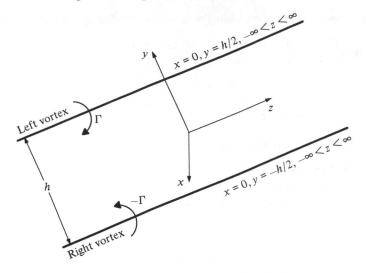

12.2-1 Configuration of a trailing vortex pair.

12.2 Long-wave co-operative instabilities

We consider the long wave instability of a pair of counter-rotating vortex filaments (Crow [1970]).[4] Axes are taken with x vertically downwards, z parallel to the undisturbed straight filaments, and y span-wise. The undisturbed pair moves downwards with velocity $\Gamma/2\pi h$, where h is the distance between them and $\pm\Gamma$ is their strengths. We refer the motion to axes moving with this speed by adding an upwards velocity $\Gamma/2\pi b$; the undisturbed motion is now steady (see Figure 12.2-1).

Now put a small disturbance on the vortices. Let ξ be a Lagrangian parameter increasing along each vortex, such that $\xi = $ constant is a fluid particle moving with the Biot–Savart law cut off speed. The disturbed vortices have equations $\mathbf{r} = \mathbf{R}_L(\xi, t)$ and $\mathbf{R}_R(\xi, t)$. Using the cut off as a

---

[4] A similar calculation in the context of liquid Helium II was carried out by Raja Gopal [1963, 1964]. Calculations of the three-dimensional instability of a staggered double row of vortices (the Karman street) were attempted by Schlayer [1928] and Rosenhead [1930].

first approximation, we obtain the equation for the left vortex

$$\frac{\partial}{\partial t} \mathbf{R}_L(\xi, t) = -\frac{\Gamma}{4\pi} \int_{[\delta]} \frac{\partial \mathbf{R}'_L}{\partial s'} \wedge \frac{\mathbf{R}_L(\xi) - \mathbf{R}_L(\xi')}{|\mathbf{R}(\xi) - \mathbf{R}_L(\xi')|^3} \, ds'$$

(self induction)

$$+ \frac{\Gamma}{4\pi} \int \frac{\partial \mathbf{R}'_R}{\partial s'} \wedge \frac{\mathbf{R}_L(\xi) - \mathbf{R}_R(\xi')}{|\mathbf{R}_L(\xi) - \mathbf{R}_R(\xi')|^3} \, ds' - \frac{\Gamma}{2\pi h} \mathbf{i}$$

(mutual induction),

(1)

and similarly for the right vortex. The cut-off parameter depends upon the internal structure as discussed in Chapter 9.

We examine infinitesimal disturbances by writing

$$\mathbf{R}_L(\xi, t) = \mathbf{i} \, x_L(t) e^{ik\xi} + \mathbf{j}(-\tfrac{1}{2} h + y_L(t) e^{ik\xi}) + \mathbf{k}(\xi + z_L(t) e^{ik\xi}),$$

(2)

where x_L, y_L, z_L are supposed small of the first order, and similarly for the right vortex. Substituting and carrying out the details of the algebraic manipulations, one obtains for the self-induction contribution to the velocity of the left filament (cf §11.2)

$$\frac{1}{4\pi} \Gamma k^2 e^{ik\xi} (\mathbf{i} \, y_L - \mathbf{j} \, x_L) \, S(ak\delta),$$

(3)

where

$$S(ak\delta) = 2 \int_{ak\delta}^{\infty} \frac{1 - \cos v}{v^3} \, dv - \int_{ak\delta}^{\infty} \frac{\sin v}{v^2} \, dv$$

$$\sim -\log ak\delta - C + \tfrac{1}{2} + O(a^2 k^2).$$

(4)

The mutual induction gives a disturbance to the left filament velocity

$$\frac{\Gamma}{2\pi h^2} [(-y_L + kh K_1(kh) y_R)\mathbf{i} + (-x_L + \{kh K_1(kh) + k^2 h^2 K_0(kh)\} x_R)\mathbf{j}$$

$$- i x_R k^3 h^3 K_1(kh)\mathbf{k}].$$

(5)

The terms in x_L and y_L are due to the displacement of the left filament in the undisturbed velocity field induced by the right filament; as discussed below, they produce the basic instability mechanism. We equate the sum of (3) and (5) (cancelling the $e^{ik\xi}$ factor) to $\mathbf{i} dx_L/dt + \mathbf{j} dy_L/dt + \mathbf{k} dz_L/dt$, and repeat the calculation for the right vortex (replace L by R, h by $-h$, and Γ by $-\Gamma$), obtaining four linear equations for the time derivatives of x_L, y_L, x_R and y_R in terms of these quantities. The perturbations z_L and z_R do not enter these equations. Also, the change in length of the filaments is second order and therefore the radius a and cut-off parameter δ can be taken as constants given by the undisturbed internal structure. The

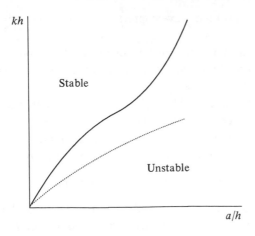

12.2-2 Stability boundary ——— and $\cdots\cdots$ maximum growth rate for symmetric modes of trailing vortex pair.

solutions are therefore proportional to $e^{\sigma t}$, where σ is the root of a quartic. We can simplify by noting that the system breaks up into two independent modes of symmetrical, $y_L = -y_R$, $x_L = x_R$, and antisymmetrical, $y_L = y_R$, $x_L = -x_R$, disturbances.

Consider the symmetrical mode. Define

$$x_S = x_L + x_R, \quad y_S = y_L - y_R. \tag{6}$$

Then

$$\frac{dx_S}{dt} = -\frac{\Gamma}{2\pi h^2}[1 + khK_1(kh) - \tfrac{1}{2}k^2h^2S(ak\delta)]y_S,$$

$$\frac{dy_S}{dt} = \frac{\Gamma}{2\pi h^2}[1 - khK_1(kh) - k^2h^2K_0(kh) + \tfrac{1}{2}k^2h^2S(ak\delta)]x_S. \tag{7}$$

Then

$$\sigma = \pm[1 + khK_1(kh) - \tfrac{1}{2}k^2h^2S(ak\delta)]^{\frac{1}{2}}$$
$$\times [1 - khK_1(kh) - k^2h^2K_0(kh) + \tfrac{1}{2}k^2h^2S(ak\delta)]^{\frac{1}{2}}. \tag{8}$$

The configuration is stable if σ is pure imaginary. Computation gives stability boundaries sketched in Figure 12.2-2. It must be kept in mind that validity requires $ka \ll 1$. The formula (8) gives a spurious band of instability for large kh when $ak = O(1)$. The antisymmetric modes are found to be stable. It can be seen from the figure that vortices of given structure and separation are unstable to sufficiently long wavelength disturbances and there is a favoured wavelength for which the growth rate is largest.

Moore [1972] has integrated the non-linear equations of motion of the cut-off approximation numerically up to the time when the vortex filaments are nearly touching.

Yuen (see Moore and Saffman [1972 Appx. D]) has investigated the linear stability using the higher approximation of (11.4.49). He finds that the stability criteria is unaltered, but there is a change in the travelling wave speed $c = -\Im \sigma / k$ of the unstable symmetric mode, which is

$$c = (a^2/h^2)(2khK_1(kh) + k^2h^2K_0(kh) - k^2h^2S(ak\delta) + 2k^2h^2\lambda)W, \tag{9}$$

where λ is the internal structure (11.4.26) and W is the mean axial flow. Thus the instability waves travel slowly relative to the fluid at infinity. The result (9) agrees with the prediction of Widnall and Bliss [1971] except for the addition of the extra term with λ, which is comparable with the others.

A physical explanation of the co-operative instability is provided by considering the three-dimensional long-wave oscillations of a vortex filament in a uniform strain $u = \epsilon y$, $v = \epsilon x$. Owing to the perturbation of the vortex, it rotates with angular velocity ω about its equilibrium position, where

$$\omega \sim \frac{\Gamma k^2}{4\pi}(\tfrac{1}{2} - C - \log ak\delta) \tag{10}$$

is the dispersion relation for long waves. The vortex moves with the fluid, so that its x-y co-ordinates have the rate of change

$$\dot{x} = (\epsilon + \omega)y, \qquad \dot{y} = (\epsilon - \omega)x. \tag{11}$$

Then the variables are proportional to $e^{\sigma t}$, where

$$\sigma^2 = \epsilon^2 - \omega^2. \tag{12}$$

The external flow destabilizes, and the self-induced motion will stabilize. The critical wavenumber k_c occurs when $\omega = \epsilon$. For $k > k_c$, there will be stability as the rotation of the filament due to self-induced velocity exceeds the rate of convection away from the hyperbolic stagnation point of the external flow. Note, however, that there is a narrow spurious instability band around $ak = e^{\frac{1}{2}-C}/\delta = O(1)$, when the formula (10) is improperly used for the rotation rate. Jimenez [1975] has studied the co-rotating vortex pair and finds stability to long-wave disturbances.

Robinson and Saffman [1982] have repeated the work of Schlayer [1928] and Rosenhead [1930] on the co-operative three-dimensional infinitesimal instability of the staggered double-array Karman vortex street (which mod-

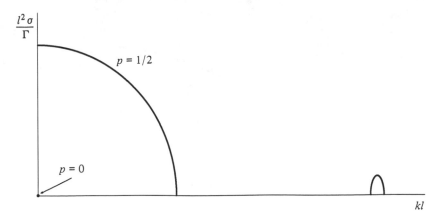

12.2-3 Growth rate of subharmonic disturbances to a single row.

els a wake) and considered also the stability of a single array (which models a mixing layer) and the stability of a symmetric double array (which models a boundary layer when only symmetric oscillations are studied). The details are straightforward but very tedious. Only the main features of the results are described here.

The single row consists of filaments parallel to the z-axis at the points $x_n = nl$, $-\infty < n < \infty$. These are disturbed so that the perturbations of the positions of the intersections with planes $z = $ constant are

$$x_n' = x'(t)e^{2\pi inp+ikz}, \quad y_n' = y'(t)e^{2\pi inp+ikz}. \tag{13}$$

The cut-off equation leads to two first-order equations for $\dot{x}$ and $\dot{y}$ with coefficients which are complicated functions of ak and kl and have to be computed numerically. The dimensionless growth rate $\sigma/\Gamma l^2$ then follows as a function of ak, kl and p. Without loss of generality, we can take $0 \le p \le \frac{1}{2}$. The disturbance is a subharmonic modulation of wavelength l/p. For given a and l, it is found that the most unstable disturbance is always two-dimensional, that is, $k = 0$, and is also the pairing mode $p = \frac{1}{2}$. The results are in qualitative agreement with the long-axial wavelength case of Pierrehumbert and Widnall's study of the three-dimensional stability of an array of Stuart vortices. Growth rates are sketched in Figure 12.2-3. Note that the spurious instability band for $ka \sim 1$ is present.

In the staggered double row, there is another parameter present: namely h, the distance between the rows. The perturbations are still of the form (13) for each of the rows, and the calculation proceeds in the same way.

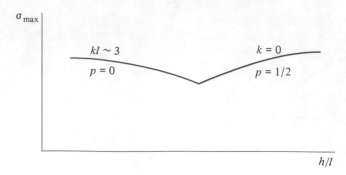

12.2-4 Most unstable disturbance to the staggered double row.

We call $\kappa = h/l$ the aspect ratio. For large aspect ratio, the results are qualitatively similar to those for the single row. The most unstable disturbances are two-dimensional and pairing; that is, $k = 0$ and $p = \frac{1}{2}$. For small aspect ratio the most unstable disturbances are three-dimensional, with $kl \approx 2$, and superharmonic; that is, $p = 0$. The critical value of κ depends upon a. For $a/l = 0.1$, the critical value is about 0.35. See Figure 12.2-4.

For the symmetric double row, the most unstable disturbance is always three-dimensional with $kh \sim 3$ and pairing with $p = \frac{1}{2}$. The dependence upon a is weak.

The stability of a vortex ring or helical vortex can be investigated as a long-wavelength co-operative instability, with different parts of the ring or helix behaving like parts of another filament. Thomson [1883] investigated perturbations described by the cut-off approximation and concluded there was stability to disturbances in which ka is small, where kR is the number of waves on the ring of radius R. The frequencies of the stable perturbations were of interest in connection with Kelvin's vortex theory of matter. There is an extensive body of theory on short-wavelength oscillations of vortex rings based on solutions of the Euler equations (e.g., Pocklington [1895b]). Widnall and Sullivan [1973] questioned Thomson's conclusions as they showed that the cut-off equations predicted instabilities in agreement with the experimentally observed vortex ring instability described by Krutzsch [1939], but this prediction was based on the spurious short-wavelength behaviour of the cut-off theory. However, Widnall, Bliss and Tsai [1974] later identified heuristically a short-wave co-operative instability, whose physics is based on the simple instability mechanism described above and depends upon the existence of non-rotating modes (see §1). Widnall and

Tsai [1977] then gave a complete theory of the instability of a uniform vortex ring. The interpretation of the short-wave co-operative instability as a parametric or mode collision instability is due to Moore and Saffman [1975b].

12.3 Short-wave co-operative instability

The considerations of the previous section suggest that stationary bending modes of a vortex column may be unstable if the vortex is in an external straining field. We now consider the mathematical discussion of this problem. Suppose we have a vortex column with boundary $r = a$ with azimuthal velocity $V(r)$, circulation Γ and axial velocity $W(r)$, and we perturb it by the addition of a uniform constant straining field. It is supposed that the strain is weak; that is,

$$\epsilon \ll \frac{\Gamma}{\pi a^2}, \tag{1}$$

so that the deformation is small. It is then assumed that a steady perturbed vortex exists, with boundary and velocities

$$r = a + \epsilon f(\theta), \quad V = V(r) + \epsilon V_1(r, \theta), \quad W = W(r) + \epsilon W_1(r, \theta). \tag{2}$$

Let us now examine the properties of infinitesimal disturbances to the perturbed column described by (2).[5] Note that the variables can still be separated in z and t, but not in θ. The (complex) boundary will have eigenfunction behaviour

$$r - a + \epsilon f(\theta) + \delta e^{im\theta + ikz + i\omega t} + \epsilon \delta e^{ikz + i\omega t} \, \text{fn}(\theta), \tag{3}$$

where δ is infinitesimal. The first infinitesimal term describes Kelvin waves on the unperturbed vortex, and the last term is an interaction. We expect that the eigenfunction problem will give

$$\omega = \omega_0(m, k) + \epsilon \omega_1 \tag{4}$$

where $\omega_0(m, k)$ is the frequency of a Kelvin mode with azimuthal wavenumber m and axial wavenumber k.

The Widnall, Bliss and Tsai [1974] argument implies that ω_1 is complex when $\omega_0(1, k) = 0$. The perturbed vortex is then unstable to disturbances of wavenumber k. We follow now the treatment of Moore and Saffman [1975b].

---

[5] Of course, for general velocity profiles the unperturbed vortex may have unstable modes, but our interest here is in stable modes that are destabilized by the perturbation.

We start with the problem of determining $f(\theta)$. This steady-flow problem is indeterminate until we specify the particular functional dependence of the vorticity ζ on the stream function Ψ. In the unperturbed column, $V(r) = -\partial\Psi(r)/\partial r$ defines $V = V(\Psi)$, and

$$\zeta = \frac{1}{r}\frac{d}{dr}(rV) = F(\Psi). \tag{5}$$

We suppose that this relation holds also in the perturbed vortex.[6] Then the stream function ψ in the perturbed vortex satisfies $\nabla^2\psi = F(\psi)$, where

$$\psi \sim \tfrac{1}{2}\epsilon r^2\cos 2\theta + \frac{\Gamma}{2\pi}\log r + O(1) \quad \text{as } r \to \infty, \quad \psi = O(r^2) \quad \text{as } r \to 0,$$

$$\tag{6}$$

assuming that the velocities are analytic at the centre. Put $\psi = \Psi(r) + \epsilon\psi_1(r,\theta)$, and expand $F(\psi) = F(\Psi) + \epsilon F'(\Psi)\psi_1$. Hence

$$\nabla^2\psi_1 = F'(\Psi)\psi_1, \tag{7}$$

where

$$F'(\Psi) = \frac{dF}{dr} \Bigg/ \frac{d\Psi}{dr} = \frac{rd^2\Omega/dr^2 + 3d\Omega/dr}{\Omega r}$$

is a known function of r and $\Omega = V/r$.

Put $\psi_1 = \tfrac{1}{2}f(r)\cos 2\theta$, where we require that $f \to r^2$ as $r \to \infty$ and $f = O(r^2)$ as $r \to 0$. Then $f(r)$ satisfies the equation

$$\frac{d^2f}{r^2} + \frac{1}{r}\frac{df}{dr} - \left(\frac{3d\Omega/dr + rd^2\Omega/dr^2}{r\Omega} + \frac{4}{r^2}\right)f = 0. \tag{8}$$

This equation has a regular singularity at $r = 0$, and the roots of the indicial equation are ± 2. Let $g(r)$ denote the solution which equals r^2 at $r = 0$. Since $\Omega \sim \Gamma/2\pi r^2$ as $r \to \infty$, it follows that $g(\infty) \sim br^2$, where b is a number to be found numerically in general. Then the perturbation to the vortex is $f(r) = g(r)/b$. Suppose $b = 0$. Then there is a steady perturbation with the $\cos 2\theta$ symmetry; that is, there is a steady 'Kirchhoff' vortex and $\omega(2,0) = 0$. In this case, the weak strain causes a two-dimensional 'resonance' and the vortex disintegrates. We suppose this is not the case. The distribution of axial flow is likewise assumed to be the same function of the stream function, so that

$$w = W(\Psi) = W_0(r) + \epsilon W'(\Psi)\psi_1. \tag{9}$$

---

[6] This is not an assumption but rather a statement that a particular class of vortex motions is being considered.

The perturbed undisturbed vortex[7] has the velocity and pressure fields

$$\bar{v}_r = \epsilon \frac{f(r)}{r} \sin 2\theta, \quad \bar{v}_\theta = V(r) + \epsilon f'(r) \cos 2\theta,$$

$$\bar{v}_z = \epsilon f(r) \cos 2\theta \, W'(\Psi), \quad \bar{p} = P(r) + \epsilon V f(r) \cos 2\theta. \tag{10}$$

We now consider disturbed velocity fields

$$v_r = \bar{v}_r + u e^{ikz+i\omega t}, \quad v_\theta = \bar{v}_\theta + v e^{ikz+i\omega t},$$

$$v_z = \bar{v}_z + w e^{ikz+i\omega t}, \quad p = \bar{p} + p e^{ikz+i\omega t}, \tag{11}$$

where u, v, w and p are functions of r and θ. Substitution into the equations of motion, linearising in the perturbation and retaining terms $O(\epsilon)$ in the disturbance leads to equations of the form

$$i\omega \mathbf{L} \mathbf{U} + \mathbf{M} \mathbf{U} = \epsilon e^{2i\theta} \mathbf{N} \mathbf{U} + \epsilon e^{-2i\theta} \mathbf{N}^* \mathbf{U}, \tag{12}$$

where we are employing a matrix notation

$$\mathbf{U} = \begin{pmatrix} u \\ v \\ w \\ p \end{pmatrix} \tag{13}$$

$$\mathbf{L} = \begin{pmatrix} 1 & 0 & 0 & 0 \\ 0 & 1 & 0 & 0 \\ 0 & 0 & 1 & 0 \\ 0 & 0 & 0 & 1 \end{pmatrix}, \tag{14}$$

and in the absence of axial velocity

$$\mathbf{M} = \begin{pmatrix} \Omega \frac{\partial}{\partial \theta} & -2\Omega & 0 & \frac{\partial}{\partial r} \\ 2\Omega + r \frac{\partial \Omega}{\partial r} & \Omega \frac{\partial}{\partial \theta} & 0 & \frac{1}{r} \frac{\partial}{\partial \theta} \\ 0 & 0 & \Omega \frac{\partial}{\partial \theta} & ik \\ \frac{\partial}{\partial r} + \frac{1}{r} & \frac{1}{r} \frac{\partial}{\partial \theta} & ik & 0 \end{pmatrix} \tag{15}$$

---

[7] We adopt the terminology that perturbations are the small but finite steady changes due to steady external effects, while disturbances are infinitesimal unforced time-dependent variations.

and

$$
\mathbf{N} = \begin{pmatrix}
\frac{if'}{r} - \frac{if}{r}^2 + \frac{if}{r}\frac{\partial}{\partial r} - \frac{f'}{2r}\frac{\partial}{\partial\theta} & \frac{f'}{r} - \frac{2f}{r^2} & 0 & 0 \\
-\frac{1}{2}f'' - \frac{1}{2}\frac{f'}{r} & -\frac{if'}{r} + \frac{if}{r}^2 + \frac{if}{r}\frac{\partial}{\partial r} - \frac{f'}{2r}\frac{\partial}{\partial\theta} & 0 & 0 \\
0 & 0 & 0 & 0
\end{pmatrix}.
$$

(16)

The equation is to be solved for the eigenvalue ω subject to homogeneous boundary conditions that $\mathbf{U}$ vanishes at ∞ and that the flow field and pressure are analytic in the neighbourhood of the axis.

The method of solution is to write

$$\mathbf{U} = \mathbf{U}_0 + \mathbf{U}_1 + \cdots,$$

(17)

$$\omega = \omega_0 + \epsilon\omega_1 + \cdots.$$

(18)

Thus to leading order

$$i\omega_0\mathbf{L}\mathbf{U}_0 + \mathbf{M}\mathbf{U}_0 = 0.$$

(19)

This equation determines the frequencies of the Kelvin waves. It was studied in §1 for the case of a uniform vortex. Here we shall continue in terms of general profiles. The $O(\epsilon)$ terms give the equation

$$i\omega_0\mathbf{L}\mathbf{U}_1 + \mathbf{M}\mathbf{U}_1 = -i\omega_1\mathbf{L}\mathbf{U}_0 + e^{2i\theta}\mathbf{N}\mathbf{U}_0 + e^{-2i\theta}\mathbf{N}^*\mathbf{U}_0.$$

(20)

In general, this inhomogeneous system of equations will not have a solution satisfying the boundary conditions and a solubility condition arises for ω_1, which measures the effect of the strain on the frequency of small oscillations.

The system (19) for the zero-order problem can be collapsed into a single equation for the disturbance pressure and if we take $p_0(r,\theta) = \tilde{p}(r)e^{im\theta}$, the equation is

$$
\frac{d^2\tilde{p}}{dr^2} + \left(\frac{1}{r} - \frac{1}{\Delta}\frac{d\Delta}{dr}\right)\frac{d\tilde{p}}{dr}
$$

$$
+ \left(\frac{2m}{r\Sigma}\left(\frac{d\Omega}{dr} - \frac{\Omega}{\Delta}\frac{d\Delta}{dr}\right) + \frac{k^2\Delta}{\Sigma^2} - \frac{m^2}{r^2}\right)\tilde{p} = 0.
$$

(21)

Here

$$\Sigma(r) = \omega_0 + m\Omega(r), \quad \text{and} \quad \Delta = 4\Omega^2 + 2\Omega r\frac{d\Omega}{dr} - \Sigma^2.$$

(22)

The boundary conditions on (21) are that $\tilde{p}$ is bounded as $r \to 0$ and vanishes as $r \to \infty$. We assume that (21) has a discrete enumerable set of real eigenvalues $\omega_0^{(n)}$, $n = 0, 1, 2, \ldots$, and consider how these are perturbed by the external strain.

Note that in the absence of axial flow, the form of the equations is such that

$$\omega_0^{(n)}(m, k) = -\omega_0^{(n)}(-m, k); \tag{23}$$

This case has a special importance, as will be seen later. When this symmetry exists, non-rotating bending-mode disturbances occur at a critical value k_c when $\omega_0^{(n)}(\pm 1, k_c) = 0$. In general, the rotation rate $\omega^{(n)}$ of a plane sinusoidal disturbance obtained by superposing the helical waves of opposite senses is

$$\omega^{(n)} = -\tfrac{1}{2}[\omega_0^{(n)}(1, k) - \omega_0^{(n)}(-1, k)]. \tag{24}$$

Take

$$\mathbf{U}_0 = \mathbf{U}(1)\, e^{i\theta}. \tag{25}$$

The dependence on k and r is implicit. We substitute into (20). The right-hand side implies that we should take

$$\mathbf{U}_1 = \mathbf{u}(1)e^{i\theta} + \mathbf{u}(-1)e^{-i\theta} + \mathbf{u}(3)e^{3i\theta}. \tag{26}$$

and equating coefficients of the various angle dependencies on using (20) gives

$$i\omega_0^{(n)}\mathbf{L}\mathbf{u}(1) + \mathbf{M}(1)\mathbf{u}(1) = -i\omega_1\mathbf{L}\mathbf{U}(1), \tag{27}$$

$$i\omega_0^{(n)}\mathbf{L}\mathbf{u}(-1) + \mathbf{M}(-1)\mathbf{u}(-1) = \mathbf{N}^*(-1)\mathbf{U}(1), \tag{28}$$

$$i\omega_0^{(n)}\mathbf{L}\mathbf{u}(3) + \mathbf{M}(3)\mathbf{u}(3) = \mathbf{N}(1)\mathbf{U}(1). \tag{29}$$

The matrices $\mathbf{M}(m)$ and $\mathbf{N}(m)$ are obtained by replacing $\partial/\partial\theta$ by im. According to the basic idea of the Fredholm alternative, (28) and (29) will have unique solutions provided $\omega_0^{(n)}$ is not an eigenvalue of the operators $i\omega\mathbf{L} + \mathbf{M}(3)$ or $i\omega\mathbf{L} + \mathbf{M}(-1)$. By construction, it is of course an eigenvalue of the left-hand side of (27). In this case, (27) has no solution unless the right hand side is orthogonal to the adjoint eigenfunction of the left-hand side.

For the present problem, an inner product of two solution vectors $\mathbf{U}$, $\mathbf{V}$ can be defined by

$$\langle \mathbf{U}, \mathbf{V} \rangle = \int_0^\infty (U_1^* V_1 + U_2^* V_2 + U_3^* V_3 + U_4^* V_4)\, dr$$

and an adjoint operator $\mathbf{A}^\dagger$ of an operator $\mathbf{A}$ by the requirement that for arbitrary vectors satisfying the boundary conditions

$$\langle \mathbf{V}, \mathbf{A}\mathbf{U} \rangle = \langle \mathbf{U}, \mathbf{A}^\dagger \mathbf{V} \rangle^*. \tag{30}$$

According to these definitions,

$$\mathbf{L}^\dagger = \begin{pmatrix} 1 & 0 & 0 & 0 \\ 0 & 1 & 0 & 0 \\ 0 & 0 & 1 & 0 \\ 0 & 0 & 0 & 0 \end{pmatrix}$$

and

$$\mathbf{M}^\dagger(m) = \begin{pmatrix} -\Omega im & 2\omega + r\frac{d\Omega}{dr} & 0 & -\frac{\partial}{\partial r} + \frac{1}{r} \\ -2\Omega & -\Omega im & 0 & -\frac{im}{r} \\ 0 & 0 & -\Omega im & -ik \\ -\frac{\partial}{\partial r} & -\frac{im}{r} & -ik & 0 \end{pmatrix}. \tag{31}$$

Since in general $\langle \mathbf{U}^\dagger(1), \mathbf{L}\mathbf{U}(1) \rangle \neq 0$, it follows from (29) that

$$\omega_1 = 0. \tag{32}$$

Thus there is in general no $O(\epsilon)$ shift in the frequency of helical waves on the perturbed vortex and, in particular, the vortex remains stable at this order.

In order for instability to occur, it is necessary for there to be a degeneracy; for example, a wavenumber k such that (23) is satisfied.[8] We shall restrict discussion to this special no axial flow case because here instability can be demonstrated without detailed calculation.[9] We suppose that the axial wavenumber $k \approx k_c$, where we anticipate that $k - k_c = \nu \epsilon$, where $\nu = O(1)$. Then

$$\mathbf{M} = \mathbf{M}_c + i(k - k_c)\mathbf{P}, \tag{33}$$

---

[8] The degeneracy need not be between the same mode. For instance, $\omega_0^{(n)}(1, k) = \omega_0^{(n')}(-1, k)$ with $n \neq n'$ or $\omega_0^{(n)}(1, k) = \omega_0^{(n')}(3, k)$ may give rise to instability.

[9] The concept of signature applicable to Hamiltonian systems may be of use here. See the application to water waves by MacKay and Saffman [1986]. The concept appears to go back to Weierstrass [1858]

where $\mathbf{M}_c$ is $\mathbf{M}$ evaluated at $k = k_c$, and

$$\mathbf{P} = \begin{pmatrix} 0 & 0 & 0 & 0 \\ 0 & 0 & 0 & 0 \\ 0 & 0 & 0 & 1 \\ 0 & 0 & 1 & 0 \end{pmatrix}. \tag{34}$$

We take the zero order disturbance to be

$$\mathbf{U}_0 = \alpha \mathbf{U}(1)e^{i\theta} + \beta \mathbf{U}(-1)e^{-i\theta}, \tag{35}$$

where α and β are complex constants and $\mathbf{U}(1)$ and $\mathbf{U}(-1)$ are the eigenfunctions for $\mathbf{M}_c(1)$ and $\mathbf{M}_c(-1)$, respectively. Then the equation for the first-order disturbance is

$$\mathbf{M}_c\mathbf{U}_1 = -i\omega_1\mathbf{L}\mathbf{U}_0 - i\nu\mathbf{P}\mathbf{U}_0 + e^{2i\theta}\mathbf{N}\mathbf{U}_0 + e^{-2i\theta}\mathbf{N}^*\mathbf{U}_0. \tag{36}$$

To solve this equation, we put

$$\mathbf{U}_1 = \mathbf{u}(3)e^{3i\theta} + \mathbf{u}(1)e^{i\theta} + \mathbf{u}(-1)e^{-i\theta} + \mathbf{u}(-3)e^{-3i\theta}. \tag{37}$$

The equations for $\mathbf{u}(\pm 3)$ have unique solutions because (ignoring possible coincidences of eigenfrequencies) the equations $\mathbf{M}_c(\pm 3)\mathbf{u} = 0$ have no non-trivial solutions. We do not consider these equations further, because they yield no information about the shift in frequency.

A solubility condition arises from the equations for $\mathbf{u}(\pm 1)$ because the adjoint equations

$$\mathbf{M}_c^\dagger(1)\mathbf{U}^\dagger(1) = 0, \qquad \mathbf{M}_c^\dagger(-1)\mathbf{U}^\dagger(-1) = 0, \tag{38}$$

have non-trivial solutions. For solubility of the equations

$$\left. \begin{aligned} \mathbf{M}_c(1)\mathbf{u}(1) &= -\alpha i\omega_1\mathbf{L}\mathbf{U}(1) - \alpha i\nu\mathbf{P}\mathbf{U}(1) + \beta\mathbf{N}(-1)\mathbf{U}(-1), \\ \mathbf{M}_c(-1)\mathbf{u}(-1) &= -\beta i\omega_1\mathbf{L}\mathbf{U}(-1) - \beta i\nu\mathbf{P}\mathbf{U}(-1) + \alpha\mathbf{N}^*(1)\mathbf{U}(1), \end{aligned} \right\} \tag{39}$$

we require that the right hand sides be orthogonal to the adjoint vectors. So,

$$\left. \begin{aligned} 0 &= -\alpha i\omega_1 L_{11} - \alpha i\nu P_{11} + \beta N_{-11}(-1), \\ 0 &= -\beta i\omega_1 L_{-1-1} - \beta i\nu P_{-1-1} + \alpha N_{1-1}^*(1), \end{aligned} \right\} \tag{40}$$

where the matrix coefficients $L_{mm'}$, and so on, are defined by

$$L_{mm'} = \langle \mathbf{U}^\dagger(m'), \ \mathbf{L}\mathbf{U}(m) \rangle, \quad m = \pm 1, \quad m' = \pm 1. \tag{41}$$

Eliminating α and β from these equations, we obtain

$$(i\omega_1 L_{11} + i\nu P_{11})(i\omega_1 L_{-1-1} + i\nu P_{-1-1}) = N_{1-1}^*(1)N_{-11}(-1). \tag{42}$$

Considerable computational effort is needed to find the actual values of ω_1. Tsai and Widnall [1976] have calculated the values of ω_1 for the uniform cored vortex, and Widnall and Tsai [1977] have calculated the eigenvalues of a uniform, thin cored vortex ring where roughly speaking the role of ϵ is played by a/R. Saffman [1978] has considered the case of non-uniform cores and estimated the unstable wavelengths for non-uniform thin cored rings. Tsai and Widnall find for the uniform core strained vortex that the most unstable disturbances are those corresponding to the zero frequency Kelvin waves. Growth rates for the other degeneracies are considerably smaller. Robinson and Saffman [1984a] have studied the finite ϵ case for a uniform core. In this case, the perturbed vortex is the Moore–Saffman elliptical vortex. For small values of the strain, agreement with the calculations of Tsai and Widnall is excellent. For larger strain, the total range of unstable wavenumbers is large and the magnitudes of the growth rates for a significant portion of the axial wavenumber space are comparable. Thus the selection mechanism at small strains evaporates. Vladimirov and Il'in [1988] have studied the three-dimensional instability of the nearly circular Kirchhoff vortex.

In the absence of axial flow, Moore and Saffman [1975b] showed that the symmetries of the problem make it possible to demonstrate the instability without having to evaluate the inner products numerically. It is easy to verify that in the absence of axial flow with $\mathbf{M}$ given by (15), $\mathbf{U}(1)$, $\mathbf{U}(-1)$, $\mathbf{U}^\dagger(1)$ and $\mathbf{U}^\dagger(-1)$ can be chosen so that

$$\mathbf{U}(1) = \mathbf{U}^\dagger(1) = \begin{pmatrix} ix_1 \\ x_2 \\ x_3 \\ x_4 \end{pmatrix} \quad \text{and} \quad \mathbf{U}(-1) = \mathbf{U}^\dagger(-1) = \begin{pmatrix} -ix_1 \\ x_2 \\ -x_3 \\ x_4 \end{pmatrix}, \tag{43}$$

where x_1, x_2, x_3 and x_4 are all real functions of r. Direct calculation shows that with an appropriate normalisation,

$$L_{11} = L_{-1-1} = 1, \quad P_{11} = -P_{-1-1} = Q, \quad N^*_{1-1}(1) = -N_{-11}(-1) = iR,$$

$$\tag{44}$$

where Q and R are real numbers. Then the frequency equation (42) can be expressed as

$$\omega_1^2 = \nu^2 Q^2 - R^2. \tag{45}$$

This demonstrates the existence of instability as long as $|\nu| < |R/Q|$ or

$$|k - k_c| < \epsilon |R/Q|. \tag{46}$$

Thus there is a narrow band of unstable axial wavenumbers centred on the wavenumber k_c corresponding to zero Kelvin wave frequency. Notice that the quantitative prediction of the plausibility argument of (§2) is not borne out unless $R^2 = 1$. That argument gives a maximum growth rate ϵ, whereas the calculation gives a maximum growth rate $\epsilon|R|$.

The effect of axial flow on the instability is an open question. The above formalism stands, but **N** contains extra terms owing to the strain-induced distortion of the axial flow field. The principal effect of axial flow is on the oscillations of the unperturbed vortex, which no longer have the symmetry (23) between oppositely propagating helical waves. Thus axial flow shifts the degeneracy to a non-zero frequency but the effect of this on the stability cannot be decided without detailed calculation.

However, if the unperturbed axial flow, $W(r)$, is of relative order ϵ, it can be treated as a perturbation. Then (33) becomes

$$\mathbf{M} = \mathbf{M}_c + i(k - k_c)\mathbf{P} + i\mathbf{T}, \tag{47}$$

where

$$\mathbf{T} = \begin{pmatrix} k_c W & 0 & 0 & 0 \\ 0 & k_c W & 0 & 0 \\ -i\frac{\partial W}{\partial r} & 0 & k_c W & 0 \\ 0 & 0 & 0 & 0 \end{pmatrix}. \tag{48}$$

It is readily verified that the matrix coefficients satisfy $T_{11} = T_{-1-1} = \epsilon S$, where S is real and $O(1)$. The calculation leading to (42) is easily repeated and gives

$$(\omega_1 + S)^2 = \nu^2 Q^2 - R^2. \tag{49}$$

Thus the instability persists for weak axial flow, the width of the band of unstable wavenumbers being unaltered. However, the most unstable disturbance now arises from helical waves with rotation rate ϵS.

An alternative interpretation of the parametric instability[10] can be given in terms of eigenvalue avoidance (see Figure 12.3-1). If the modes which cross have opposite signatures, then for finite values of the parameter a bubble of instability in which $(\Im\omega)^2 \propto (k - k_c)^2 - \epsilon^2$ occurs.

---

[10] So called because the instability arises from the variation of a parameter ϵ which controls the perturbation from a stable state.

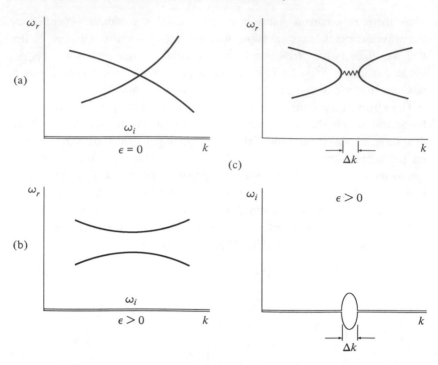

12.3-1 Instability associated with eigenvalue crossing. (a) Unperturbed. (b) Avoidance. (c) Instability.

12.4   Ultra short-wave co-operative instability

The studies reported in the previous section indicated the existence of unstable disturbances of perturbed vortices when the axial wavenumber of the disturbances satisfies conditions on the Kelvin waves of the un-perturbed vortex. For uniform vortices without axial flow, details can be worked out explicitly, and suggest that in general there are an infinite number of such wavenumbers with a point of accumulation at infinity. Moreover, the calculations suggest that the growth rates depend weakly on the wavenumbers and as a consequence it is suggested that for very small wavelengths there is likely to be a continuum of unstable modes. This has been confirmed numerically by Pierrehumbert [1986], and analytical support has been given by Bayly [1986]. Further studies have been carried out by Landman and Saffman [1987], who discussed the effect of viscosity, and Waleffe [1990]. Here, we follow the treatment of Bayly as extended by Landman and Saffman.

We consider ultra short wavelength disturbances to a perturbed vortex. We assume that when the wavelength of the disturbance is small, its evolution will be controlled by the flow near the centre of the perturbed vortex where the stream function has the form

$$\psi = -\tfrac{1}{2}\gamma(x^2 + y^2) - \tfrac{1}{2}\epsilon(y^2 - x^2). \tag{1}$$

The first term on the right-hand side describes solid-body rotation with angular velocity one-half the value of the vorticity on the axis, and the second term describes the perturbation which has altered the circular streamlines to ellipses. Co-ordinates have been chosen with the z-axis parallel to the vortex axis.

In a matrix notation, we can write the undisturbed perturbed velocity field as

$$\mathbf{u} = \mathbf{Ar}, \quad \text{where } \mathbf{u} = \begin{pmatrix} u \\ v \\ w \end{pmatrix}, \quad \mathbf{r} = \begin{pmatrix} x \\ y \\ z \end{pmatrix}, \quad \mathbf{A} = \begin{pmatrix} 0 & -\gamma - \epsilon & 0 \\ \gamma - \epsilon & 0 & 0 \\ 0 & 0 & 0 \end{pmatrix}. \tag{2}$$

We now consider a disturbance to the velocity and pressure fields of the form

$$\mathbf{u}' = \mathbf{v}(t)e^{i\mathbf{k}(t)\cdot\mathbf{r}}, \quad p' = p(t)e^{\mathbf{k}(t)\cdot\mathbf{r}}. \tag{3}$$

It is easy to verify that this is an exact solution of the Euler equations (i.e., no linearisation in the disturbance is necessary) if

$$\dot{\mathbf{k}} = -\mathbf{A}^T\mathbf{k} \tag{4}$$

and

$$\dot{\mathbf{v}} = \left(\frac{2\mathbf{k}^T\mathbf{k}}{|\mathbf{k}|^2} - \mathbf{I} \right) \mathbf{v}. \tag{5}$$

Equation (4) can be integrated in closed form with solution

$$\mathbf{k} = k_0 (\sin\theta \cos\Omega t, \ \alpha\sin\theta\sin\Omega t, \ \cos\theta), \tag{6}$$

where θ and k_0 are arbitrary constants of integration and

$$\alpha = \sqrt{\frac{\gamma + \epsilon}{\gamma - \epsilon}}, \quad \Omega = \sqrt{\gamma^2 - \epsilon^2}. \tag{7}$$

Thus (5) is of the form

$$\dot{\mathbf{v}} = \mathbf{Q}(t)\mathbf{v}, \tag{8}$$

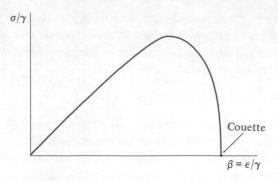

12.4-1 Growth rate dependence upon strain for ultra-short wavelength disturbances.

where $\mathbf{Q}$ is periodic in time with period $2\pi/\Omega$. Hence by Floquet theory,

$$\mathbf{v} = e^{\sigma t}\mathbf{P}(t), \qquad \sigma = \sigma(\alpha, \Omega), \qquad (9)$$

where $\mathbf{P}(t)$ is periodic. The flow is unstable if $\Re\sigma$ is not equal to zero. The growth rate has to be calculated numerically. A plot of $\Re\sigma$ is shown in Figure 12.4-1. The flow stabilizes as $\epsilon \to \gamma$ as it approaches Couette flow. Landman and Saffman show that an exact solution remains in the presence of viscosity, and moreover there are still unstable modes, however large the viscosity may be.

13

EFFECTS OF VISCOSITY

13.1 Viscous cores

The simplest viscous vortex is the Lamb–Oseen vortex (Lamb [1932 §334a], Batchelor [1967 §4.5]). This is a two-dimensional flow with circular symmetry in which the streamlines are circles around the axis and the vorticity is a function of radial distance r and time t and parallel to the axis. It arises as an exact solution of the Navier–Stokes equations for the initial condition

$$\omega(r, 0) = \Gamma_0 \, \delta(x) \, \delta(y). \tag{1}$$

The Navier–Stokes equations reduce to the single equation for the vorticity $\omega(r, t)$

$$\frac{\partial \omega}{\partial t} = \nu \nabla^2 \omega. \tag{2}$$

The exact solution[1] is

$$\omega = \frac{\Gamma_0}{4\pi \nu t} e^{-r^2/4\nu t}, \quad v = \frac{\Gamma}{2\pi r} = \frac{\Gamma_0}{2\pi r} \left(1 - e^{-r^2/4\nu t}\right),$$
$$\Gamma = \Gamma_0 \left(1 - e^{-r^2/4\nu t}\right), \tag{3}$$

where v is the tangential velocity around the axis and Γ is the circulation around a circle of radius a. Profiles of tangential velocity and circulation are sketched in Figure 13.1-1. Note that the tangent to the circulation curve at the radius r_1 of maximum tangential velocity v_1 passes through

---

[1] There are various ways of solving the problem. Taking a Laplace transform in time, or a Fourier expansion in x and y, or assuming the appropriate functional form implied by dimensional analysis, all work equally well.

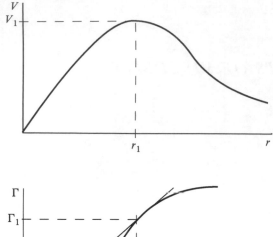

13.1-1 Sketch of profile for Lamb–Oseen vortex.

the origin. For the distribution (3),

$$\Gamma_1 = 0.716\Gamma_0, \qquad r_1 = 2.24\sqrt{\nu t}. \tag{4}$$

The Lamb–Oseen vortex can be regarded as the 'desingularisation' of the rectilinear line vortex, in which the vorticity has a delta-function singularity. Studies of inviscid roll-up of a two-dimensional vortex sheet give vortices in which $v \propto r^{-n}$ as $r \to 0$ where n is a real number, positive or negative. For example, the case $n = \frac{1}{2}$ arises from the roll-up of the vortex produced at the edge of a sharp plate in uniform motion. This non-analytic behaviour can also be removed by viscosity. The Navier–Stokes equations reduce to

$$\frac{\partial v}{\partial t} = \nu \left(\frac{\partial^2 v}{\partial r^2} + \frac{1}{r} \frac{\partial v}{\partial r} - \frac{v}{r^2} \right). \tag{5}$$

We are interested in solutions such that $v \sim \beta/r^n$ as $r/\sqrt{\nu t} \to \infty$, and v is analytic at $r = 0$. The singular behaviour is then smoothed out in a boundary layer core of radius $O(\sqrt{\nu t})$.

A solution to this problem was given by Kirde [1962]. Dimensional analysis suggests that we write

$$v = \frac{\beta}{(\nu t)^{n/2}} F(\eta), \qquad \eta = -\frac{r^2}{4\nu t}. \tag{6}$$

Substituting into (5) gives

$$\eta F'' + (1 - \eta)F' - \frac{n}{2}\left(1 + \frac{1}{\eta}\right)F = 0. \tag{7}$$

The solution of this equation is

$$F(\eta) = 2^{-n}\Gamma\left(\frac{3}{2} - \frac{n}{2}\right)(-\eta)^{\frac{1}{2}} M(\tfrac{1}{2} + \tfrac{1}{2}n; 2; \eta), \quad -\infty < \eta < 0, \tag{8}$$

where M is the confluent hypergeometric function (Abramowitz and Stegun [1964]). For the case $n = \frac{1}{2}$, the maximum occurs when $\eta = -2.10$ and $r_1 = 2.92\sqrt{\nu t}$. This type of viscous core structure is appropriate for vortices formed by roll-up of vortex sheets.

When a laminar vortex becomes unstable, the possibility always exists that the fluid motion will become turbulent. The structure of a turbulent line vortex is a problem of considerable interest, but the status of the present understanding is unclear. The problem is to determine the structure of average quantities, the average being an ensemble or azimuthal or axial average, as functions of r and t. Hoffman and Joubert [1963] have argued that there is a law similar to the law of the wall for a turbulent boundary layer in which the mean azimuthal velocity depends only upon r, r_1 and Γ_1. In this layer, which extends from about r_1 to a radius small compared with r_0 (see §2.1), the mean circulation $\overline{\Gamma}$ behaves like

$$\overline{\Gamma} = \Gamma_1\left(\log\frac{r}{r_1} + 1\right), \qquad r_1 \leq r \ll r_0. \tag{9}$$

For $r \ll r_1$, it is speculated that the core is laminar in solid-body rotation (see Vladimirov and Tarasov [1979]).

Govindaraju and Saffman [1971] obtained the strange result that there is likely to be an overshoot of circulation in a turbulent line vortex. Assuming that the mean properties are axisymmetric and independent of axial co-ordinate, the evolution equation for the mean azimuthal velocity is

$$\frac{\partial \overline{v}}{\partial t} = \nu\left(\frac{\partial^2 \overline{v}}{\partial r^2} + \frac{1}{r}\frac{\partial \overline{v}}{\partial r} - \frac{\overline{v}}{r^2}\right) - \frac{1}{r^2}\frac{\partial}{\partial r}\overline{r^2 u'v'}, \tag{10}$$

where u', v' are the fluctuations of radial and azimuthal velocity. The last term in (10) is the Reynolds stress. Multiplying (10) by r^2 and integrating,

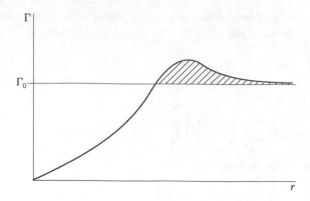

13.1-2 Overshoot of mean circulation for turbulent vortex.

we obtain

$$\int_0^\infty r^2 \frac{\partial \bar{v}}{\partial t}\, dr = \nu \left[\frac{\partial}{\partial r} \left(r^3 \frac{\partial}{\partial r} \frac{\bar{v}}{r} \right) \right]_0^\infty - \left[\overline{u'v'}r^2 \right]_0^\infty = -\frac{\nu \Gamma_0}{\pi}, \tag{11}$$

since $\bar{v} \sim \Gamma_0/2\pi r$ and the Reynolds stress vanishes as $r \to \infty$. It follows from (11) that

$$\frac{\partial}{\partial t} \int_0^\infty r(\Gamma_0 - \Gamma)\, dr = 2\nu \Gamma_0 \tag{12}$$

and hence

$$\int_0^\infty r(\Gamma_0 - \Gamma)\, dr = 2\nu \Gamma_0 t + \text{constant}. \tag{13}$$

Suppose now that the turbulent vortex is self-similar; that is,

$$\Gamma = \Gamma_0 f(\eta), \quad \eta = r/r_1, \quad r_1 = b\sqrt{\Gamma_0 t}. \tag{14}$$

Then substitution into (12) gives

$$\int_0^\infty (1 - f)\eta\, d\eta = \frac{2\nu}{b\Gamma_0} \to 0 \quad \text{as} \quad \frac{\nu}{\Gamma_0} \to 0. \tag{15}$$

Analyticity requires that $f \propto \eta^2$ as $r \to 0$, and the consequence is that for large Reynolds number, there must be values of f for which $f > 1$. In other words, there has to be an overshoot of circulation.

The assumption of self-similarity is not in fact necessary. It suffices that for fixed r, the mean circulation Γ should tend to zero faster than the rate at which it does for the Lamb–Oseen vortex; it then follows from (13) that there must exist values of r for which $\Gamma_0 - \Gamma$ is negative. A circulation profile like that shown in Figure 13.1-2 is expected. The mean vorticity for large r is negative, and the Rayleigh criterion suggests axisymmetrical instability in the outer part of the vortex with turbulent Taylor cells appearing.

13.2 Decay of trailing vortices

According to inviscid theory, a vortex sheet emanates from the trailing edge of a lifting wing and rolls up under the action of its self-induced velocity field. The initial strength of the vortex sheet is determined by the shape of the wing. The steady three-dimensional problem is intractable, and it is customary to replace it by a conceptually and numerically simpler unsteady two-dimensional problem where t replaces z/U and the initial strength of the the sheet is given by lifting surface theory (when solutions are available) or by the further approximation of lifting-line theory. As discussed in §6.3 and §8.4, vortex sheets roll up into spirals. The number of spirals depends upon the structure of the sheet, and the presence of wing flaps, for example, may cause several spirals to appear. However, the vorticity of like sign tends to accumulate into two spirals of opposite strength with axes along the vorticity shed from the wing tips. This pair of trailing vortices forms the far wake of the wing, and its properties are of considerable interest because of potential hazard to following aircraft. The co-operative long-wavelength (Crow) instability of the vortex pair is discussed in §12.2. Here, we are concerned with the internal structure of the trailing vortices and the effect of viscosity on their evolution, neglecting their interaction and making the further assumption that the roll-up has made the velocity distributions axisymmetric.

Thus we suppose that at a distance z downstream of the wing,[2] the vortex has developed an axisymmetric form. In cylindrical polar co-ordinates (r, θ, z) fixed with respect to the wing, the velocity components are taken to be $(u, v, U + w)$, with u radial, v azimuthal and w the excess speed in the wake. U is the speed at infinity, that is, the speed of the wing in a fixed frame. All components are assumed to depend only on r and z; that is, the flow is steady and $\partial/\partial\theta = 0$ (see Figure 13.2-1).

The equations of motion are

---

[2] The distance depends on the wing loading distribution, i.e., the actual initial vortex sheet properties.

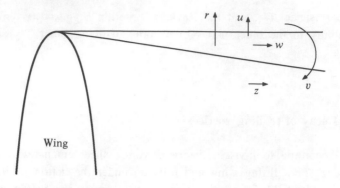

13.2-1 Co-ordinate system for a trailing vortex.

continuity:

$$\frac{1}{r}\frac{\partial}{\partial r}(ur) + \frac{\partial w}{\partial z} = 0,\tag{1}$$

azimuthal momentum:

$$U\frac{\partial v}{\partial z} = \nu\left(\frac{\partial^2 v}{\partial r^2} + \frac{1}{r}\frac{\partial v}{\partial r} - \frac{v}{r^2}\right)$$
$$+ \nu\left\langle\frac{\partial^2 v}{\partial z^2}\right\rangle - \left[\frac{1}{r^2}\frac{\partial}{\partial r}(uvr^2) + \frac{\partial}{\partial z}(vw)\right],\tag{2}$$

radial momentum:

$$-\frac{v^2}{r} + \frac{\partial p}{\partial r} = \nu\left(\frac{\partial^2 u}{\partial r^2} + \frac{1}{r}\frac{\partial u}{\partial r} - \frac{u}{r^2}\right)$$
$$+ \nu\left\langle\frac{\partial^2 u}{\partial z^2}\right\rangle - \left[\frac{1}{r}\frac{\partial}{\partial r}(u^2 r) + \frac{\partial}{\partial z}(uw)\right],\tag{3}$$

axial momentum:

$$U\frac{\partial w}{\partial z} = -\left\{\frac{\partial p}{\partial z}\right\} + \nu\left(\frac{\partial^2 w}{\partial r^2} + \frac{1}{r}\frac{\partial w}{\partial r}\right)$$
$$+ \nu\left\langle\frac{\partial^2 w}{\partial z^2}\right\rangle - \left[\frac{1}{r}\frac{\partial}{\partial r}(uwr) + \frac{\partial}{\partial z}(w^2)\right].\tag{4}$$

Terms are grouped in brackets according to their orders of magnitude.
If there is no viscosity, these equations have the Bernoulli integral

$$p + \tfrac{1}{2}(U+w)^2 + \tfrac{1}{2}u^2 + \tfrac{1}{2}v^2 = \text{constant on streamlines}$$
$$= p_\infty + \tfrac{1}{2}U^2,\tag{5}$$

as all streamlines come from the same state (assuming no closed streamlines), or alternatively noting that $\mathbf{u} \times \omega = 0$ as the vortices are
free.

Batchelor (1964) deduced the dynamical necessity of axial flow relative
to the free stream in the core. The pressure in the core is lower than that
upstream of the wing because of the need to balance the centrifugal force.
Thus, in the absence of viscous effects, fluid particles on a streamline
originating ahead of the wing and entering the core are accelerated in
the downstream direction. A formal demonstration that $w > 0$ comes
from making the inviscid, light-loading approximation in which the right-
hand sides of equations (2), (3) and (4) are neglected. Then $v = v(r)$,
$w = w(r)$, $u = 0$. We combine the reduced form of the Bernoulli equation
$w^2 + 2Uw = -v^2 - 2p$ with the radial momentum equation $\partial p/\partial r = v^2/r$ to
give

$$\frac{d}{dr}(w^2 + 2Uw) = -\frac{dv^2}{dr} - 2\frac{v^2}{r} = \frac{1}{r^2}\frac{d}{dr}(v^2 r^2) < 0, \qquad (6)$$

since $v^2 r^2$ will increase with r if all the vorticity is of the same sign. Since
$w \to 0$ as $r \to \infty$, we have that

$$w^2 + 2Uw > 0 \qquad (7)$$

and we expect $w > 0$ and accelerated flow in the core. (An estimate of the
range of validity of this result can be made as follows. Viscous diffusion
is negligible if $\nu t \ll r_0^2$, where r_0 is the radius of the vortex and t is
identified with z/U. For a wing of span b, it is expected that $r_0 \sim b/5$
(Saffman [1974]), and this leads to $z/c \ll (Uc/\nu)A_R^2/25$, where c is the
chord and $A_R = b/c$ is the aspect ratio. The neglect of the non-linear terms
would seem to be valid after the roll-up is completed. The time for roll-up
is roughly r_0^2/Γ, where $\Gamma \sim Uc\alpha$. Here, α is the angle of attack. Then the
downstream distance to roll up is $b^2/25\alpha c$, which has to be small compared
with $Ub^2/25\nu$. There will then be a range of validity if $(Uc/\nu)\alpha \gg 1$.)

There may be a deceleration in the near wake owing to entrainment of
the viscous boundary layer into the core. This can be represented by a loss
of head ΔH. The addition of $-\Delta H$ to the right-hand side of the reduced
Bernoulli equation can lead to $w < 0$.

To study the effects of viscosity on the evolution of the core, we introduce the viscous light-loading approximation. This is a boundary layer
type approximation, with $\partial/\partial z \ll \partial r$ and $u \ll w \ll U$. We neglect the
terms in (2), (3) and (4) in brackets and angle brackets. All terms on the
right-hand side of the radial momentum equation are also neglected. This

leaves us with the equations

$$U\frac{\partial v}{\partial z} = \nu\left(\frac{\partial^2 v}{\partial r^2} + \frac{1}{r}\frac{\partial v}{\partial r} - \frac{v}{r^2}\right),\tag{8}$$

$$-\frac{v^2}{r} + \frac{\partial p}{\partial r} = 0,\tag{9}$$

$$U\frac{\partial w}{\partial z} = -\frac{\partial p}{\partial z} + \nu\left(\frac{\partial^2 w}{\partial r^2} + \frac{1}{r}\frac{\partial w}{\partial r}\right).\tag{10}$$

These equations are to be solved with initial conditions on $z = 0$ for $v(r,0)$ and $w(r,0)$. Further, $w \sim 0$ and $v \sim \Gamma/2\pi r$ as $r \to \infty$.

It follows from (9) and (10), on carrying out integrations by parts, that

$$U\frac{d}{dz}\int_0^\infty wr\,dr = -\frac{d}{dz}\int_0^\infty pr\,dr = \frac{d}{dz}\int_0^\infty \frac{1}{2}r^2\frac{\partial p}{\partial r}\,dr$$

$$= \frac{1}{2}\frac{d}{dz}\int_0^\infty rv^2\,dr.\tag{11}$$

Now (8) can be written, after multiplying by vr,

$$\frac{\partial}{\partial z}Urv^2 = -\nu\left(\frac{v^2}{r} + r\left(\frac{\partial v}{\partial r}\right)^2\right) + \nu\frac{\partial}{\partial r}\left(rv\frac{\partial v}{\partial r}\right).\tag{12}$$

Hence

$$\frac{d}{dz}\int_0^\infty rv^2\,dr < 0, \quad\text{and consequently}\quad \frac{d}{dz}\int_0^\infty rw\,dr < 0.\tag{13}$$

Thus the axial flux decays under the action of viscosity, and (11) can be interpreted physically as transfer of induced drag from kinetic energy of transverse motion to the axial momentum flux deficit.

Batchelor [1964] has studied the far-field asymptotic solution for $z \to \infty$. The asymptotic solution of (8) is taken as

$$v = \frac{\Gamma}{2\pi r}\left(1 - e^{-Ur^2/4\nu z}\right).\tag{14}$$

The pressure is

$$p = -\frac{\Gamma^2 U}{32\pi^2\nu z}P(\zeta), \quad\text{where}\quad P(\zeta) = \int_\zeta^\infty \frac{(1-e^{-\xi})^2}{\xi^2}\,d\xi \quad\text{and}\quad \zeta = \frac{Ur^2}{4\nu z}.\tag{15}$$

The equation for w then integrates to give[3]

$$w = -\frac{\Gamma^2 U}{32\pi^2\nu z}e^{-\zeta}\left[\log\frac{Uz}{\nu} - \int_0^\zeta \left(1 - e^\xi \xi + Pe^\xi\right)\,d\xi + K\right],\tag{16}$$

[3] The calculation is not entirely straightforward.

where K is an arbitrary constant. Note that w is negative on the axis.

The intermediate region between the completion of roll-up and the far field where the Batchelor solution holds has been investigated by Moore and Saffman [1973] (see also Baker et al. [1974]). The central part of the core is studied, where the roll-up produces an azimuthal distribution

$$v = \beta r^{-n}, \tag{17}$$

where β and n depend upon the shape of the wing but n is expected to lie between 0 and 1. The special case of elliptic loading gives $n = \frac{1}{2}$.[4] The inviscid light-loading approximation then gives

$$p \sim -\frac{1}{2n} \frac{\beta^2}{r^{2n}} \quad \text{as } r \to 0. \tag{18}$$

Then the Bernoulli equation gives

$$w \sim \frac{\beta^2}{2U} \left(\frac{1}{n} - 1 \right) r^{-2n} \quad \text{as } r \to 0. \tag{19}$$

It is noted that $w > 0$ in accordance with Batchelor's general argument, but $w \to \infty$ as $r \to 0$, and the need to consider viscous effects is obvious.

The appropriate equations for the viscous modification of the central core are the viscous light-loading approximation with initial conditions

$$v(r,0) = \frac{\beta}{r^n}, \qquad w(r,0) = \frac{\beta^2}{2U} \left(\frac{1}{n} - 1 \right) \frac{1}{r^{2n}}. \tag{20}$$

The solution is now straightforward, using the similarity variables of Kirde [1962]. It is convenient to put $z = Ut$ and then $\eta = -r^2/4\nu t$. Comparing with the results of §1, we find that

$$v = \frac{\beta}{(\nu t)^{\frac{1}{2n}}} V_n(\eta), \tag{21}$$

where

$$V_n(\eta) = 2^{-n} \Gamma(\frac{3}{2} - \frac{1}{2} n)(-\eta)^{\frac{1}{2}} M(\frac{1}{2} + \frac{1}{2} n; 2; \eta). \tag{22}$$

M denotes the confluent hypergeometric function (Abramowitz and Stegun [1964]). For the case $n = \frac{1}{2}$, the maximum of $V_{1/2}(\eta)$ occurs at $\eta = -2.10$, giving a radius r_1 of maximum azimuthal velocity

$$r_1 = 2.92 \left(\frac{\nu z}{U} \right)^{\frac{1}{2}} = 2.92 c \left(\frac{z}{c} \right)^{\frac{1}{2}} \left(\frac{Uc}{\nu} \right)^{\frac{1}{2}}. \tag{23}$$

---

[4] The corresponding tip loading is proportional to x^{1-n} where x is the distance from the tip.

If the core radius is defined as the radius at which v takes its maximum value, experimentally observed core radii can be expected to be small. The pressure is obtained by substituting (21) into (9) and integrating. We find

$$p = -\frac{\beta^2}{(\nu t)^n} P_n(\eta),$$ (24)

where

$$P_n(\eta) = -\tfrac{1}{2} \int_{-\infty}^{\eta} \eta^{-1} V_n^2(\eta) \, d\eta.$$ (25)

Then the axial velocity is

$$w = \frac{\beta^2}{U(\nu t)^n} W_n(\eta)$$ (26)

and (10) for w becomes

$$\eta \frac{d^2 W_n}{d\eta^2} + (1 - \eta) \frac{dW_n}{d\eta} - nW_n = -nP_n - \eta \frac{dP_n}{\eta}.$$ (27)

The initial and boundary conditions require

$$W_n \sim 2^{-1-2n} \left(\frac{1}{n} - 1 \right) (-\eta)^{-n} \quad \text{as } \eta \to -\infty.$$ (28)

The differential equation can be reduced to quadrature by the method of variation of parameters, but it seems better to proceed numerically. A particular integral W_I of (27) is defined by the requirement that $W_I(0) = 0$. Because the equation is singular, this defines W_I uniquely. The asymptotics of (27) show that

$$W_I(\eta) \sim \gamma(-\eta)^{-n} \quad \text{as } \eta \to -\infty,$$ (29)

and γ can be found by numerical integration. The general solution of (27) is

$$W_n(\eta) = W_I(\eta) + \mu M(n; 1; \eta).$$ (30)

Since $M(n; 1; \eta) \sim (-\eta)^{-n}/\Gamma(1 - n)$ as $\eta \to -\infty$, we must choose μ so that

$$\frac{\mu}{\Gamma(1 - n)} + \gamma = 2^{-1-2n} \left(\frac{1}{n} - 1 \right).$$ (31)

Since $W_n(0) = \mu$, the axial velocity on the core axis is determined. The values of μ are sketched in Figure 13.2-2. It is positive for $n < 0.44$. This is consistent with the known result for conical flow (Stewartson and Hall [1963]), which corresponds to $n = 0$. The negative values for $n > 0.44$ are consistent with Batchelor's case, which corresponds to $n = 1$. Note that the

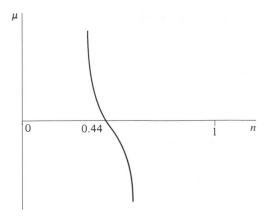

13.2-2 Dependence of axial velocity parameter μ on core structure.

limiting cases $n = 0$ and $n = 1$ require special treatment; logarithmic terms need to be included. This accounts for the apparent singular behaviour of μ as $n \to 0$ or 1. For a comparison with experiments, see Moore and Saffman [1973] and Baker et al. [1974].

These results imply that if a balloon or similar object is released into the trailing vortex wake of an aircraft, an observer on the ground could see the balloon either following the aircraft or moving in the opposite direction (until wake diffusion becomes dominant) depending upon the shape of the wing.

In using (17), we are in effect neglecting the detailed spiral structure of the inviscid vortex and replacing the flow field which has discrete jumps by a smoothed-out distribution. This is, in fact, not a mathematical artifice but a valid procedure because, as was pointed out by Maskell [1964], the turns are thickened by viscous diffusion and the spiral structure disappears. From the analysis of §8.5, the equation of the spiral is

$$r = \left(\frac{\beta t}{\theta - \theta_\infty} \right)^{1/(n+1)}, \tag{32}$$

which will hold out to a radius of the central portion

$$a_0(t) \propto (\beta t)^{1/(n+1)} \tag{33}$$

until a_0 becomes comparable with the dimensions of the sheet from which the core is formed, and the roll-up is complete and a_0 ceases to grow. It follows from (32) that the radial distance between successive turns is $2\pi r^{n+2}/\{(n+1)\beta t\}$. This distance is small compared with the distance $(\nu t)^{\frac{1}{2}}$

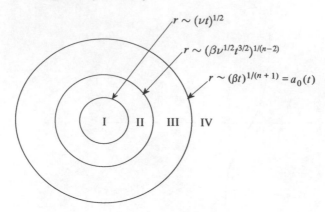

$r \sim (\nu t)^{1/2}$

$r \sim (\beta \nu^{1/2} t^{3/2})^{1/(n-2)}$

$r \sim (\beta t)^{1/(n+1)} = a_0(t)$

13.2-3 Sketch of vortex structure. I, viscous inner region of radius $O(\nu t)^{\frac{1}{2}}$. II, smoothed-out spiral; the velocity distribution is essentially inviscid. III, tightly wound inviscid spiral. IV, external region containing unrolled-up part of vortex sheet.

over which viscosity smears the sheet (and which is also the radius of the viscosity-dominated central core region) if

$$r \ll (\beta \nu^{\frac{1}{2}} t^{\frac{3}{2}})^{1/(n+2)} = a_s. \tag{34}$$

The radius a_s is larger than that of the viscous core when this is small compared with a_0 given by (33). This picture of the vortex is summarized in Figure 13.2-3.

13.3   Burgers vortices

It is known[5] that solutions of the equations for a solenoidal vector field $\omega(\mathbf{x}, t)$

$$\frac{\partial \omega_i}{\partial t} + U_j \frac{\partial \omega_i}{\partial x_j} = \omega_j \frac{\partial U_i}{\partial x_j} + \nu \nabla^2 \omega_i \tag{1}$$

(the summation convention is employed) can be expressed as integrals in closed form when

$$U_i = \alpha_{ij} x_j, \tag{2}$$

[5] See Burgers [1948]. The results have been rediscovered frequently. Alexandrou [1963] employed solutions of the general equation to investigate the behaviour of magnetic fields in linear straining fields.

where the coefficients α_{ij}, subject to $\alpha_{ii} = 0$ so that the imposed straining field is incompressible, can be arbitrary functions of time. This is also true for the scalar diffusion equation for a concentration $s(\mathbf{x}, t)$

$$\frac{\partial s}{\partial t} + U_j \frac{\partial s}{\partial x_j} = \nu \nabla^2 s. \tag{3}$$

Townsend [1951] gives the Green's function for (3) in the case that α_{ij} is diagonal with coefficients α, β, γ. It is

$$s = s_m(t) \exp\left[-\frac{1}{2}\frac{(x - \xi e^{\int \alpha\, dt})^2}{a^2} - \frac{1}{2}\frac{(y - \eta e^{\int \beta\, dt})^2}{b^2} - \frac{1}{2}\frac{(z - \zeta e^{\int \gamma\, dt})^2}{c^2}\right],$$

$$\tag{4}$$

where

$$\frac{da^2}{dt} - 2\alpha a^2 = 2\nu, \qquad \frac{db^2}{dt} - 2\beta b^2 = 2\nu, \qquad \frac{dc^2}{dt} - 2\gamma c^2 = 2\nu \quad . \tag{5}$$

and

$$s_m = \frac{(2\pi)^{3/2}}{abc}. \tag{6}$$

If the strain rates are constant in time,

$$a^2 = \frac{\nu}{\alpha}(e^{2\alpha t} - 1), \qquad b^2 = \frac{\nu}{\beta}(e^{2\beta t} - 1), \qquad c^2 = \frac{\nu}{\gamma}(e^{2\gamma t} - 1). \tag{7}$$

As noted by Burgers and Townsend, these give exact solutions of the Navier–Stokes equations if the vorticity is unidirectional, parallel to the z-axis say and independent of z, and if the vorticity field is either one-dimensional (i.e., $\omega = \omega(x, t)$) or the vorticity field and imposed strain are axisymmetric. In the first case,

$$\boldsymbol{\omega} = (0, 0, \omega(x, t)), \qquad U = \alpha x, \qquad V = \beta z, \qquad W = \gamma z. \tag{8}$$

In the second case,

$$\boldsymbol{\omega} = (0, 0, \omega(r, t)), \qquad U = -\tfrac{1}{2}\gamma x, \qquad V = -\tfrac{1}{2}\gamma y, \qquad W = \gamma z, \tag{9}$$

where $r = \sqrt{x^2 + y^2}$. The velocity induced by the vorticity lies in the xy-plane, with components u, v, say, which are independent of z. It is immediately apparent that the self-induced convection $u\partial\omega/\partial x + v\partial\omega/\partial y$ is zero and there is no self-induced stretching, so that the solution of the appropriate form of (1) is also a solution of the Navier–Stokes equation with the full non-linear vorticity stretching and convection terms included.

In the one-dimensional case, the equation for ω is

$$\frac{\partial \omega}{\partial t} + \alpha x \frac{\partial \omega}{\partial x} = \gamma \omega + \nu \frac{\partial^2 \omega}{\partial x^2}. \tag{10}$$

If $\gamma > 0$ and is constant, and $\alpha = -\gamma$, this equation has the steady solution

$$\omega = \omega_m e^{-\gamma x^2/2\nu} \tag{11}$$

in which the diffusion of vorticity in the x-direction is balanced by the convection of the vortex lines towards the $x = 0$ plane and the stretching in the z-direction by the imposed strain.

In the axisymmetric case, the steady solution is

$$\omega = \omega_m e^{-\gamma(x^2+y^2)/4\nu}. \tag{12}$$

Robinson and Saffman [1984b] have investigated the existence of steady Burgers vortices which are two-dimensional but not axisymmetric. That is, solutions of the equations

$$(-\alpha x + u)\frac{\partial \omega}{\partial x} + (-\beta y + v)\frac{\partial \omega}{\partial y} = (\alpha + \beta)\omega + \nu\nabla^2\omega, \tag{13}$$

$$\nabla^2\psi = -\omega, \quad u = \frac{\partial\psi}{\partial y}, \quad v = -\frac{\partial\psi}{\partial x}. \tag{14}$$

Solutions to these equations are referred to as fields of stretched vorticity. As just described, simple closed-form solutions exist with $\alpha = \beta > 0$ and $\alpha > 0, \beta = 0$. The equations can be conveniently non-dimensionalised by scaling the time on $(\alpha + \beta)/2$ and lengths on $[2\nu(\alpha + \beta)]^{1/2}$. In addition, the magnitude of the non-dimensional vorticity and stream function can be scaled on the Reynolds number $R = \iint \omega \, dx \, dy/(2\pi\nu)$, so that the circulation is fixed in the scaled form and the Reynolds number enters as a parameter in the equations. In the scaled variables, (14) remains unaltered and (13) becomes

$$\omega_{xx} + \omega_{yy} + [(1 + \epsilon)x - Ru]\omega_x + [(1 - \epsilon)y - Rv]\omega_y + 2\omega = 0, \tag{15}$$

where

$$\epsilon = \frac{\alpha - \beta}{\alpha + \beta}, \quad \iint \omega \, dx \, dy = 2\pi. \tag{16}$$

Without loss of generality, we can take $\alpha \geq \beta$ and anticipate that a necessary condition for solutions to exist is that $\alpha + \beta > 0$. The axisymmetric case is the limit $\epsilon = 0$, and the one-dimensional case is $\epsilon = 1$.

For $R = 0$, a simple closed-form solution exists

$$\omega = \sqrt{1 - \epsilon^2} \exp[-(1 + \epsilon)x^2/2 - (1 - \epsilon)y^2/2], \tag{17}$$

for which the streamlines are ellipses. Robinson and Saffman demonstrate formally the existence of expansions in powers of R and ϵ, and calculate solutions numerically up to $R = 100$ in the range $0 \le \epsilon < 100$. There is no evidence to suggest non-existence for arbitrary R and $\epsilon \le 1$, and indeed the solutions appear to be unique. An interesting feature of the solutions is that for all ϵ, the streamlines become more and more circular as R increases. The dominance of non-linear terms at large R requires that the velocity is orthogonal to the gradient of the vorticity, which produces a circular shape. Robinson and Saffman also study the linear stability of the axisymmetric Burgers vortex to two-dimensional disturbances and conclude that the axisymmetric vortex is linearly stable to this class of disturbance. Stability to three-dimensional disturbances is an open question.

Unsteady one-dimensional exact solutions can be studied using (10), for which the general initial value problem solution can be written (using Townsend's Green's function (4) and integrating with respect to η and ζ, or finding it directly)

$$\omega(x,t) = s_0(t) \frac{e^{\gamma t}}{2\pi a} \int_{-\infty}^{\infty} \omega_0(\xi) \exp\left[-\frac{1}{2} \frac{(x - \xi e^{\alpha t})^2}{a^2} \right] d\xi \qquad (18)$$

where $\omega_0(x) = \omega(x,0)$ and

$$\frac{\dot{s}_0}{s_0} = \gamma - \frac{\nu}{a^2}, \qquad s_0 \sim 1/\sqrt{4\pi\nu t} \quad \text{as } t \to 0. \qquad (19)$$

as required for the proper normalisation of the Green's function. Note also that $d(as_0)/dt = (\gamma + \alpha) as_0$, which is a direct consequence of equation (10).

It can be shown that for the case $\alpha + \gamma = 0$, $\gamma > 0$, (18) tends to the steady solution (11) with $\omega_m = s_0(\infty) \int \omega_0(\xi) d\xi$,[6] $s_0(\infty) = \sqrt{-2\pi\alpha/\nu}$.

Buntine and Pullin [1989] have studied a case in which the initial vorticity distribution is not one-dimensional but consists of two equal and opposite circular vortices and $\beta = -(\gamma + \alpha) > 0$, so that the vortices are pulled out in the y-direction. Equation (13) was integrated numerically and results for the decay of circulation $\Gamma(t)$ were found to be modelled quite well by a one-dimensional model as follows.[7]

Consider the exact one-dimensional flow, in which the initial distribution of vorticity is two strips of equal and opposite strengths at $x = \pm X_0$. Use the vorticity in the strip between $y = \pm\frac{1}{2} Y_0$ to model the two vortices.

---

[6] Kambe [1983] has discussed various cases in which $\gamma = 0$ and the imposed strain is purely two-dimensional and there is no vortex line stretching.

[7] The following paragraph is based on ideas of Prof D. I. Pullin.

The circulation of the vortices is

$$\Gamma_0 = Y_0 \int_0^\infty \omega_0(x)\,dx. \tag{20}$$

The centroids $\pm X_0$ can be taken as the y-impulse divided by the circulation. After time t, the span-wise extent of the strip becomes $Y = e^{\beta t}Y_0$. The circulation of the vortex becomes

$$\Gamma(t) = Y_0 e^{\beta t} \int_0^\infty \omega(x, t)\,dx. \tag{21}$$

Now it can be shown by asymptotic evaluation of the Green's function that

$$\omega(x, t) \sim \frac{1}{\sqrt{2\pi}} \left(\frac{-\alpha}{\nu}\right)^{3/2} e^{(\gamma+2\alpha)t} x e^{\alpha x^2/2\nu} \int_{-\infty}^\infty x' \omega_0(x')\,dx'. \tag{22}$$

It follows that

$$\frac{\Gamma(t)}{\Gamma_0} \sim \left(\frac{-2X_0^2 \alpha}{\pi\nu}\right)^{1/2}. \tag{23}$$

These predictions of a simple model that the decay of the circulation is independent of the stretching and depends on the rate at which the vortices are squashed together by the converging strain are consistent with the numerical calculations of Buntine and Pullin.

Lundgren [1982] has shown that two-dimensional strained vortices which are also stretched by an axisymmetric irrotational strain correspond to two-dimensional strained but unstretched vortices. When ω is parallel to the z-axis and depends only on (x, y, t) and the external irrotational strain is $-\frac{1}{2}\gamma, -\frac{1}{2}\gamma, \gamma$, the vorticity equation becomes (in cylindrical polar coordinates)

$$\frac{\partial \omega}{\partial t} + \left(u_r - \tfrac{1}{2}\gamma r\right)\frac{\partial \omega}{\partial \theta} + \frac{u_\theta}{r}\frac{\partial \omega}{\partial \theta} = \gamma\omega + \nu\nabla^2\omega. \tag{24}$$

The continuity equation for the induced velocity (u_r, u_θ) takes the two-dimensional form and there exists a stream function $\psi(r, \theta)$ such that

$$u_r = \frac{1}{r}\frac{\partial \psi}{\partial \theta}, \quad u_\theta = -\frac{\partial \psi}{\partial r}, \quad \text{where } \omega = -\nabla^2\psi. \tag{25}$$

Lundgren points out that solutions of this set of equations are related to solutions of a strictly two-dimensional flow, that is, the same equations with $\gamma = 0$. He defines new variables by

$$R = \sqrt{S(t)}r, \quad T = \int_0^t S(t')\,dt',$$

$$\Psi(R, \theta, t) = \psi(r, \theta, t), \quad \Omega(R, \theta, T) = \omega(r, \theta, t), \tag{26}$$

where

$$S(t) = \exp\left(\int_0^t \gamma(t')\,dt'\right) \tag{27}$$

can be interpreted as the stretch factor; it is the length of an element of the vortex at time t which had unit length at $t = 0$. Then it is readily verified that Ψ and Ω satisfy the two-dimensional, no external strain, Navier–Stokes equations

$$\frac{\partial \Omega}{\partial T} + \frac{1}{R}\frac{\partial \Psi}{\partial \theta}\frac{\partial \Omega}{\partial R} - \frac{\partial \Psi}{\partial R}\frac{1}{R}\frac{\partial \Omega}{\partial \theta} = \nu \nabla_R^2 \Omega. \tag{28}$$

Thus a two-dimensional solution $\Omega(R, \theta, T)$ generates a three-dimensional axisymmetrically stretched vortex

$$\omega(r, \theta, t) = S(t)\Omega\left(\sqrt{S(t)}\,r, \theta, \int_0^t S(t')\,dt'\right). \tag{29}$$

Lundgren gives the example of an axially symmetric flow, for which Ω satisfies the two-dimensional axisymmetric heat equation which has the fundamental solution

$$\Omega = (1/T)\exp(-R^2/4\nu T). \tag{30}$$

When γ is a positive constant, $S(t) = e^{\gamma t}$ and

$$\omega = \frac{\gamma}{1 - e^{-\gamma t}}\exp\left(-\frac{\gamma^2}{4\nu[1 - e^{-\gamma t}]}\right). \tag{31}$$

The axisymmetric steady Burgers vortex is the limit as $t \to \infty$.[8]

We mention here that Neu [1984] showed that the unsteady equations obtained by Kida [1981a] for the inviscid motion of two-dimensional uniform vortex patches in uniform two-dimensional strain (§9.3) can be generalized to include uniform stretching. Kida's equations for the major and minor axes and orientation of the ellipse in the straining field $u = \alpha x$, $v = \beta y$, can be written

$$\dot{a} = (\alpha \cos^2 \theta + \beta \sin^2 \theta)a, \quad \dot{b} = (\alpha \sin^2 \theta + \beta \cos^2 \theta)b, \tag{32}$$

and

$$\dot{\theta} = \frac{\omega ab}{(a+b)^2} - \frac{1}{2}(\alpha - \beta)\frac{a^2 + b^2}{a^2 - b^2}\sin 2\theta, \tag{33}$$

---

[8] Buntine and Pullin [1989] point out that a stretched-vortex problem can be mapped into an equivalent fully two-dimensional flow when the stretching is not axisymmetric, but in this case there is a time-dependent uniform irrotational strain in the equivalent flow. They also give examples of transformations of flows with time-dependent uniform strains into flows with constant uniform strains.

with

$$\alpha = -\beta. \tag{34}$$

Neu shows that (32) and (33) remain true when (34) is not satisfied; that is, there is a z-direction stretching $\gamma = \beta + \alpha$. In this class of solutions, the cross-sectional area of the patch does not remain constant, and its relative rate of change is equal to γ. The extension, unlike Lundgren's transformation, requires that the flow is inviscid and that the vorticity is uniform.

14

MISCELLANEOUS TOPICS

14.1   Minimum induced drag

The force on a lifting body moving with uniform speed through an inviscid fluid can be interpreted in terms of the vortex force acting on the bound vorticity (see §3.1). For a three-dimensional body, the vortex lines cannot all be closed inside the body, because then the total drag (consisting of lift perpendicular plus drag antiparallel to the direction of motion) is zero. The bound vorticity continues into free vortex lines extending downstream in an inviscid wake. The free vorticity induces a velocity on the body which is equivalent to a change in the angle of attack, which alters the direction of the lift and thereby causes an effective drag equal to the component of the modified lift in the direction of flight (see Durand [1934], Glauert [1948]) which is called the induced drag.

A 'light-loading' approximation can be employed to calculate the induced drag by assuming that the free vortex lines are parallel to the undisturbed flow and the velocity that they induce is given by the Biot–Savart formula for a semi-infinite line vortex. When the bound vorticity is approximated further by a straight vortex filament of strength $\Gamma(s)$, $-\frac{1}{2}b \leq s \leq \frac{1}{2}b$, where b is the span, and the wake is consequently a semi-infinite vortex sheet of strength $d\Gamma/ds$, the downwards induced velocity $v(s)$ at station s is

$$v(s) = \frac{1}{4\pi} \int_{-\frac{1}{2}b}^{\frac{1}{2}b} \frac{d\Gamma(s')}{ds'} \frac{ds'}{s - s'}. \tag{1}$$

Using the Kutta formula for the lift on an element of the bound vortex, it follows that the induced drag is

$$D_i = \int_{-\frac{1}{2}b}^{\frac{1}{2}b} v(s)\Gamma(s)\,ds. \tag{2}$$

This approach is known as Prandtl's lifting-line theory (for details, see Batchelor [1967 p. 583] or Glauert [1948 chap. XI]). It can be shown that the induced drag is a minimum for given lift when

$$\Gamma(s) = \Gamma_0 \left(1 - \frac{4s^2}{b^2} \right)^{1/2}. \tag{3}$$

This is called the case of elliptical loading and would be the result of elementary airfoil theory for a wing of constant section and elliptical plan form.

An alternative approach to the calculation of the induced drag is Karman and Burgers's use of energy conservation (see Durand [1934]). Suppose the direction of flight is the z-axis. The inviscid wake in a plane $z = $ constant (a Trefftz plane) is approximated by a vorticity distribution $\omega(x, y)$ which gives rise to a stream function $\psi(x, y)$ for the two-dimensional motion in the plane. The kinetic energy E of the quasi-two-dimensional motion is

$$E = \tfrac{1}{2} \iint \omega \psi \, dx \, dy. \tag{4}$$

Since the wake is increasing in length, work must be done to provide its increasing kinetic energy. This is provided by the induced drag D_i, and hence

$$D_i = E/U, \tag{5}$$

where U is the speed of the body. Since the kinetic energy of a plane vortex sheet, $-\tfrac{1}{2} b \leq y \leq \tfrac{1}{2} b$, $x = 0$, is

$$\frac{1}{2\pi} \iint \kappa(y) \, \kappa(y') \log |y - y'| \, dy \, dy', \tag{6}$$

the two expressions give the same answer on writing $\kappa(y) = d\Gamma(y)/dy$. However, the energy approach is conceptually more satisfactory as the integral (4) is invariant under the light-loading approximation. That is, it does not vary with z, and the roll-up of the vorticity field under its self-induced motion can therefore be neglected (the kinetic energy of the axial flow produced by roll-up discussed in §13.2 is of higher order). This is not immediately obvious from the calculation of effective angle of attack.

Suppose now that the wing can be designed to give a required distribution of vorticity in a Trefftz plane. We consider the question of what distribution gives minimum induced drag given the total lift and the fact that the total vorticity must be zero; that is, ω satisfies

$$\iint \omega y \, dx \, dy = L, \qquad \iint \omega \, dx \, dy = 0, \tag{7}$$

where L is given. We also suppose that the vorticity lies inside a contour C of finite area. It was shown by Munk [1919] that given C (because it is intuitively clear that the induced drag can always be decreased by making the contour larger) the drag is a minimum when the vorticity is confined to a vortex sheet on the boundary of C whose strength is such that the induced velocity $u(x, y)$ in the Trefftz plane is vertical and constant inside C. (Winglets on modern aircraft reduce induced drag without increasing the span.)

It is easy to show that these are necessary conditions for stationarity. Denoting Lagrange multipliers by λ and μ, we require that ψ have the property that the first variation

$$\delta \left(-\frac{1}{2} \iint \psi \nabla^2 \psi \, dx \, dy + \lambda \iint y \nabla^2 \psi \, dx \, dy + \mu \iint \nabla^2 \psi \, dx \, dy \right) = 0. \tag{8}$$

Integrating by parts, we obtain the condition that

$$-\iint \delta\psi \, \nabla^2 \psi \, dx \, dy - \frac{1}{2} \oint \psi \frac{\partial}{\partial n} \delta\psi \, ds + \frac{1}{2} \oint \frac{\partial \psi}{\partial n} \delta\psi \, ds$$
$$+ \lambda \oint y \frac{\partial}{\partial n} \delta\psi \, ds - \lambda \oint \delta\psi \frac{dy}{ds} \, ds + \mu \oint \frac{\partial}{\partial n} \delta\psi \, ds = 0 \tag{9}$$

for arbitrary $\delta\psi$. It follows immediately that $\nabla^2 \psi = 0$, inside C, and hence there is no vorticity inside C and the boundary is a vortex sheet. Further, we find that on the boundary,

$$-\tfrac{1}{2}\psi + \lambda y + \mu = 0, \qquad \frac{1}{2}\frac{\partial \psi}{\partial n} - \lambda \frac{y}{s} = 0. \tag{10}$$

Then $u = \partial\psi/\partial y = 2\lambda = \text{constant}$, and $v = 0$. The second equation of (10) states that the tangential velocity on the boundary is $u \, dx/ds$, which is automatically satisfied. Since the kinetic energy is positive definite, it follows that the energy is a minimum.

14.2 Kelvin's variational principle

Consider two-dimensional inviscid incompressible unbounded flow. We are interested in states that are steady relative to some translating rotating frame of reference. The linear and angular impulses and kinetic energy are

$$I_x = \int \omega y \, dS, \quad I_y = -\int \omega x \, dS, \quad A = -\tfrac{1}{2} \int \omega r^2 \, dS, \quad T = \tfrac{1}{2} \int \omega \psi \, dS,$$

$$(1)$$

respectively, where $\nabla^2 \psi = -\omega$. These quantities are conserved during the motion. Kelvin [1875] showed that the steady solutions are the critical points of

$$H = T - UI_x - VI_y - \Omega A, \tag{2}$$

where U, V and Ω are Lagrange multipliers, with respect to area-preserving displacements which conserve vorticity, that is, variations $\delta \omega$ such that

$$\delta \omega = -\mathbf{s} \cdot \nabla \omega, \quad \text{where } \operatorname{div} \mathbf{s} = 0. \tag{3}$$

We can take $\mathbf{s} = (\partial \epsilon / \partial y, -\partial \epsilon / \partial x)$, where ϵ is an arbitrary function of x and y.

Taking the variation, we have

$$\delta H = \int (\psi - Uy + Vx + \tfrac{1}{2}\Omega r^2)\,\delta \omega \, dS. \tag{4}$$

Substituting (3) and integrating by parts, we have

$$\delta H = \int \left[\frac{\partial}{\partial y}(\psi - Uy + Vx + \tfrac{1}{2}\omega r^2)\,\frac{\partial \omega}{\partial x} \right.$$
$$\left. - \frac{\partial}{\partial x}(\psi - Uy + Vx + \tfrac{1}{2}\omega r^2)\,\frac{\partial \omega}{\partial y} \right] \epsilon \, dS. \tag{5}$$

Thus when H is stationary,

$$(\psi_y - U + \Omega y)\,\partial \omega / \partial x + (-\psi_x - V - \omega x)\,\partial \omega / \partial y = 0. \tag{6}$$

The motion is then steady in a frame translating with velocity (U, V) and rotating with angular velocity Ω. Kelvin mentions that if the second variation is definite (i.e., the equilibrium state is a maximum or minimum), then the equilibrium is stable. Kelvin's variational principle was used by Saffman and Szeto [1980] (see also §9.6) to discuss the stability of the co-rotating vortex pair. In this case, stable equilibria of uniform vortex patches of given area, S, would be found by minimizing $T - \Omega A$.

The use of variational methods for the calculation of equilibrium configurations is rendered difficult by the absence of simple and accurate formulae for the calculation of the kinetic energy. For the case of a uniform vortex patch, one requires the stream function inside the patch and its integral over the area S of the patch. The following formula is sometimes helpful and can be verified directly:

$$\int_S \psi \, dx \, dy = \tfrac{1}{2} \oint_{\partial S} \psi(x \, dy - y \, dx) - \tfrac{1}{4} \oint_{\partial S} (x^2 + y^2) \left(\frac{\partial \psi}{\partial x} dy - \frac{\partial \psi}{\partial y} dx \right)$$

$$+ \int_S (x^2 + y^2) \nabla^2 \psi \, dx \, dy. \tag{7}$$

This form has the advantage that it requires just the values of the tangential and normal components of the velocity on the boundary of the patch. $\nabla^2 \psi$ is either zero or minus the vorticity, depending upon whether ψ is generated by external vorticity or the constant vorticity in the patch.

It can be mentioned here that the use of variational methods or Newton's method to find numerical solutions for the steady shapes of vortex patches can encounter serious counting difficulties since the number of equations may not equal the number of unknowns. These problems can be caused by hidden or non-apparent kinematic symmetries or identities. Examples are the vanishing of

$$\int \mathbf{r} \cdot \mathbf{q} \, dS = \oint \tfrac{1}{2} r^2 (\mathbf{q} \cdot \mathbf{n}) \, ds \tag{8}$$

when $\operatorname{div} \mathbf{q} = 0$ and $\mathbf{q} = \Omega \times \mathbf{r} + k s$ on the boundary, as the right-hand side of (8) is then $-\Omega/8 \oint dr^4 = 0$. Another kinematic constraint for the calculation of vortex pairs is provided by the result that for a bounded distribution of vorticity[1]

$$\int \omega (\mathbf{r} \cdot \mathbf{q}) \, dS = 0. \tag{9}$$

This follows from the identity

$$-(\mathbf{r} \cdot \mathbf{q})\omega = (x\psi_y - y\psi_x)\nabla^2 \psi$$

$$= \frac{\partial}{\partial x} (x\psi_x\psi_y + \tfrac{1}{2} y\psi_y^2 - \tfrac{1}{2} y\psi_x^2) + \frac{\partial}{\partial y} (-\tfrac{1}{2} x\psi_x^2 + \tfrac{1}{2} x\psi_y^2 - y\psi_x\psi_y)$$

Hence,

$$\int \omega (\mathbf{r} \cdot \mathbf{q}) \, dS = \oint (x\psi_x\psi_y + \tfrac{1}{2} y\psi_y^2 - \tfrac{1}{2} y\psi_x^2) \, dy$$

$$+ \oint (y\psi_x\psi_y + \tfrac{1}{2} x\psi_x^2 - \tfrac{1}{2} x\psi_y^2) \, dx.$$

At infinity, the dominant contributions to the contour integral come from $\psi_x = -(\Gamma/2\pi r)\cos\theta$, $\psi_y = -(\Gamma/2\pi r)\sin\theta$, and the integrals vanish when

---

[1] Derived also by Dritschel [1985].

taken around a circle of large radius. For examples on the use of the kinematic constraints, see Meiron, Saffman and Schatzman [1984] and Kamm [1987].

An alternative variational principle has been given by Arnol'd [1965 1969][2] for a class of steady two-dimensional inviscid flows with the property that the Euler equation integral $\omega = \Omega(\psi)$ can be inverted to give $\psi = \Psi(\omega)$. A necessary condition is that $\nabla\omega \neq 0$ inside the domain D of the flow.[3] This excludes vortex patches and vortical flows in simply connected domains. Consider a flow inside a fixed simple closed contour C_0, with internal fixed simple closed boundaries C_i $(i = 1,\ldots,N)$. The unsteady evolution of the flow is given by[4]

$$\frac{\partial \omega}{\partial t} = -\nabla\psi \times \nabla\omega, \quad \nabla^2\psi = -\omega, \quad \frac{d\Gamma_i}{dt} = 0, \tag{10}$$

where $\Gamma_i = \oint_{C_i} \nabla\psi \times d\mathbf{s}$ is the circulation around the ith contour. In a steady equilibrium with $\omega = \omega_e$, it follows, of course, that $\nabla\omega \parallel \nabla\psi$.

We search now for a functional H which is a constant of the motion such that the equilibrium state is a critical point of H; that is, $\delta H = 0$ for infinitesimal variations of the vorticity. There are the following constants of the motion:

$$E = \int_D \tfrac{1}{2}(\nabla\psi)^2 \, dS, \quad F = \int_D \Phi(\omega) \, dS, \quad \Gamma_i. \tag{11}$$

Here, Φ is any function.

Suppose now that the dependence of the vorticity on the stream function can be inverted; that is, $\psi_e = \Psi(\omega_e)$ is a single-valued function over the range of ω. Then define

$$\Phi(\omega) = -\int^{\omega} \psi(\omega') \, d\omega', \quad \Psi_i = \psi_e \big|_{C_i}. \tag{12}$$

The equilibrium is then a critical point of

$$H = E + F - \sum_{i=0}^{N} \Gamma_i \Psi_i. \tag{13}$$

A proof is as follows:

$$\delta H = \delta \int [\tfrac{1}{2}(\nabla\psi)^2 + \Phi(\omega)] \, dS - \sum \Psi_i \, \delta\Gamma_i$$

---

[2] The treatment here is based on notes of Dr R. S. MacKay.

[3] In the case of parallel flow, this condition means no points of inflection.

[4] For two-dimensional vectors $\mathbf{a}$, $\mathbf{b}$, $\mathbf{a} \times \mathbf{b} = a_1 b_2 - a_2 b_1$.

$$= \int [\nabla \Psi \cdot \nabla \delta \psi + \Psi \nabla^2 \delta \psi] \, dS - \Psi_i \, \delta \Gamma_i, \quad \text{since } \Phi' \, \delta \omega = \Psi \nabla^2 \delta \psi,$$

$$= \int \text{div}(\Psi \cdot \nabla \delta \psi) \, dS - \Psi_i \, \delta \Gamma_i$$

$$= 0 \quad \text{since } \delta \Gamma_i = \oint \nabla \, \delta \psi \times d\mathbf{s}. \tag{14}$$

To find the second variation $\delta^2 H$, we write $\psi = \psi_e + \tilde{\psi}$, $\omega = \omega_e + \tilde{\omega}$. Then

$$\delta^2 H = \int [\tfrac{1}{2} (\nabla \tilde{\psi})^2 + \tfrac{1}{2} \Phi''(\omega_e) \tilde{\omega}^2] \, dS. \tag{15}$$

If $\Phi'' > 0$ in the domain, then $\delta^2 H$ is positive definite. Stability would then follow in a finite number of degrees of freedom system, But in a system with an infinite number of degrees of freedom, more work is required to establish stability. Suppose that the equilibrium flow is such that

$$0 < c \leq -\frac{d\Psi}{d\omega} \leq C < \infty. \tag{16}$$

Then

$$\int [|\nabla \tilde{\psi}|^2 + c \tilde{\omega}^2] \, dS \leq \int [|\nabla \tilde{\psi}_0|^2 + C \tilde{\omega}_0^2], \, dS \tag{17}$$

where $(\tilde{\psi}, \tilde{\omega})$ are the perturbations resulting from initial perturbations $(\tilde{\psi}_0, \tilde{\omega}_0)$. This proves stability in the energy-enstrophy norm. To prove (17), consider

$$H_2 = \int \left[\tfrac{1}{2} (\nabla \tilde{\psi})^2 + \Phi(\omega_e + \tilde{\omega}) - \Phi(\omega_e) - \Phi'(\omega_e) \tilde{\omega} \right] \, dS. \tag{18}$$

Now H_2 is a constant of the flow. This follows on writing $\tfrac{1}{2} (\nabla \tilde{\psi})^2 = \tfrac{1}{2} [\nabla(\psi_e + \tilde{\psi})]^2 - \nabla \psi_e \cdot \nabla \tilde{\psi} - \tfrac{1}{2} (\nabla \psi_e)^2]$ and remembering that the energy and functions of the vorticity and constants are conserved, and that the terms linear in $\tilde{\psi}$ and $\tilde{\omega}$ integrate to contour integrals equal to $\sum (\Psi_i)_e \, \delta \tilde{\Gamma}_i$, which are constant. Equation (16) puts bounds on Φ'', and hence

$$\frac{c}{2} \tilde{\omega}^2 \leq \Phi(\omega_e + \tilde{\omega}) - \Phi(\omega_e) - \Phi'(\omega_e) \tilde{\omega} \leq \frac{C}{2} \tilde{\omega}^2. \tag{19}$$

The result now follows, since

$$\int \left[\tfrac{1}{2} |\nabla \tilde{\psi}|^2 + \frac{c}{2} \tilde{\omega}^2 \right] \, dS \leq H_2(\tilde{\psi}(t)) = H_2(\tilde{\psi}_0) \leq \int \left[\tfrac{1}{2} |\nabla \tilde{\psi}_0|^2 + \frac{C}{2} \tilde{\omega}_0^2 \right] \, dS. \tag{20}$$

Note that the condition (16) is equivalent to the non-existence of a point of inflexion for channel flow, and for flow in an annulus for which the azimuthal velocity $v = -d\Psi/dr$ and $\omega = (1/r)\,d(vr)/dr$, it is equivalent to $d(vr)^2/dr > 0$, which is Rayleigh's criterion for the stability of flow between rotating cylinders.

14.3   Hamiltonian dynamics of vortex patch moments

It is well known and has been demonstrated by a number of authors that the motion of an incompressible inviscid fluid is a Hamiltonian system.[5] In particular, this is true for uniform two-dimensional vortex patches, and any functional F of the patches will evolve according to the equation

$$\frac{dF}{dt} = \{F, H\}, \tag{1}$$

where H is the Hamiltonian and $\{*, *\}$ is the Poisson bracket of two functionals. For N uniform vortex patches, $\alpha = 1, \ldots, N$, each of area S_α containing constant vorticity ω_α, the Hamiltonian is

$$H = \frac{1}{2} \sum_{\alpha=1}^{N} \omega_\alpha \int_{S_\alpha} \psi_\alpha(x, y)\, dx\, dy, \tag{2}$$

where ψ_α is the stream function inside the patch S_α given by

$$\psi_\alpha(x, y) = -\frac{1}{2\pi} \sum_{\beta=1}^{N} \omega_\beta \int_{S_\beta} \log|\mathbf{r} - \mathbf{r}'|\, dx'\, dy'. \tag{3}$$

If the patches are contained within a rigid boundary, the logarithmic Green's function would be replaced by another appropriate Green's function. Also, the area integrals in (1) and (2) can be replaced by contour integrals around the boundaries ∂S_α using the contour dynamics integration by parts approach and the formula (2.7) for the area integral of the stream function.

The Poisson bracket is calculated as follows. First given F, we construct a function ψ_F defined on the boundaries ∂S_α by

$$\lim_{\epsilon \to 0} \frac{F(\epsilon) - F(0)}{\epsilon} = \sum_{\alpha=1}^{N} \omega_\alpha \int_{\partial S_\alpha} \psi_F\, d\psi. \tag{4}$$

In this formula, ψ is an arbitrary smooth function defined on the domain, which defines an incompressible velocity field by $(u, v) = (\psi_y, -\psi_x)$. Then

---

[5] This section is based on work and notes by Dr A. Rouhi, who followed Marsden and Weinstein's [1983] treatment of Hamiltonian dynamics.

the left-hand side of (4) defines the rate of change of the functional F as the patches are convected by the flow defined by ψ, and ψ_F is to be chosen (independent of ψ) so that the relation (4) holds. It is a consequence of the abstract theory that ψ_F exists, and the method of calculation will be illustrated for the case of the moments. Then the Poisson bracket of functionals F and G is

$$\{F, G\} = \sum_{\alpha=1}^{N} \omega_\alpha \oint_{\partial S_\alpha} \psi_F \, d\psi_G, \tag{5}$$

where ψ_G is defined in the same way. This Poisson bracket is skew symmetric and satisfies the Jacobi identity. We now consider the functionals defined by the moments of the patches. For the patch S_α, we define for non-negative integers n and m the functional

$$K_{n,m}^\alpha = \int_{S_\alpha} x^n y^m \, dx \, dy. \tag{6}$$

The rate of change of this functional on the flow defined by ψ is

$$\frac{dF}{d\epsilon} = \int_{\partial S_\alpha} x^n y^m \, q_n \, ds = \int_{\partial S_\alpha} x^n y^m \, d\psi, \tag{7}$$

since $q_n \, ds = d\psi$, where q_n denotes the component of the flow normal to the boundary. Comparing with (4), we see that

$$\psi_F(x, y) = \frac{1}{\omega_\alpha} x^n y^m \quad \text{for } (x, y) \in \partial S_\alpha \tag{8}$$

and zero elsewhere. Then from the definition (5) of the Poisson bracket, we find

$$\{K_{n,m}^\alpha, K_{k,l}^\beta\} = \frac{\delta_{\alpha\beta}}{\omega_\alpha} \int_{\partial S_\alpha} (x^n y^m) \, d(x^k y^l) = \frac{\delta_{\alpha\beta}}{\omega_\alpha} (nl - mk) \, K_{n+k-1,m+l-1}^\alpha, \tag{9}$$

on converting the contour integrals to area integrals. It is readily confirmed that (9) satisfies the required properties of skew symmetry and the Jacobi identity. Note that the lowest-order moment $K_{0,0}^\alpha$ is the area of the α patch and its conservation is ensured by the fact that $K_{0,0}^\alpha = S_\alpha$ is a 'Casimir'; that is, it has vanishing bracket with any function of the moments.

It is clear that if the distance between the patches is large compared with their size and the patches are nearly circular, then it will be possible to expand the Hamiltonian as a series in the moments. The Hamiltonian equations then become an infinite system of ordinary differential equations, and it is natural to consider the truncation of this system. We call $n + m$ the order of the moment $K_{n,m}^\alpha$, and consider describing the evolution of the patches by ignoring moments of order greater than some number M. This, however, fails and there is a subtle and as yet unsolved closure problem because the truncated system fails to satisfy the Jacobi identities. Suppose, for instance, that $m = 3$, and consider the Jacobi identity

$$\{K_{3,0}^\alpha, \{K_{0,3}^\alpha, K_{0,1}^\alpha\}\} + \{K_{0,3}^\alpha, \{K_{3,0}^\alpha, K_{0,1}^\alpha\}\} + \{K_{0,1}^\alpha, \{K_{3,0}^\alpha, K_{0,3}^\alpha\}\} = 0. \tag{10}$$

The first term vanishes from (9). The last term would be dropped in the truncation because $\{K_{3,0}^\alpha, K_{0,3}^\alpha\}$ is of order 4. But the middle term is $18K_{1,2}^\alpha/\omega_\alpha^2$ which is not in general zero.

The case $M = 2$ is an important exception, and is connected to the properties of the exact KMSK (Kirchhoff, Moore, Saffman, Kida) unsteady elliptical patch solutions of the Euler equations. The brackets of moments of order 2 or less are

$$\{K_{0,0}^\alpha, H\} = 0 \quad \text{for any } H, \quad \{K_{1,0}^\alpha, K_{0,1}^\alpha\} = \frac{K_{0,0}^\alpha}{\omega_\alpha} = \frac{S_\alpha}{\omega_\alpha}, \tag{11}$$

$$\{K_{1,0}^\alpha, K_{1,1}^\alpha\} = \frac{K_{1,0}^\alpha}{\omega_\alpha}, \quad \{K_{1,0}^\alpha, K_{0,2}^\alpha\} = \frac{2K_{0,1}^\alpha}{\omega_\alpha},$$

$$\{K_{0,1}^\alpha, K_{1,1}^\alpha\} = -\frac{K_{0,1}^\alpha}{\omega_\alpha}, \quad \{K_{0,1}^\alpha, K_{2,0}^\alpha\} = -2\frac{K_{1,0}^\alpha}{\omega_\alpha},$$

$$\{K_{2,0}^\alpha, K_{0,2}^\alpha\} = 4\frac{K_{1,1}^\alpha}{\omega_\alpha}, \quad \{K_{2,0}^\alpha, K_{1,1}^\alpha\} = 2\frac{K_{2,0}^\alpha}{\omega_\alpha}, \quad \{K_{0,2}^\alpha, K_{1,1}^\alpha\} = -2\frac{K_{0,2}^\alpha}{\omega_\alpha},$$

where all other brackets involving just functionals of order 2 or less vanish, as do those between moments of different patches.

It proves to be more convenient to use the moments relative to the centroids of the patches. We define in the usual way for two-dimensional motion the centroids

$$(X_\alpha, Y_\alpha) = \frac{1}{S_\alpha} (K_{1,0}^\alpha, K_{0,1}^\alpha). \tag{12}$$

Second-order moments (n+m=2) relative to the centroid are

$$J_{2,0}^\alpha = K_{2,0}^\alpha - S_\alpha X_\alpha^2,$$
$$J_{1,1}^\alpha = K_{1,1}^\alpha - S_\alpha X_\alpha Y_\alpha, \tag{13}$$
$$J_{0,2}^\alpha = K_{0,2}^\alpha - S_\alpha Y_\alpha^2.$$

The relations (11) then become

$$\{S_\alpha, H\} = 0 \quad \text{for any } H, \quad \{X_\alpha, Y_\alpha\} = \frac{1}{S_\alpha \omega_\alpha},$$

$$\{J_{2,0}^\alpha, J_{0,2}^\alpha\} = 4\frac{J_{1,1}^\alpha}{\omega_\alpha}, \quad \{J_{2,0}^\alpha, J_{1,1}^\alpha\} = 2\frac{J_{2,0}^\alpha}{\omega_\alpha}, \quad \{J_{0,2}^\alpha, J_{1,1}^\alpha\} = -2\frac{J_{0,2}^\alpha}{\omega_\alpha}. \tag{14}$$

Note that the brackets of X_α and Y_α with the second-order moments vanish. It is also a consequence of these relations that the quantity

$$C_\alpha = J_{2,0}^\alpha J_{0,2}^\alpha - (J_{1,1}^\alpha)^2 \tag{15}$$

has zero bracket with all quantities S_α, X_α, Y_α, $J_{n,m}^\alpha$ $(n+m=2)$. That is, C_α behaves like a Casimir for the second-order truncated system. It will be shown below that this is another way of expressing the conservation of patch area.

Let us now consider the case of a single elliptical patch in a uniform straining field. The dynamics of these KMSK vortices were discussed in §9.3, where it was shown that they are exact solutions of the Euler equations. For a single patch, we can drop the label α. The area S and centroid co-ordinates X, Y are three variables. For an ellipse with major axis $2a$ and minor axis $2b$, two further variables are the ratio of axis length $\lambda = a/b$ and the angle between the major axis and the x-axis, which is denoted by ϕ. These are related to the second-order moments by (Melander, Zabusky and Styczek [1986])

$$J_{2,0} = \frac{S^2}{4\pi\lambda}(\lambda^2 + (1 - \lambda^2)\sin^2\phi), \tag{16}$$

$$J_{0,2} = \frac{S^2}{4\pi\lambda}(\lambda^2 + (1 - \lambda^2)\cos^2\phi), \tag{17}$$

$$J_{1,1} = -\frac{S^2}{8\pi\lambda}(\lambda^2 + (1 - \lambda^2)\sin 2\phi). \tag{18}$$

There appears to be a counting conflict because the ellipse is described completely by five variables λ, ϕ, S, X, Y, whereas the second-order truncation involves six variables, S, X, Y, $J_{2,0}$, $J_{1,1}$, $J_{0,2}$, but it can be shown from (16)–(18) that for an ellipse the moments are not independent and that

$$J_{2,0}J_{0,2} - J_{1,1}^2 = \frac{S^4}{16\pi}. \tag{19}$$

The left-hand side of (19) is precisely the Casimir (15), which is conserved for an ellipse by the conservation of area.

The Hamiltonian is given by (2), with an additional term due to the external uniform straining field. Since the stream function inside the ellipse is a polynomial of degree 2 in x and y with coefficients that are functions of S, λ and ϕ, it follows that the Hamiltonian is completely determined by the moments of order 2 and less. Directly, we write $H_T = H_S + H_E$, where the suffices refer to the total, self-induced and external contributions, respectively. The stream function can be written

$$\psi = -\frac{\omega}{2\pi} \int_S \log |\mathbf{r} - \mathbf{r}'| \, dx' \, dy' + \epsilon xy \tag{20}$$

for a constant straining field $(\epsilon x, -\epsilon y)$. (Modifications to deal with unsteady fields are immediate.) The integrals of ψ_S and ψ_E over the patch can be evaluated in closed form to give[6]

$$H_S = -\frac{\omega^2 S^2}{8\pi} \left[\log \frac{(\lambda + 1)^2}{\lambda} + \log \frac{A}{4\pi} + \frac{1}{4} \right] \tag{21}$$

and

$$H_E = \epsilon \omega J_{1,1} + \epsilon \omega SXY. \tag{22}$$

In terms of the moments, we can express (21) as

$$H_S = -\frac{\omega^2 S^2}{8\pi} \log \left(2 + \frac{4\pi}{S^2} (J_{2,0} + J_{0,2}) \right), \tag{23}$$

where the constant term and function only of A on the right-hand side are neglected since they are Casimirs and have no effect on the dynamics.

The total Hamiltonian is obtained by adding (22) and (23). Applying the evolution equation (1), we first obtain the trivial results

$$\dot{A} = \{A, H_T\} = 0, \quad \dot{X} = \{X, H_T\} = \epsilon X, \quad \dot{Y} = \{Y, H_T\} = -\epsilon Y. \tag{24}$$

These equations express the conservation of area and the convection of the centroid by the local external velocity. The second-order moments have non-trivial evolutions. We have

$$
\begin{aligned}
\dot{J}_{2,0} &= \{J_{2,0}, H_T\} \\
&= \{J_{2,0}, J_{0,2}\} \frac{\partial H_T}{\partial J_{0,2}} + \{J_{2,0}, J_{1,1}\} \frac{\partial H_T}{\partial J_{1,1}} \\
&= \frac{-\omega S^2 J_{1,1}}{S^2 + 2\pi(J_{2,0} + J_{0,2})} + 2\epsilon J_{2,0}.
\end{aligned}
\tag{25}
$$

---

[6] There is a subtlety here with regard to the contribution from the straining field, and also for a contribution from a constant uniform velocity. The existence of a strain implies a singularity at infinity, which should be incorporated into the Hamiltonian. In principle, this could be handled by regarding the strain as being due to the limit of vortices at infinity. There is then a contribution to the integral over the vortices at infinity as the vortices at a finite distance move. The reciprocal theorems for harmonic fields remove the necessity of explicitly calculating this contribution, as they imply that it is sufficient to multiply the integral over the vortices by a factor of 2.

Similarly,

$$\dot{J}_{0,2} = \frac{\omega S^2 J_{1,1}}{S^2 + 2\pi(J_{2,0} + J_{0,2})} - 2\epsilon J_{0,2}, \tag{26}$$

$$\dot{J}_{1,1} = \frac{\omega S^2 (J_{2,0} - J_{0,2})}{2S^2 + 4\pi(J_{2,0} + J_{0,2})} . \tag{27}$$

These equations are not independent since any can be derived from the other two using the vanishing of C defined by (15).

Expressed in terms of λ and ϕ, we obtain

$$\dot{\lambda} = 2\epsilon\lambda\cos 2\phi, \quad \dot{\phi} = \frac{\omega\lambda}{(1+\lambda)^2} + \epsilon\frac{1+\lambda^2}{1-\lambda^2}\sin 2\phi, \tag{28}$$

which are precisely the equations given by Kida for the evolution of the ellipse parameters. The case $\epsilon = 0$ gives the Kirchhoff solution, and the steady solutions are the Moore–Saffman steady states.

An alternative procedure is to employ the centroid and ellipse parameters directly. The bracket $\{X, Y\}$ is given in (12). Using the moment-parameter relations (16)-(18) and moment brackets (14), we find that

$$\{\phi, \lambda\} = \frac{8\pi}{\omega S^2} \frac{\lambda^2}{1-\lambda^2} . \tag{29}$$

The total Hamiltonian in terms of the ellipse parameters is

$$H_T = -\frac{\omega^2 S^2}{8\pi}\log\frac{(1+\lambda)^2}{\lambda} - \frac{\epsilon\omega S^2}{8\pi\lambda}(1-\lambda^2)\sin 2\phi + \epsilon XY. \tag{30}$$

These brackets and Hamiltonian lead again to the evolution equation (28).[7]

We can examine the elliptical vortex model for the motion and structure of patches in terms of the Hamiltonian formulation described here.[8] The vortex patches are defined by the variables S_α, X_α, Y_α, ϕ_α, λ_α. The relations between the second-order moments and their brackets are unaltered. Hence, we have

$$\{X_\alpha, Y_\beta\} = \frac{\delta_{\alpha\beta}}{\omega_\alpha S_\alpha}, \quad \{\phi_\alpha, \lambda_\beta\} = \frac{8\pi\delta_{\alpha\beta}}{\omega_\alpha S_\alpha^2}\frac{\lambda_\alpha^2}{1-\lambda_\alpha^2}, \tag{31}$$

---

[7] Note that the argument given here is incomplete, in that it assumes and does not prove that the shape remains elliptical. To close the argument, it needs to be shown, for example, that the relations between the higher-order moments and those of order 2 or less are invariant. (This remark comes from an anonymous referee.)

[8] Saffman and Szeto [1980 1981] showed that the approximation of taking a patch as an ellipse in the uniform straining field defined by the value of the strain at the centroid produced by the other elliptically approximated patches was quite accurate. A refined set of equations using Kida's unsteady solution for elliptical patches was developed by Melander, Zabusky and Styczek [1986] and shown to have Hamiltonian form.

with the remaining brackets vanishing. The total Hamiltonian H_T is decomposed into the sum of the self energies

$$H_S = -\frac{1}{4\pi} \sum_\alpha \omega_\alpha^2 \int_{S_\alpha} \int_{S'_\alpha} \log |\mathbf{r} - \mathbf{r}'| \, dx \, dy \, dx' \, dy' \tag{32}$$

and interaction energies

$$H_I = -\frac{1}{4\pi} \sum_\alpha \sum_{\alpha \neq \beta} \omega_\alpha \omega_\beta \int_{S_\alpha} \int_{S_\beta} \log |\mathbf{r} - \mathbf{r}'| \, dx \, dy \, dx' \, dy'. \tag{33}$$

These integrals are now evaluated assuming that the patches are elliptical and expanding the logarithmic term in the interaction energy by a Taylor expansion of

$$\log |\mathbf{r} - \mathbf{r}'| = \log |\mathbf{R}_\alpha - \mathbf{R}_\beta + (\mathbf{r} - \mathbf{R}_\alpha) - (\mathbf{r}' - \mathbf{R}_\beta)|, \tag{34}$$

where $|\mathbf{r} - \mathbf{R}_\alpha|$ and $|\mathbf{r}' - \mathbf{R}_\beta|$ are small compared with $|\mathbf{R}_\alpha - \mathbf{R}_\beta|$ since the patches are well separated. Neglecting third and higher moments, and expressing second moments in terms of the variables, we obtain

$$H_S = -\frac{1}{8\pi} \sum_\alpha \omega_\alpha^2 S_\alpha^2 \log \frac{(1 + \lambda_\alpha)^2}{\lambda_\alpha} \tag{35}$$

and

$$H_I = -\frac{1}{4\pi} \sum_\alpha \sum_{\alpha \neq \beta} \omega_\alpha \omega_\beta S_\alpha S_\beta$$
$$\times \left[\log R_{\alpha\beta} - \frac{S_\alpha}{4\pi R_{\alpha\beta}^2} \frac{\lambda_\alpha^2 - 1}{\lambda_\alpha} \cos 2(\phi_\alpha - \theta_{\alpha\beta}) \right], \tag{36}$$

where

$$(X_\alpha - X_\beta, \ Y_\alpha - Y_\beta) = R_{\alpha\beta}(\cos \theta_{\alpha\beta}, \ \sin \theta_{\alpha\beta}). \tag{37}$$

The Hamiltonian given by (35) and (36) and the Poisson brackets (31) enable the calculation of the rates of change using (1).[9] The results are

$$\dot{\lambda}_\alpha = \lambda_\alpha \sum_{\alpha \neq \beta} \frac{\omega_\beta S_\beta}{\pi R_{\alpha\beta}^2} \sin 2(\theta_{\alpha\beta} - \phi_\alpha), \tag{38}$$

$$\dot{\phi}_\alpha = \frac{\omega_\alpha \lambda_\alpha}{(1 + \lambda_\alpha)^2} + \frac{1 + \lambda_\alpha^2}{1 - \lambda_\alpha^2} \sum_{\alpha \neq \beta} \frac{\omega_\beta S_\beta}{2\pi R_{\alpha\beta}^2} \cos 2(\theta_{\alpha\beta} - \phi_\alpha), \tag{39}$$

---

[9] Remember that the rules for evaluation of products and functions in the Poisson brackets are $\{F, GH\} = \{F, G\}H + \{F, H\}G$, and $\{F, f(G)\} = \{F, G\}f'(G)$.

$$\dot{X}_\alpha = -\frac{1}{2\pi} \sum_{\alpha \neq \beta} \frac{\omega_\beta S_\beta}{R_{\alpha\beta}} \left[\sin\theta_{\alpha\beta} - \frac{1}{4\pi R_{\alpha\beta}^2} \left(S_\beta \frac{1 - \lambda_\beta^2}{\lambda_\beta} \sin(3\theta_{\alpha\beta} - 2\phi_\beta) \right. \right.$$

$$\left. \left. + S_\alpha \frac{1 - \lambda_\alpha^2}{\lambda_\alpha} \sin(3\theta_{\alpha\beta} - 2\phi_\beta) \right) \right], \tag{40}$$

$$\dot{Y}_\alpha = \frac{1}{2\pi} \sum_{\alpha \neq \beta} \frac{\omega_\beta S_\beta}{R_{\alpha\beta}} \left[\cos\theta_{\alpha\beta} - \frac{1}{4\pi R_{\alpha\beta}^2} \left(S_\beta \frac{1 - \lambda_\beta^2}{\lambda_\beta} \cos(3\theta_{\alpha\beta} - 2\phi_\beta) \right. \right.$$

$$\left. \left. + S_\alpha \frac{1 - \lambda_\alpha^2}{\lambda_\alpha} \cos(3\theta_{\alpha\beta} - 2\phi_\beta) \right) \right], \tag{41}$$

These equations agree with those derived by a direct expansion of the equations of motion by Melander et al. An advantage in having equations of motion in Hamiltonian form is that methods of symplectic integration may then be employed (see, for example, Pullin and Saffman [1991] where symplectic integration was shown to be more accurate for long time calculations of the motion of four point vortices than standard methods).

14.4 Vortex breakdown

Vortex filaments or tubes that occur in experimental situations are usually thought of as being slender; that is, $\partial/\partial s \ll 1/a$, where s is distance along the filament and a is the core radius. It was discovered experimentally (e.g., Elle [1960], Lambourne and Bryer [1961], see Batchelor [1967 plate 22]) that there exist circumstances in which this condition is violated. The core exhibits a rapid change, usually to a rather different flow state. This phenomenon is called vortex breakdown, or vortex bursting as the core appears to burst or increase its size in a distance of order a. Numerical simulations at finite Reynolds number have also found this behaviour. The breakdown takes the form of the appearance of a symmetrical bubble on the axis, or non-axisymmetric spiralling is observed. We discuss here the question of the extent to which the phenomenon can be understood or predicted in terms of inviscid incompressible vortex dynamics.

Various theories have been proposed and the matter is controversial (e.g., see Hall [1972] for a critical discussion of various theories). It is suggested here that there is a close association with the (in general unsolved) problem of existence and uniqueness for solutions of the steady incompressible Euler equations. Consider flow in a volume V, assumed to be cylindrical, with length large compared with radius. The cylindrical walls are assumed impermeable and are stream surfaces, and the inlet and outlet are at the ends. Intuition suggests that for a solution to be unique (or at least isolated[10]) an appropriate class of boundary conditions should be consistent with three requirements. The first is that $\mathbf{u} \cdot \mathbf{n}$ and ω should be given at the inlet, where $\mathbf{u} \cdot \mathbf{n} < 0$ ($\mathbf{n}$ being the outward normal); the second is that $\mathbf{u} \cdot \mathbf{n} > 0$ be given at the outlet; and the third requirement is the constraint that $\mathbf{u} \neq 0$ anywhere inside the tube. This last condition ensures that there are no closed streamlines and that all fluid is in the tube for only a finite time. Such a solution, if it exists, will be called a regular solution. Now it is expected that the existence of regular solutions is limited to a subclass of boundary conditions. If the imposed boundary conditions are not in this class, then a regular flow does not exist and further information is needed to specify the flow and make it unique, such as the Bernoulli constant in regions of closed streamlines. Note that the boundary conditions which allow regular solutions do not necessarily imply a unique (or isolated) solution in the class of all flows, since, for example, infinitesimal vortex rings could be added to a regular solution without changing the boundary conditions. We now assert that if a slender, quasi-parallel geometry does not admit regular solutions, then vortex breakdown will occur.[11]

According to these ideas, the criterion for breakdown is the non-existence of regular flows and the mathematical problem is to determine for a given geometry the conditions on the inlet velocity and vorticity for non-existence and demonstrate, as is needed for the results to be useful, that the dependence on the outlet flow profiles is weak. In general, for flows with similar flow profiles, the criteria can be expected to depend on a characteristic velocity U, a characteristic vorticity Ω and a radius a, through a Rossby number $R = U/\Omega a$. Breakdown is then expected when $R < R_c$, where the critical Rossby number R_c depends on the entire geometry.[12] A local

---

[10] That is, no other solution exists which is arbitrarily close and satisfies the same boundary conditions.

[11] Hall [1972] argued for vortex breakdown being associated with failure of a slender flow approximation and the appearance of longitudinal gradients comparable with the transverse ones, in analogy to boundary layer separation. Trigub [1985] has identified vortex breakdown with the existence of folds or limit points for the family of quasi-parallel flows.

[12] Compare the Taylor–Proudman theorem of rotating fluids, where small Rossby number is the criterion for a rigidity of the flow which does not allow a variation in the direction parallel to the axis of rotation. See Greenspan [1968].

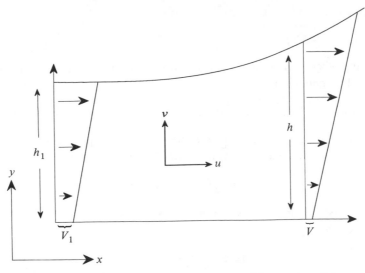

14.4-1 Inlet and outlet velocity profiles for a constant vorticity, quasi-parallel, two-dimensional flow through a channel of varying width.

Rossby number dependent only on the inlet conditions and independent of downstream geometry is not expected to exist in general.[13]

Two separate but not altogether unrelated phenomena may signal the occurrence of breakdown. The first is the appearance of stagnation points in the flow. Then unsteadiness or inflow data at the downstream outlet will in general be required to fix the flow, either of which can lead to violations of the slenderness requirement. The second is the appearance of a fold or limit point in the dependence of a family of solutions on the Rossby number so that no solution of the assumed form exists.

These cases can be illustrated by simple examples. Consider first a two-dimensional flow in a channel of slowly varying width (see Figure 14.4-1).[14] We make a quasi-parallel approximation supposing that $v \ll u$ and $\partial/\partial x \ll \partial/\partial y$. Then in terms of the stream function $\psi(x, y)$, the longitudinal velocity $u = \partial\psi/\partial y$ and vorticity $\omega = -\partial^2\psi/\partial y^2$. At the inlet u and ω are supposed to be given functions of y, which implies a functional

---

[13] At finite Reynolds number, a dependence of R_c on Reynolds number is to be expected, of course. See Spall, Gatski and Grosch [1987], who emphasized the relevance of the Rossby number as a criterion for vortex breakdown.

[14] The suggestion that vortex breakdown phenomena can be studied in a two-dimensional flow is due to Prof D. I. Pullin (private communication).

dependence

$$\omega = \Omega(\psi). \tag{1}$$

At any other station, the stream function is determined by the equation and boundary conditions

$$d^2\psi/dy^2 = -\Omega(\psi), \qquad \psi = 0 \quad \text{at } y = 0, \qquad \psi = Q \quad \text{at } y = h(x), \tag{2}$$

where Q is the flux, determined at the inlet. A simple example is constant shear $\Omega = -\kappa = \text{constant}$, with $\kappa > 0$. The inlet flow is

$$u = \kappa y + V_1, \quad \psi = \tfrac{1}{2}\kappa y^2 + V_1 y, \quad Q = \tfrac{1}{2}\kappa h_1^2 + V_1 h_1. \tag{3}$$

The quasi-parallel approximation removes the freedom to fix the outlet profile which is

$$u = \kappa y + V, \quad \text{where } \tfrac{1}{2}\kappa h^2 + Vh = \tfrac{1}{2}\kappa h_1^2 + V_1 h_1. \tag{4}$$

This equation determines the outlet velocity V at $y = 0$ in terms of the width h.

Defining a Rossby number by

$$R_1 = \frac{V_1}{\kappa h_1}, \tag{5}$$

we see that regular solutions exist; that is, $V > 0$, if

$$R_1 > \frac{1}{2}\left(\frac{h^2}{h_1^2} - 1\right) = R_c. \tag{6}$$

The implication is that breakdown does not occur for flow in an expanding channel if the Rossby number is sufficiently high. Figure 14.4-2 shows the breakdown region and the region of existence of regular flows in the R_1-h/h_1 plane. For $R < R_c$, there is no regular solution (for given geometry defined by h/h_1), and either inflow conditions must be given at the outlet, or an unsteady problem should be considered. These conclusions are based on the use of a quasi-parallel approximation, but it is not expected that using the full Euler equations and dropping the slenderness approximations would change the ideas significantly.

For a general two-dimensional profile, the relation between the velocity V on $y = 0$ and a Rossby number will follow from the equation

$$h = \int_0^Q \frac{d\psi}{\sqrt{V^2 - 2\int_0^\psi \Omega \, d\psi}}. \tag{7}$$

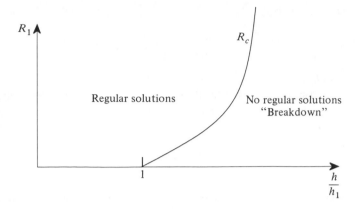

14.4-2 Regions of existence and non-existence of regular solutions in the inlet Rossby number – expansion factor plane.

For an example where the failure of regular solutions to exist is due to the presence of a fold or limit point rather than the occurrence of a stagnation point, we refer to Batchelor's [1967 §7.5] discussion of steady axisymmetric flow with swirl. The flow is determined by the Bragg–Hawthorne (usually called Squire–Long) equation (3.13.9) for the stream function ψ. We again use the quasi-parallel approximation appropriate for slender flow, neglecting transverse derivatives and velocities, and the equation takes the form (employing cylindrical polar co-ordinates (r, θ, z) with velocity components (u, v, w))

$$\frac{\partial^2 \psi}{\partial r^2} - \frac{1}{r} \frac{\partial \psi}{\partial r} = r^2 \frac{dH}{d\psi} - C \frac{dC}{d\psi}, \tag{8}$$

where $H(\psi)$ is the Bernoulli constant, and $2\pi C(\psi)$ is the azimuthal circulation round a circle of radius r. The axial velocity w is related to ψ by $w = (1/r)\partial \psi/\partial r$. In the quasi-parallel approximation, $dp/dr = v^2/r = C^2/r^3$, and

$$H = \tfrac{1}{2}(v^2 + w^2) + \int^r \frac{C^2}{r^3} \, dr = \tfrac{1}{2}w^2 + \int^r \frac{C}{r^2} \frac{dC}{dr} \, dr. \tag{9}$$

We consider the case of an isolated vortex with constant vorticity in a uniform stream. Upstream, the velocity is

$$w = U_1, \quad v = \kappa r, \quad \text{for } r < a,$$
$$w = U_1, \quad v = \kappa a^2/r, \quad \text{for } r > a. \tag{10}$$

These are equivalent to

$$C = \frac{2\kappa}{U_1}\psi, \quad H = \tfrac{1}{2}U_1^2 + \frac{2\kappa^2}{U_1}\psi, \quad \text{for } \psi < \tfrac{1}{2}U_1 a^2, \tag{11}$$

and $C = \kappa a^2$, $H = \tfrac{1}{2}U_1^2 + \kappa^2 a^2$, for $\psi > \tfrac{1}{2}U_1 a^2$.

We now suppose that the external flow is decelerated. At a station where the external velocity is U ($< U_1$), the stream function is determined by the equation

$$\frac{d^2\psi}{dr^2} - \frac{1}{r}\frac{d\psi}{dr} = \frac{2\kappa^2 r^2}{U_1} - \frac{4\kappa^2}{U_1^2}\psi, \tag{12}$$

and the new core radius b follows from the boundary conditions (in addition to analyticity on $r = 0$)

$$\psi = \frac{1}{2}U_1 a^2, \quad \frac{1}{r}\frac{\partial\psi}{\partial r} = U, \quad \text{when } r = b. \tag{13}$$

The solution is

$$\psi = \tfrac{1}{2}U_1 r^2 + Ar J_1(kr), \tag{14}$$

where $k = 2\kappa/U_1$, $A = \tfrac{1}{2}U_1(a^2 - b^2)/bJ_1(kb)$ and b is determined in terms of U and the inlet conditions by the equation (Batchelor [1967 eq. (7.5.24)]

$$\frac{U}{U_1} = 1 + \left(\frac{a^2}{b^2} - 1\right)\frac{\tfrac{1}{2}kbJ_0(kb)}{J_1(kb)}. \tag{15}$$

The velocity on the axis is

$$W = U_1 - \frac{U_1 - U}{J_0(kb)}. \tag{16}$$

We define an inlet Rossby number by

$$R_1 = \frac{U_1}{\kappa a} = \frac{2}{ka}. \tag{17}$$

We now consider the solution of (15) as a function of inlet Rossby number for given values of the (geometrical) external factor U/U_1. There are an infinite number of solutions of this equation. However, we are interested in the existence of regular solutions, not the presence of non-regular solutions. Therefore our main concern is the primary branch of solutions for which $W \to U_1$ and $b/a \to \sqrt{U_1/U}$ as $R_1 \to \infty$. A sketch of this solution showing the dependence of b and W on R_1 for $U/U_1 = \tfrac{1}{2}$ is shown in Figure 14.4-3. Note the existence of a fold at $R_1 = R_c = 2.5325$. We interpret the non-existence of regular solutions on the primary branch for $R_1 < R_c$ as signalling breakdown.

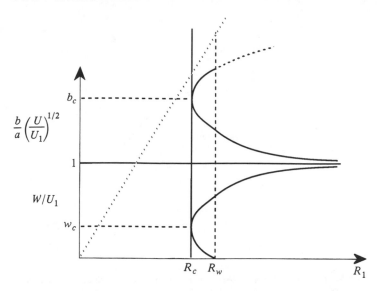

14.4-3 Sketch of dependence of vortex core radius b on R_1 for a fixed value of the deceleration factor U/U_1. ——— is regular solution on the primary branch; - - - - is non-regular extension of the primary branch; · · · · · · denote flows with significant internal structure not in general regular. Also shown is minimum axial velocity on primary branch. For $U/U_1 = \frac{1}{2}$, $R_c = 2.5325$, $(b/a)_c = 1.8167$, $W_c/U_1 = 0.0875$. Minimum axial velocity goes to zero at $R_w = 2.5655$.

At the fold, the minimum axial velocity occurs on the centreline, where its value is given by (16) and is found to be positive ($W_c/U_1 = 0.0875$), so that breakdown in this case is not associated with the occurrence of a stagnation point. The velocity W on the axis goes to zero on the upper branch at $R_1 = R_w = 2.5655$, and the primary branch continues back to infinity as a family of non-regular solutions with reverse flow, asymptoting $b/a \sim \frac{1}{2} R_1 \eta$, where η is the smallest root of

$$\frac{\eta}{2} \frac{J_0(\eta)}{J_1(\eta)} = 1 - \frac{U}{U_1}. \tag{18}$$

The infinity of other solutions start at the origin in the b vs R_1 plane and have this same asymptotic behaviour, with η being one of the other roots of (18). These solutions have rapidly varying internal structure and are not in general regular. They cannot be reached continuously from the primary branch and are not expected to be of physical significance.

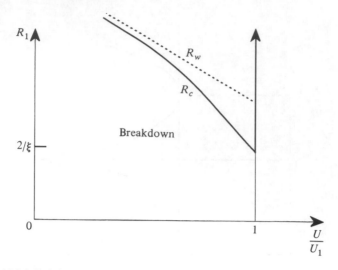

14.4-4 Variation of R_c with U/U_1 on the fold. Inlet conditions are subcritical for $R_1 < 2/\xi = 0.83$. $\cdots\cdots$ shows value of R_w on upper branch for stagnation after passing the fold.

If the flow is accelerated (i.e., $U > U_1$), then there are regular flows for all Rossby numbers and vortex breakdown is not expected.

In Figure 14.4-4, we sketch the regions of existence of regular solutions in an R_1–U/U_1 plane.[15] and show the location of the fold. It is found for this particular flow that the minimum axial velocity W is positive everywhere on the boundary of the regions, so that breakdown is due to non-existence rather than the appearance of stagnation flow. It is also found that for the existence of regular flows when the external flow is decelerated it is necessary that the inlet Rossby number be greater than $2/\xi$, where ξ is the smallest root of the Bessel function J_0. As $U \rightarrow U_1$, the nose of the primary branch goes towards $(b/a)_c = 1$ and $R_1 \rightarrow 2/\xi$, with the upper branch moving upwards towards its asymptotic behaviour deduced from (18). Also shown is the value of the inlet Rossby number R_w for which stagnation occurs.

The non-existence of decelerated regular flows when the inlet Rossby number is too small is related to the criterion of supercriticality introduced into the vortex breakdown context by Squire [1960] and Benjamin ([1962]. The speed of the slowest long wave (i.e., that going upstream fastest) on the top-hat inlet profile (relative to the frame of reference in which the

---

[15] Mr J. D. Buntine assisted with these calculations.

inlet flow is steady) is $c = -2\kappa a/\xi + U_1$, see (11.3.8). The inlet flow is therefore supercritical, in the sense that the flow cannot support standing waves if

$$c > 0, \quad \text{i.e., } R_1 > 2/\xi = 0.83. \tag{19}$$

Thus a sufficient condition for breakdown of a decelerated flow is that the inlet conditions are subcritical; that is, $c < 0$. It can be argued on general grounds that the flow profile at the fold must be critical. This is because the difference of neighbouring solutions on different sides of the limit point is a non-propagating eigenfunction of infinite wavelength; thus a standing wave exists.[16] But it should be emphasised that this condition is not necessary and that breakdown can also be associated with the appearance of stagnation in the flow. Waves of all speeds exist in the above example of stagnation in a two-dimensional flow, and the concept of criticality is not relevant there.

If the region of non-slenderness is of finite extent when a regular flow is not possible, the question arises of the existence of a slender, quasi-parallel state downstream of the burst. If H, C and the flow outside the core are unaltered in the burst region and the flow is steady, then conservation of mass, momentum and energy imply that the flows are identical. Therefore energy dissipation, momentum loss or unsteadiness are required for the quasi-parallel states to be different. Referring to Figure 14.4-4, we see that for $R_c < R_1 < R_w$, two regular flows exist for given deceleration, and it can be shown that these have different momentum flux. Thus if momentum conservation can be violated, there is the possibility in principle of a jump between states, but there is no clear reason why this should occur. There is, indeed, no good reason why the flow downstream of the breakdown should be quasi-parallel, and the calculation of this flow is expected to be a global problem dependent on the downstream boundary conditions.

A further possibility, not explored here, is that the flow with a burst region is unstable to linear or non-linear, non-axisymmetric disturbances and that this is responsible for the spiral-type bursting that is also observed.

The relation between the steady or quasi-steady vortex breakdown discussed above and the occurrence of bursting in evolving vortex filaments, commonly visible to the naked eye in the motion of trailing vortices and in the behaviour of unsteady vortex rings, is unclear. This bursting has recently been observed experimentally in filaments which thread turbulent

---

[16] This has been verified numerically for the case of the inlet top-hat profile by V. Mudkavi (private communication).

flow (Douady, Couder and Brachet [1991]) and may be significant in the generation of the small-scale components of high Reynolds number turbulence.

EPILOGUE

For God's Sake
Let us sit upon the ground and tell sad stories
Of vortex filaments.
How some have been ill-posed, some singular,
Some poisoned by their self induction, some core size killed,
Some haunted by the mathematics they have involved.

All murderous.

For within the swirling motion that rounds the mortal circulation
Of a vortex
Keeps futility his court,
And there the non-linearity sits
Scoffing at his state and grinning at his theories
Allowing him a breath, a little scene to linearize, compute and fill with
    approximations
And then at last he comes and with a little inconsistency bores through
    the costly hopes and

Farewell ...

Shakespeare, *Richard II*, Act 3, Scene 2

'Translated' by H. C. Yuen

REFERENCES

Abramowitz, M. & Stegun, I. 1964. *Handbook of mathematical functions*. Dover.

Abrashkin, A. A. & Yakubovich, E. I. 1984. Planar rotational flows of an ideal fluid. *Sov. Phys. Dokl.* 29, 370–1.

Ahlborn, C. G. F. 1902. Uber den Mechanismus des Hydrodynamischen Widerstandes. *Hamburg Nat. ver Abhandlung* 1–59.

Akhmetov, D. G., Vladimirov, V. A., Il'in, K. I., Makarenko, B. G., Nikulin, V. V. & Tarasov, B. F. 1988. *Hydrodynamics of vortical flows*. Lavrentiev Inst. Hydrodynamics, Siberian Division, Acad. Sci. USSR.

Alexander, R. C. 1971. Family of similarity flows with vortex sheets. *Phys. Fluids* 14, 231–9.

Alexandrou, N. 1963. Ph. D. Thesis, King's College, London University.

Anton, L. 1939. Ausbildung eines Wirbels an der Kante einer Platte. *Ing. Arch.* 10, 411–27.

Aref, H. 1979. Motion of three vortices. *Phys. Fluids* 22, 393–400.

Aref, H. 1983. Integrable, chaotic and turbulent vortex motion in two-dimensional flows. *Ann. Rev. Fluid Mech.* 15, 345–89.

Aref, H., Kadtke, J. B., Zawadzki, I., Campbell, L. J. & Eckhardt, B. 1988. Point vortex dynamics: recent results and open problems. *Fluid Dynamics Res.* 3, 63–74.

Arnol'd, V. I. 1965. Conditions for nonlinear stability of stationary plane curvilinear flows of an ideal fluid. *Sov. Math. Dokl.* 6, 773–7.

Arnol'd, V. I. 1969. On an a priori estimate in the theory of hydrodynamical stability. *AMS Translations* 79, 267–9.

Baker, G. R. 1980. Energetics of a linear array of hollow vortices of finite cross-section. *J. Fluid Mech.* 99, 97–100.

Baker, G. R. 1990. A study of the numerical stability of the method of contour dynamics. *Phil. Trans. Roy. Soc.* A333, 391–400.

Baker, G. R., Barker, S. J., Bofah, K. K. & Saffman, P. G. 1974. Laser anemometer measurements of trailing vortices in water. *J. Fluid Mech.* 65, 325–36.

Baker, G. R., Saffman, P. G. & Sheffield, J. S. 1976. Structure of a linear array of hollow vortices of finite cross section. *J. Fluid Mech.* 74, 469–76.

Batchelor, G. K. 1964. Axial flow in trailing line vortices. *J. Fluid Mech.* 20, 645–658.

Batchelor, G. K. 1967. *An introduction to fluid mechanics.* Cambridge Univ. Press.

Bayly, B. J. 1986. Three-dimensional instability of elliptical flow. *Phys. Rev. Lett.* 57, 2160–2163.

Bénard, H. 1908. Formation de centres de giration à l'arrière d'un obstacle en mouvement. *Comptes Rendus* 147, 839–42.

Benjamin, T. B. 1962. Theory of the vortex breakdown phenomenon. *J. Fluid Mech.* 14, 65–84.

Benjamin, T. B. 1976. The alliance of practical and analytical insights into the nonlinear problems of fluid mechanics. *Lecture Notes in Mathematics* 503 (ed. P. Germain & B. Nayrolle), 8–29. Springer.

Benjamin, T. R. & Ellis, A. T. 1990. Self propulsion of asymetrically vibrating bubbles. *J. Fluid Mech.* 212, 65–80.

Betz, A. 1932. Verhalten von Wirbelsystemen. *Z.A.M.M.* 12, 164–74. (Also: 1933. Behavior of vortex systems. *N.A.C.A.* T.N. 713.)

Betz, A. 1950. Wie ensteht ein Wirbel in einer wenig zähen Flüssigkeit? *Die Naturwissenschaften* 9, 193–96.

Birkhoff, G. & Fisher, J. 1959. 'Do vortex sheets roll up?' *Rend. Circ. Mat. Palermo* 8, 77–90.

Birkhoff, G. 1962. *Helmholtz and Taylor Instability.* Proceedings of Symposium in Applied Mathematics, vol. XIII, Am. Math. Soc.

Blendermann, W. 1969. Der Spiralwirbel am translatorisch bewegten Kreisbogenprofil; Struktur, Bewegung und Reaktion. *Schiffstechnik* 16, 3–14.

Bragg, S. L. & Hawthorne, W. R. 1950. Some exact solutions of the flow through annular cascade actuator discs. *J. Aero. Sci.* 17, 243–49.

Buntine, J. D. & Pullin, D. I. 1989. Merger and cancellation of strained vortices. *J. Fluid Mech.* 205, 263–95.

Burbea, J. 1982. On patches of uniform vorticity in a plane of irrotational flow. *Arch. Ret. Mech. Anal.* 77, 349–58.

Burgers, J. M. 1948. A mathematical model illustrating the theory of turbulence. *Adv. Appl. Mech.* 1, 171–99.

Caflisch, R. E. & Orellana, O. F. 1986. Long time existence for a slightly perturbed vortex sheet. *Comm. Pure Appl. Math.* 39, 807–38.

Chandrasekhar, S. 1961. *Hydrodynamic and hydromagnetic stability.* Dover.

Chemin, J-Y. 1991. Existence globale pour le problème des poches de tourbillon. *C.R. Acad. Sci. Paris,* t. 312, 803–6.

Christiansen, J. P. & Zabusky, N. J. 1973. Instability, coalescence and fission of finite-area vortex structures. *J. Fluid Mech.* 61, 219–43.

Coles, D. 1967. A note on Taylor instability in circular Couette flow. *Trans A.S.M.E. J. App. Math.* 34, 529–34.

Crow, S. C. 1970. Stability theory for a pair of trailing vortices. *AIAA J.* 8, 2172–9.

Da Rios, L. S. 1906. Sul Moto d'un liquido indefinito con un filetto vorticoso di forma qualunque. *Rend. Circ. Mat. Palermo.* 22, 117–135.

Deem, G. S. & Zabusky, N. J. 1978, Vortex Waves: stationary 'V states', interactions, recurrence and breaking. *Phys. Rev. Lett.* 40, 859–62.

Dhanak, M. R. 1981. Interaction between a vortex filament and an approaching rigid sphere. *J. Fluid Mech.* 110, 129–47.

Dhanak, M. R. 1992. Stability of a regular polygon of finite vortices. *J. Fluid Mech.* 234, 297–316.

Dhanak, M. R. & de Bernardinis, B. 1981. The evolution of an elliptic vortex ring. *J. Fluid Mech.* 109, 189–216.

Domm, U. 1956. Uber die Wirbelstrassen von geringster Instabilität. *Z.A.M.M.* 36, 367–71.

Donnelly, R. J. & Roberts, P. H. 1974. Superfluid mechanics. *Ann. Rev. Fluid Mech.* 6, 179–225.

Douady, S., Couder, Y. & Brachet, M. E. 1991. Direct observation of the intermittency of intense vorticity filaments in turbulence. *Phys. Rev. Lett.* 67, 983–6.

Drazin, P. G. & Reid, W. H. 1981. *Hydrodynamic stability.* Cambridge Univ. Press.

Dritschel, D. G. 1985. The stability and energetics of co-rotating uniform vortices. *J. Fluid Mech.* 157, 95–134.

Dritschel, D. G. 1988. The repeated filamentation of two-dimensional vorticity interfaces. *J. Fluid Mech.* 194, 511–47.

Durand, W. F. 1934. *Aerodynamic theory.* Stanford.

Dyson, F. W. 1893. Potential of an anchor ring, part II. *Phil. Trans. Roy. Soc.* A184, 1041–1106.

Elle, B. J. 1960. On the breakdown at high incidences of the leading edge vortex on Delta wings. *J. Roy. Aero. Soc.* 64, 491–3

Fine, K. S., Driscoll, F. F., Malmberg, J. H. & Mitchell, T. B. 1991. Measurements of symmetric vortex merger. *Phys. Rev. Lett.* 67, 588–91.

Föppl, L. 1913. Wirbelbewegung hinter einem Kreiszylinder. *Sitz. K. Bäyr Akad. Wiss.* 1, 7–18. (Also: 1983. Vortex motion behind a circular cylinder. *N.A.S.A.* T.M. 77015.)

Fraenkel, L. E. 1970. On steady vortex rings of small cross-section in an ideal fluid. *Proc. Roy. Soc. Lond.* A316, 29–62.

Fraenkel, L. E. 1972. Examples of steady vortex rings of small cross-section in an ideal fluid. *J. Fluid Mech.* 51, 119–135.

Fraenkel, L. E. & Burger, M. S. 1974. A global theory of steady vortex rings in an ideal fluid. *Acta Math.* 132, 13–51.

Fukumoto, Y. & Miyazaki, T. 1991. Three-dimensional distortions of a vortex filament with axial velocity. *J. Fluid Mech.* 222, 396–416.

Gaier, D. 1964. *Konstruktive Methoden der konformen Abbildung.* Springer Tracts in Natural Philosophy, vol. 3.

Glauert, H. 1948. *The elements of aerofoil and airscrew theory,* 2d ed. Cambridge Univ. Press.

Goldstein, S. 1960. *Lectures in fluid mechanics.* Interscience.

Goldstein, S. 1965. *Modern developments in fluid dynamics.* Dover.

Goodman, J., Hou, T. Y. & Lowengrub, J. 1990. Convergence of the point-vortex method for the 2-D Euler equations. *Comm. Pure & Applied Math.* 43, 415–30.

Govindaraju, S. P. & Saffman, P. G. 1971. Flow in a turbulent trailing vortex. *Phys. Fluids* 14, 2074–80.

Greenspan, H. P. 1968. *The theory of rotating fluids.* Cambridge Univ. Press.

Guiraud, J. P. & Zeytounian, R. Kh. 1977. A double-scaled investigation of the asymptotic structure of rolled-up vortex sheets. *J. Fluid Mech.* 79, 93–112.

Hall, H. E. 1958. An experimental and theoretical study of torsional oscillations in uniformly rotating liquid helium II. *Proc. Roy. Soc.* A245, 183.

Hall, M. G. 1972. Vortex breakdown. *Ann. Rev. Fluid Mech.* 4, 195–218.

Hama, F. R. 1962. Streaklines in a perturbed shear flow. *Phys. Fluids* 5, 644–50.

Hama, F. R. 1988. Genesis of the LIA. *Fluid Dynamics Res.* 3, 149–50.

Harvey, J. K. & Perry, F. J. 1971. Flowfield produced by trailing vortices in the vicinity of the ground. *A.I.A.A.J.* 9, 1659–60.

Hasimoto, H. 1971. Motion of a vortex filament and its relation to elastica. *J. Phys. Soc. Japan* 31, 293–4.

Hasimoto, H. 1972. A soliton on a vortex filament. *J. Fluid Mech.* 51, 477–85.

Hasimoto, H. 1988. Elementary aspects of vortex motion. *Fluid Dynamics Res.* 3, 1–12.

Havelock, T. H. 1931. The stability of motion of rectilinear vortices in ring formation. *Phil. Mag.* (Ser. 7) 1, 617–33.

Helmholtz, H. 1858. Uber Integrale der hydrodynamischen Gleichungen welche den Wirbelbewegungen entsprechen. *Crelles J.* 55, 25.

Hicks, W. M. 1885. Researches on the theory of vortex rings – part II. *Phil. Trans. Roy. Soc.* A176, 725–80.

Hicks, W. M. 1899. Researches in vortex motion. III. On spiral or gyrostatic vortex aggregates. *Phil. Trans. Roy. Soc.* A192, 33–101.

Hill, F. M. 1975a. A numerical study of the descent of a vortex pair in a stable stratified atmosphere. *J. Fluid Mech.* 71, 1–13.

Hill, F. M. 1975b. Single hollow vortex in a strain field. Ph.D. Thesis, Imperial College, London.

Hill, M. J. M. 1894. On a spherical vortex. *Phil. Trans. Roy. Soc.* A185, 213–45.

Hirasaki, G. J. & Hellums, J. D. 1970, Boundary conditions on the vector and scalar potentials in viscous three-dimensional hydrodynamics. *Qu. App. Math.* 28, 293–6.

Hocking, L. M. 1964. Instability of a non-uniform vortex sheet. *J. Fluid Mech.* 18 177–86.

Hocking, L. M. 1965. Note on the instability of a non-uniform vortex sheet. *J. Fluid Mech.* 21 333–36.

Hoffmann, E. R. & Joubert, P. N. 1963. Turbulent line vortices. *J. Fluid Mech.* 16, 395–411.

Jimenez, J. 1975. Stability of a pair of co-rotating vortices. *Phys. Fluids* 18, 1580–1.

Jimenez, J. 1987. On the linear stability of the inviscid Karman vortex street. *J. Fluid Mech.* 178, 177–94.

Jimenez, J. 1988. Linear stability of a non-symmetric, inviscid Karman street of small uniform vortices. *J. Fluid Mech.* 189, 337–48.

Joyce, G. & Montgomery, D. 1972. Simulation of the negative temperature instability for line vortices. *Phys. Lett.* 39A, 371–2.

Joyce, G. & Montgomery, D. 1973. Negative temperature states for the two-dimensional guiding center plasma. *Plasma Physics* 10, 107–21.

Kaden, H. 1931. Aufwicklung einer unstabilen Unstetigkeitsfläche. *Ing. Arch.* 2, 140–68.

Kambe, T. 1983. A class of exact solutions of two-dimensional viscous flow. *J. Phys. Soc. Japan* 52, 834–41.

Kamm, J. R. 1987. Shape and stability of two-dimensional vortex regions. Ph.D. Thesis, Caltech.

Karman, T. von 1911, 1912. Uber den Mechanismus des Widerstands, den ein bewegter Korper in einer Flüssigkeit erfährt. Göttingen Nachrichten, *Math. Phys. Kl.* 12, 509–46, 13, 547–56.

Karman, T. von & Burgers, J. M. 1934. *General aerodynamic theory, perfect fluids. Aerodynamic theory* (ed. Durand), vol. II. Stanford.

Karman, T. von & Rubach, H. L. 1913. Uber den Mechanismus des Flüssigkeits- und Luftwidertands. *Phys. Z.* 13, 49–59.

Karweit, M. 1975. Motion of a vortex pair approaching an opening in a boundary. *Phys. Fluids* 18, 1604–6.

Kasper, W. A. 1979. *The Kasper wing*. Meheen Corporation, Denver.

Keady, G. 1985. Asymptotic estimates for symmetric vortex streets. *J. Austral. Math. Soc.* Ser. B26, 487–502.

Kelvin, Lord 1867a. On vortex atoms. *Phil. Mag.* 34, 15–24.

Kelvin, Lord 1867b. The translatory velocity of a circular vortex ring. *Phil. Mag.* 33, 511–12.

Kelvin, Lord 1868. On vortex motion. *Trans. Royal Soc. Edinburgh* 25, 217–60. (Also: *Collected works*, vol. IV, 13–65.)

Kelvin, Lord 1875. Vortex Statics. *Collected works*, vol. IV, 115–28.

Kelvin, Lord 1878. Floating magnets (illustrating vortex systems). *Collected works*, vol. IV, 135–40.

Kelvin, Lord 1880. Vibrations of a columnar vortex. *Phil. Mag.* 10, 155–68. (Also: *Collected works*, vol. IV, 152–65.)

Kelvin, Lord 1881. On the average pressure due to impulse of vortex rings on a solid. *Nature* 24, 47. (Also: *Collected works*, vol. IV, 188.)

Kida, S. 1975. Statistics of the system of line vortices. *J. Phys. Soc. Japan* 39, 1395–1404.

Kida, S. 1981a. Motion of an elliptic vortex in a uniform shear flow. *J. Phys. Soc. Japan* 50, 3517–20.

Kida, S. 1981b. A vortex filament moving without change of form. *J. Fluid Mech.* 112, 397–409.

Kida, S. 1982. Stabilizing effects of finite core on Karman vortex street. *J. Fluid Mech.* 122, 487–504.

Kirde, K. 1962. Untersuchengen über die zeitliche Weiterentwicklung eines Wirbels mit vorgegebener Anfangsverteilung. Ing. Arch 31, 385–404.

Kiya, M. & Arie, M. 1979. Helmholtz instability of a vortex sheet in uniform shear flow. *Phys. Fluids* 22, 378–9.

Klein, F. 1910. Uber die Bildung von Wirbeln in reibungslosen Flüssigkeiten. *Zeit. für Math. u. Physik* 59, 259–62.

Kochin, N. E., Kibel, I. A. & Roze, N. V. 1964. *Theoretical hydrodynamics.* Interscience.

Kreiss, H. O. & Lorenz, J. 1989. *Initial boundary value problems.* Academic Press.

Krutzsch, C. H. 1939. Uber an experimentall Beobachten an Wirbelringen bei ihrer translatorischen Bewegung in wirklichen Flussigkeiten. *Ann. Phys.* 35, 497–523.

Küchemann, D. 1965. Report on the I.U.T.A.M. Symposium on concentrated vortex motions in fluids. *J. Fluid Mech.* 21, 1–20.

Küchemann, D. & Weber, J. 1965. Vortex motions, *Z.A.M.M.* 45, 457–74.

Lamb, G. 1976. Solitons and the motion of helical curves. *Phys. Rev. Lett.* 37, 235–7.

Lamb, H. 1932. *Hydrodynamics*, 6th ed. Cambridge Univ. Press.

Lambourne, N. C. & Bryer, D. W. 1961. *The bursting of leading edge vortices – some observations and discussions of the phenomenon.* Aero. Res. Council RM-3282.

Landau, L. D. & Lifshitz, E. M. 1958. *Statistical physics.* Pergamon.

Landau, L. D. & Lifshitz, E. M. 1959. *Fluid mechanics.* Pergamon.

Landman, M. J. & Saffman, P. G. 1987. The three-dimensional instability of strained vortices in a viscous fluid. *Phys. Fluids* 30, 2339–42.

Larmor, J. 1889. On the images of vortices in a spherical vessel. *Quart. J. Math.* 23, 338.

Lessen, M., Singh, P. J. & Paillet, F. 1974. The stability of a trailing vortex. Part 1, Inviscid theory. *J. Fluid Mech.* 63, 753–63,

Lichtenstein, L. 1925. Uber einige Existenzprobleme der Hydrodynamik. *Math. Zeit.* 23, 89–154.

Lighthill, M. J. 1956. The image system of a vortex element in a rigid sphere. *Proc. Camb. Phil. Soc.* 52, 317–21.

Lighthill, M. J. 1962. Physical interpretation of the mathematical theory of wave generation by wind. *J. Fluid Mech.* 14, 385–98.

Lighthill, M. J. 1963. *Laminar boundary layers* (ed. L. Rosenhead), chap. II. Oxford Univ. Press.

Lighthill, M. J. 1973. On the Weis-Fogh mechanism of lift generation. *J. Fluid Mech.* 60, 1–17.

Lin, C. C. 1941. On the motion of vortices in two-dimensions. I. Existence of the Kirchhoff–Routh function. *Proc. Nat. Acad. Sci.* 27, 570–77.

Lin, C. C. 1943. *On the motion of vortices in two-dimensions*, Univ. Toronto App. Math. Series, no. 5. Toronto Univ. Press.

Love, A. E. H. 1893. On the stability of certain vortex motions. *Proc. Lond. Math. Soc.* 25, 18–42.

Love, A. E. H. 1927. *A treatise on the mathematical theory of elasticity*, 4h ed. Cambridge Univ. Press.

Lugt, H. J. 1979. *Wirbelströmung in Natur und Technik*, Braun.

Lundgren, T. S. 1982. Strained spiral vortex model for turbulent fine structure. *Phys. Fluids* 25, 2193–203.

MacKay, R. S. 1987. Instability of vortex streets. *Dyn. Stab. Sys.* 2, 55–71.

MacKay, R. S. & Saffman, P. G. 1986. Stability of water waves. *Proc. Roy. Soc.* A406, 115–25.

Mangler, K. W. & Weber, J. 1967. The flow field near the center of a rolled up vortex sheet. *J. Fluid Mech.* 30, 177–96.

Marsden, J.E. & Weinstein, A. 1983. Coadjoint orbits, vortices and Clebsch variables for incompressible fluids. *Physica* D7, 305–23.

Maskell, E.C. 1964, Some recent developments in the study of edge vortices, *Proc. 3rd. Cong. Int. Council Aero. Sci.* 737–50. Spartan Books.

Meiron, D.I., Baker, G.R. & Orszag, S.A. 1982. Analytic structure of vortex sheet dynamics, Part I, Kelvin–Helmholtz instability. *J. Fluid Mech.* 114, 283–98.

Meiron, D.I., Saffman, P.G. & Schatzman, J.C. 1984. The linear two-dimensional stability of inviscid vortex streets of finite-cored vortices. *J. Fluid Mech.* 147, 187–212.

Melander, M.V., Zabusky, N.J. & Styczek, A.S. 1986. A moment model for vortex interactions of the two-dimensional Euler equations, Part I., Computational validation of a Hamiltonian elliptical representation. *J. Fluid Mech.* 167, 95–115.

Miloh, T. & Shlien, D.J. 1977. Passage of a vortex ring through a circular aperture in an infinite plane, *Phys. Fluids* 20, 1219–27.

Moffatt, H.K. 1969. The degree of knottedness of tangled vortex lines. *J. Fluid Mech.* 35, 117–29.

Moffatt, H.K. 1986. On the existence of localized rotational disturbances which propagate without change of structure in an inviscid fluid. *J. Fluid Mech.* 173, 289–302.

Moffatt, H.K. 1990. Structure and stability of solutions of the Euler equations: a lagrangian approach. *Phil. Trans. Roy. Soc.* A333, 321–42.

Moffatt, H.K. & Moore, D.W. 1978. The response of Hill's spherical vortex to a small axisymmetric disturbance. *J. Fluid Mech.* 87, 749–60.

Moore, D.W. 1972. Finite amplitude waves on aircraft trailing vortices. *Aero. Quarterly* 23, 307–14.

Moore, D.W. 1974. A numerical study of the roll-up of a finite vortex sheet, *J. Fluid Mech.* 63, 225–35.

Moore, D.W. 1975. The rolling up of a semi-infinite vortex sheet. *Proc. Roy. Soc.* A345, 417–30.

Moore, D.W. 1976. The stability of an evolving two-dimensional vortex sheet. *Mathematika* 23, 35–44.

Moore, D.W. 1978. The equation of motion of a vortex layer of small thickness. *Stud. App. Math.* 58 119–40.

Moore, D.W. 1979. The spontaneous appearance of a singularity in the shape of an evolving vortex sheet. *Proc. Roy. Soc.* A365, 105–19.

Moore, D.W. 1980. The velocity of a vortex ring with a thin core of elliptical cross section. *Proc. Roy. Soc.* A370, 407–15.

Moore, D.W. 1981. On the point vortex method. *SIAM J. Sci. Stat. Computing* 2, 65–84.

Moore, D.W. 1984. Numerical and analytical aspects of Helmholtz instability. *Th. App. Mechanics*, Eds. Nirodson & Olhoff, North-Holland. Proc. XVI IUTAM, 629–33.

Moore, D.W. 1985. The effect of compressibility on the speed of propagation of a vortex ring. *Proc. Roy. Soc.* A397, 87–97.

Moore, D. W. & Griffith-Jones, R. 1974. The stability of an expanding circular vortex sheet. *Mathematika* 21, 128–33.

Moore, D. W. & Saffman, P. G. 1971. Structure of a line vortex in an imposed strain. *Aircraft wake turbulence* (eds. Olsen, Goldburg, Rogers) 339–54. Plenum.

Moore, D. W., & Saffman, P. G. 1972. The motion of a vortex filament with axial flow. *Phil. Trans. Roy. Soc.* A272, 403–29.

Moore, D. W. & Saffman, P. G. 1973. Axial flow in laminar trailing vortices. *Proc. Roy. Soc.* A.333, 491–508.

Moore, D. W. & Saffman, P. G. 1974. A note on the stability of a vortex ring of small cross-section. *Proc. Roy. Soc.* A338, 535–7.

Moore, D. W. & Saffman, P. G. 1975a. The density of organized vortices in a turbulent mixing layer, *J. Fluid Mech.* 69, 465–73.

Moore, D. W. & Saffman, P. G. 1975b. The instability of a straight vortex filament in a strain field. *Proc. Roy. Soc.* A346, 413–25.

Moore, D. W., Saffman, P. G. & Tanveer, S. 1988. The calculation of some Batchelor flows: the Sadovskii vortex and rotational corner flow. *Phys. Fluids* 31, 978–90.

Munk, M. 1919. Isoperimetrische Aufgaben aus der Theorie des Fluges. Inaug.-Dissertation, Göttingen.

Neu, J. C. 1984. The dynamics of a columnar vortex in an imposed strain. *Phys. Fluids* 27, 2397–402.

Norbury, J. 1972. A steady vortex ring close to Hill's spherical vortex. *Proc. Camb. Phil. Soc.* 72, 253–84.

Norbury, J. 1973. A family of steady vortex rings. *J. Fluid Mech.* 57, 417–31.

Novikov, E. A. 1976. Dynamics and statistics of a system of vortices, *Sov. Phys. J.E.T.P.* 41, 937–43.

Oshima, Y. 1978. The game of passing-through of a pair of vortex rings. *J. Phys. Soc. Japan* 45, 660–4.

Onsager, L. 1949. Statistical hydrodynamics. *Nuovo Cimento* 6 (Supplement), 279–87.

Peace, A. J. & Riley, N. 1983. A viscous vortex pair in ground effect. *J. Fluid Mech.* 129, 409–26.

Pierrehumbert, R. T. 1980. A family of steady, translating vortex pairs with distributed vorticity. *J. Fluid Mech.* 99, 129–44.

Pierrehumbert, R. T. 1986. A universal short wave instability of two-dimensional eddies in an inviscid fluid. *Phys. Rev. Lett.* 57, 2157–9.

Pierrehumbert, R. T. & Widnall, S. E. 1981. The structure of organized vortices in a free shear layer. *J. Fluid Mech.* 102, 301–13.

Pierrehumbert, R. T. & Widnall, S. E. 1982. The two- and three-dimensional instabilities of a spatially periodic shear layer. *J. Fluid Mech.* 114, 59–82.

Pocklington, H. C. 1895a. The configuration of a pair of equal and opposite hollow straight vortices of finite cross-section, moving steadily through fluid. *Proc. Cambridge Philos. Soc.* 8, 178–87.

Pocklington, H. C. 1895b. The complete system of periods of a hollow vortex ring. *Phil. Trans. R. Soc.* A186, 603–19.

Poincaré, H. 1893. *Théorie des tourbillons.* Gauthier-Villars.

Pointin, Y. B. & Lundgren, T. S. 1976. Statistical mechanics of two-dimensional vortices in a bounded container. *Phys. Fluids* 10, 1459–70.

Polvani, L. M., Flierl, G. R. & Zabusky, N. J. 1989. Filamentation of unstable vortex structures via separatrix crossing: A quantitative estimate of onset time. *Phys. Fluids* A1, 181–84.

Prandtl, L. 1918. Tragflügeltheorie I. *Mitteilungen, Nachrichten Ges. Wiss. Göttingen, Math-Phys. Kl.* 151–77. (Also: *Gesammelte Abhandlungen*, vol. 1, 322–45.)

Prandtl, L. 1919. Tragflügeltheorie II. *Mitteilungen, Nachrichten. Ges. Wiss. Göttingen. Math-Phys. Kl.* 107–37. (Also 1920 *N.A.C.A.* T.N. 10 Calculation of core of rolled up vortex by energy.)

Prandtl, L. 1922. *Uber die Entstehung von Wirbeln in der Idealen Flüssigkeit. Vorträge aus dem Gebiete der Hydro- und Aerodynamik* (eds. von Karman and Levi Civita). Springer. (Also, *Gesammelte Abhandlungen*, vol. 2, 697.)

Pullin, D. I. 1978. The large-scale structure of unsteady self-similar rolled-up vortex sheets. *J. Fluid Mech.* 88, 401–30.

Pullin, D. I. 1981. The nonlinear behaviour of a constant vorticity layer at a wall. *J. Fluid Mech.* 108, 401–21.

Pullin, D. I. 1989. On similarity flows containing two-branched vortex sheets. *Mathematical aspects of vortex dynamics* (ed. R. Caflisch), 97–106. S.I.A.M.

Pullin, D. I. 1991. Contour dynamics methods. *Ann. Rev. Fluid Mech.* 24, 84–115.

Pullin, D. I., Jacobs, P. A., Grimshaw, R. H. J. & Saffman, P. G. 1990. Instability and filamentation of finite-amplitude waves on vortex layers of finite thickness. *J. Fluid Mech.* 209, 359–85.

Pullin, D. I. & Moore, D. W. 1990. Remark on a result of D. Dritschel. *Phys. Fluids* A2, 1039–41.

Pullin, D. I. & Phillips, W. R. C. 1981. On a generalization of Kaden's problem. *J. Fluid Mech.* 104, 45–53.

Pullin, D. I. & Saffman, P. G. 1991. Long-time symplectic integration: the example of four-vortex motion. *Proc. Roy. Soc.* A432, 481–94.

Raja Gopal, E. S. 1963. Motion and stability of vortices in a finite channel: application to liquid Helium II. *Ann. Phys.* 25, 196–220.

Raja Gopal, E. S. 1964. Oscillations of quantized vortices in rotating liquid Helium II. *Ann. Phys.* 29, 350–65.

Rayleigh, Lord 1880. On the stability or instability of certain fluid motions. *Proc. Lond. Math. Soc.* 11, 57–75.

Rayleigh, Lord 1945. *Theory of Sound.* Dover.

Ricca, R. L. 1991. Rediscovery of Da Rios equations. *Nature* 352, 561–62.

Roberts, K. V. & Christiansen, J. P. 1972. Topics in computational fluid mechanics. *Comput. Phys. Commun.* 3 (suppl.). 14–32.

Roberts, P. H. & Donnelly, R. J. 1970. Dynamics of vortex rings. *Physics Letters* 31A, 137–38.

Robinson, A. C. & Saffman, P. G. 1982. Three-dimension stability of vortex arrays. *J. Fluid Mech.* 125, 411–27.

Robinson, A. C. & Saffman, P. G. 1984a. Three-dimensional stability of an elliptical vortex in a straining field. *J. Fluid Mech.* 142, 451–66.

Robinson, A. C. & Saffman, P. G. 1984b. Stability and structure of stretched vortices. *Stud. App. Math.* 70, 163–81.

Rosenbluth, M. N. & Simon, R. A. 1990. Algebraic instability of hollow electron columns and cylindrical vortices. *Phys. Rev. Lett.* 64, 649–52.

Rosenhead, L. 1930. The spread of vorticity in the wake behind a cylinder. *Proc. Roy. Soc.* A127, 590–612.

Rott, N. 1956. Diffraction of a weak shock with vortex generation. *J. Fluid Mech.* 1, 111–28.

Routh, E. J. 1881. Some applications of conjugate functions. *Proc. Lond. Math. Soc.* 12, 73–89.

Roy, M. 1972. Stationnarité et stabilisation de tourbillons rectilignes en ecoulement plan. *C.R. Acad. Sci.* 274, 1659–62.

Sadovskii, V. S. 1971. Vortex regions in a potential stream with a jump of Bernoulli's constant at the boundary. *App. Math. Mech.* 35, 773–79.

Saffman, P. G. 1967. The self propulsion of a deformable body in a perfect fluid. *J. Fluid Mech.* 28, 385–89.

Saffman, P. G. 1970. The velocity of viscous vortex rings. *Stud. App. Math.* 49, 371–80.

Saffman, P. G. 1974. The structure and decay of trailing vortices. *Archives of Mechanics* 26, 423–39.

Saffman, P. G. 1978. The number of waves on unstable vortex rings. *J. Fluid Mech.* 84, 625–39.

Saffman, P. G. 1979. The approach of a vortex pair to a plane surface in inviscid fluid. *J. Fluid Mech.* 92, 497–503.

Saffman, P. G. 1981. Dynamics of vorticity. *J. Fluid Mech.* 106, 49–58.

Saffman, P. G. 1985. Perspectives in vortex dynamics. Perspectives in fluid mechanics. p. 91–102. *Lecture Notes in Physics* 320 (ed. D. Coles), Springer-Verlag.

Saffman, P. G. 1988. The stability of vortex arrays to two- and three-dimensional disturbances. *Fluid Dynamics Res.* 3, 13–21.

Saffman, P. G. 1990. A model of vortex reconnection. *J. Fluid Mech.* 212, 395–402.

Saffman, P. G. 1991. Approach of a vortex pair to a rigid free surface in viscous fluid. *Phys. Fluids* A3, 984–5.

Saffman, P. G. & Baker, G. R. 1979. Vortex interactions. *Ann. Rev. Fluid Mech.* 11, 95–122.

Saffman, P. G. & Meiron, D. I. 1986. Difficulties with three-dimensional weak solutions for inviscid incompressible flows. *Phys. Fluids* 29, 2373–75.

Saffman, P. G. & Schatzman, J. C. 1982a. Stability of a vortex street of finite vortices. *J. Fluid Mech.* 117, 171–85.

Saffman, P. G. & Schatzman, J. C. 1982b. An inviscid model for the vortex street wake. *J. Fluid Mech.* 122, 467–86.

Saffman, P. G. & Sheffield, J. S. 1977. Flow over a wing with an attached free vortex. *Stud. App. Math.* 57, 107–17.

Saffman, P. G. & Szeto, R. 1980. Equilibrium shapes of a pair of equal uniform vortices. *Phys. Fluids* 23, 2339–42.

Saffman, P. G. & Szeto, R. 1981. Structure of a linear array of uniform vortices. *Stud. App. Math.* 65, 223–48.

Saffman, P. G. & Tanveer, S. 1982. The touching pair of equal and opposite uniform vortices. *Phys. Fluids* 25, 1929–30.

Saffman, P. G. & Tanveer, S. 1984a. Vortex induced lift on two-dimensional low speed wings. *Stud. App. Math.* 71, 65–78.

Saffman, P. G. & Tanveer, S. 1984b. Prandtl–Batchelor flow past a flat plate with a forward facing flap. *J. Fluid Mech.* 143, 351–65.

Schwartz, L. W. 1981. A semi-analytic approach to the self-induced motion of vortex sheets. *J. Fluid Mech.* 111, 475–90.

Schlayer, K. 1928. Uber die Stabilitat der Karmanschen Wirbelstrasse gegenuber beliebigen Storungen in drei Dimensionen. *ZAMM* 8, 352–72.

Sedov, L. J. 1965. *Two-dimensional problems in hydrodynamics and aerodynamics.* Interscience.

Serrin, J. 1959. Mathematical principles of classical fluid mechanics. *Handbuch der Physik* 8, 125–263. Springer Verlag.

Shariff, K. & Leonard, A. 1992. Vortex rings. *Ann. Rev. Fluid Mech.* 24, 235–79.

Sheffield, J. S. 1977. Trajectories of an ideal vortex pair near an orifice. *Phys. Fluids* 20, 543–45.

Sheffield, J. S. 1978. Topics in vortex motion. Ph.D. Thesis, Caltech.

Shelley, M. J. & Baker, G. R. 1990. On the connection between thin vortex layers and vortex sheets. *J. Fluid Mech.* 215, 161–94.

Smith, J. H. B. 1968. Improved calculations of leading-edge separation from slender, thin, delta wings. *Proc. Roy. Soc.* A306, 67–90.

Smith, J. H. B. 1986. Vortex flows in aerodynamics. *Ann. Rev. Fluid Mech.* 18, 221–42.

Smith, J. H. B. 1982. The representation of planar separated flow by regions of uniform vorticity. *Vortex Motion* (eds. H. Hornung & E.-A. Müller), 157–72. Vieweg.

Smith, J. H. B. & Clark, R. W. 1975. Non-existence of stationery vortices behind a two-dimensional normal plate. *AIAA. J.* 13, 1114–15.

Spall, R. E., Gatski, T. B. & Grosch, C. E. 1987. A criterion for vortex breakdown. *Phys. Fluids* 30, 3434–40.

Squire, H. B. 1960. Analysis of the 'vortex breakdown' phenomenon. Aero. Dept. Imperial Coll. London, Rep. 102. Also *Misz. Ang. Mech. Festschrift W. Tollmien* (ed. M. Schäfer) 306–12. Akad. Verlag, Berlin.

Spreiter, J. R. & Sachs, A. H. 1951. The rolling up of the trailing vortex and its effect on the downwash behind wings. *J. Aero. Sci.* 18, 21–32.

Stewartson, K. & Hall, M. G. 1963. The inner viscous solution for the core of a leading edge vortex. *J. Fluid Mech.* 15, 306–18.

Stuart, J. T. 1967. On finite amplitude oscillations in laminar mixing layers. *J. Fluid Mech.* 29, 417–40.

Su, C. H. 1979. Motion of a fluid with constant vorticity in a simply connected region. *Phys. Fluids* 22, 2032–33.

Sunderland, L. D. 1976. What happened to the Kasper wing? *Sport Aviation* 25, 30–35.

Synge, J. L. & Lin, C. C. 1943. On a statistical model of isotropic turbulence. *Trans. Roy. Soc., Canada* 37, 45–79.

Tait, P. G. 1867. Translation of 'On integrals of the hydrodynamical equations which express vortex-motion' by H. Helmholtz. *Phil. Mag.* 33, 485–512.

Tanveer, S. 1983. Topics in 2-D separated vortex flows. Ph.D. Thesis, Caltech.

Tanveer, S. 1985. Vortex induced lift on a flat plate with a curved forward flacing flap. *Stud. App. Math.* 72, 173–87.

Tanveer, S. 1986. A steadily translating pair of equal and opposite vortices with vortex sheets on their boundaries. *Stud. App. Math.* 74, 139–54.

Taylor, G. I. 1928. The forces on a body placed in a curved or converging stream of fluid. *Proc. Roy. Soc.* A120, 260–83.

Taylor, G. I. 1953. Formation of a vortex ring by giving an impulse to a circular disk and then dissolving it away. *J. App. Phys.* 24, 104.

Thomson, J. J. 1883. *A treatise on the motion of vortex rings.* Macmillan.

Thwaites, B. (ed). 1960. *Incompressible aerodynamics.* Oxford Univ. Press.

Tkachenko, V. K. 1966. Stability of vortex lattices. *Sov. Phys. J.E.T.P.* 23, 1049–56.

Townsend, A. A. 1951. On the fine-scale structure of turbulence. *Proc. Roy. Soc.* A208, 534–42.

Trigub, V. N. 1985. The problem of breakdown of a vortex line. *PMM, U.S.S.R.* 49, 166–71.

Truesdell, C. 1954. *The kinematics of vorticity.* Indiana Univ. Press.

Tsai, C-Y. & Widnall, S. E. 1976. The stability of short waves on a straight vortex filament in a weak externally imposed strain field. *J. Fluid Mech.* 73, 721–33.

Villat, H. 1930. *Leçons sur la théorie des tourbillons.* Gauthier-Villers.

Vladimirov, V. A. & Il'in, K. 1988. Three-dimensional instability of an elliptical Kirchhoff vortex. *Fluid Dynamics* 3, 356–60.

Vladimirov, V. A. & Tarasov, F. F. 1979. Structure of turbulence near the core of a vortex ring. *Sov. Phys. Dokl.* 24, 254–56.

Wagner, H. 1925. Uber die Entstehung des dynamischen Auftriebes von Tragflügeln. *Z.A.M.M.* 5, 17–35.

Walker, P. B. 1931. *Experimental verification of Wagner's work.* Tech. Rep. A.R.C. R. & M., no. 1402.

Waleffe, F. 1990. On the three-dimensional instability of strained vortices. *Phys. Fluids* 2A, 76–80.

Wan, Y. H. & Pulvirenti, M. 1985. Nonlinear stability of circular vortex patches. *Commun. Math. Phys.* 99, 435–50.

Weierstrass, K. 1858. Uber ein die homogenen Funktionen zweiten Grades betroffendes Theorem, nebst Anwendung desselben auf die Theorie der kleinen Schwingungen. In *Weierstrass: mathematische Werke*, vol. 1, 233–46. (Berlin 1894; reprinted in 1967 by Olms & Johnson.)

Weis-Fogh, T. 1973. Quick estimates of flight fitness in hovering animals, including novel mechanisms for lift production. *J. Exptl. Biology* 59, 169–230.

Whitham, G. B. 1963. The Navier–Stokes equations of motion. I, *Laminar boundary layers* (ed. L. Rosenhead), chap. III. Oxford Univ. Press.

Widnall, S. E. 1985. *Three-dimensional instability of vortices in separated flows.* Int. symp. sep. flow. Norwegian Institute of Technology. Trondheim.

Widnall, S. E. & Bliss, D. B. 1971. Slender body analysis of the motion and stability of a vortex filament containing an axial flow. *J. Fluid Mech.* 50, 335–53.

Widnall, S. E., Bliss, D. B. & Tsai, C-Y. 1974. The instability of short waves on a vortex ring. *J. Fluid Mech.* 66, 35–47.

Widnall, S. E., Bliss, D. B. & Zalay, A. 1971. Theoretical and experimental study of the stability of a vortex pair. *Aircraft wake turbulence* (eds. Olsen, Goldburg, Rogers), 305–38. Plenum.

Widnall, S. E. & Sullivan, J. 1973. on the stability of vortex rings. *Proc. Roy. Soc.* A332, 335–53.

Widnall, S. E. & Tsai, C-Y. 1977. The instability of the thin vortex ring of constant vorticity. *Phil. Trans. Roy. Soc.* A287, 273–305.

Yamada, H. & Matsui, T. 1978. Preliminary study of mutual slip-through of a pair of vortices. *Phys. Fluids* 21, 292–94.

Zabusky, N. J., Hughes, M. H. & Roberts, K. V. 1979. Contour dynamics for the Euler equations in two-dimensions. *J. Comput. Phys.* 30, 96–106.

Zarodny, S. J. & Greenberg, M. D. 1973. On a vortex sheet approach to the numerical calculation of water waves, *J. Comput. Phys.* 11, 440–46.

INDEX